T0377692

Growth Factors in Mammalian Development

Editors

I. Y. Rosenblum, Ph.D.
Head
Section of Pharmacology and Toxicology
Department of Medical Affairs
Warner-Lambert Company
Morris Plains, New Jersey
and
Adjunct Professor
Department of Pharmacology and Toxicology
Philadelphia College of Pharmacy and Science
Philadelphia, Pennsylvania

and

Susan Heyner, Ph.D.
Director
Obstetrics and Gynecology Research Laboratories
Albert Einstein Medical Center
and
Professor
Department of Obstetrics and Gynecology
and
Department of Anatomy
School of Medicine
Temple University
Philadelphia, Pennsylvania

CRC Press, Inc.
Boca Raton, Florida

Library of Congress Cataloging in Publication Data

Growth factors in mammalian development / editors, I.Y. Rosenblum and
Susan Heyner.
    p. cm.
  Includes bibliographies and index.
  ISBN 0-8493-4540-5
  1. Growth factors--Physiological effect. 2. Embryology--Mammals.
I. Rosenblum, Irwin Y., 1942- . II. Heyner, Susan.
  [DNLM: 1. Embryo--physiology. 2. Fetal Development. 3. Growth
Substances--physiology. QS 604 G884]
QP552.G76G76 1989
 599'.031--dc20
DNLM/DLC
for Library of Congress   89-15881
                                    CIP

Direct all inquiries to CRC Press, Inc., 2000 Corporate Blvd., N.W., Boca Raton, Florida, 33431.

International Standard Book Number 0-8493-4540-5

Library of Congress Card Number 89-15881
Printed in the United States

PREFACE

The aim of this book is to provide a review and discussion of the mechanisms underlying early mammalian growth and development, with a primary focus on growth factors and their signaling mechanisms. The majority of the chapters focus on regulatory factors of importance in mammalian development, although certain topics, for example, insulin and the insulin-like growth factors IGF-I and IGF-II, have been expanded to include several nonmammalian vertebrate species and invertebrates. The overall emphasis is on the early, i.e., pre- and periimplantation, stages of development. However, the recent explosion of research literature in the area of fetal development could not be ignored. The regulatory roles of several peptides in fetal development, for example, insulin and the IGFs, have been documented amply in the literature. Hence, where appropriate, this new information on fetal aspects has been integrated with the preimplantation data.

Embryogenesis, which begins with fertilization, involves the activation of a developmental program that ultimately governs such critical events as cellular proliferation, differentiation, and commitment.[1] The temporal sequences of the morphological and molecular events during the first cycle of mouse embryogenesis have been described in detail, and reviewed recently.[2] These studies provide evidence to show that, following ovulation, development is controlled at two levels. The first level is oocyte-derived and probably relates to the final phases of oocyte maturation. A second control system becomes activated following sperm penetration and functions to trigger embryogenesis. There is little doubt that the driving force behind early development is encoded in the genome. The embryonic genome becomes activated at different times according to the species. The mouse, which is precocious in this respect, demonstrates activation of the embryonic genome at the two-cell stage, while the pig embryonic genome is activated at the four-cell stage, the sheep at the eight-cell stage, and the human between the four- and eight-cell stages.[3] Upon activation of the embryonic genome, remaining nontranscribed maternal RNA becomes inactivated, and subsequent development is dependent upon embryo-encoded information. Chapter 1 provides a background to the genetic events that are known to accompany normal development in the early mouse embryo. In addition, experimental data regarding chromosomal imprinting and information obtained from transgenic animals is reviewed.

Mammalian embryos do not demonstrate a substantial increase in mass while undergoing the first few cleavages in the oviduct, and this may be one of the reasons that has contributed to their relative ease of culture *in vitro*. However, once the embryo implants, two key events occur that set the stage for fetal development: a dramatic increase in the growth rate of the embryo and the allocation of cells to embryonic lineages. The earliest postimplantational differentiation events involve a small group of pluripotent stem cells, the primitive ectoderm, from which all the cell types in the fetus and extraembryonic tissues are derived. Fetal development is initiated during the period of implantation, and a number of studies have made clear that this is the time at which the embryo is most vulnerable. Many domestic species, as well as the human, show a high rate of periimplantation mortality. In the mouse, it has been demonstrated experimentally that embryos which lag *in vitro* demonstrate implantation failure more frequently than siblings at a more advanced stage of development. These data suggest that the rate of embryonic cellular proliferation is of great importance once the blastocyst enters the uterine cavity.

An understanding of the mechanisms underlying mammalian pre- and periimplantation development is fundamentally important for several reasons; not least is the possibility of improving the emerging technologies of *in vitro* fertilization, gene therapy, and cryopreservation. Present knowledge of early mammalian development has been gained largely through studies *in vitro*.

In addition to genetic determinants, embryonic growth and development in most mammalian species require a continuing supply of energy substrates, peptide growth factors, and the participation of key hormones and hormonal signaling systems. The metabolic requirements of early mammalian embryos have been studied predominantly in two species: the mouse and the rabbit. The availability of numerous genetically defined strains of mice has contributed significantly to the use of this species to examine the regulatory mechanisms underlying early embryonic development. It is known from early studies that energy substrate requirements change significantly during the preimplantation period of the mouse, and Chapter 2 provides a review of the carbohydrate metabolism during this developmental period. New data obtained using sensitive techniques that permit the study of single embryos and allow kinetic analysis are described.

Growth and development of all multicellular organisms depend upon the operation of complex communication networks that transfer and convey information between and within cells, as well as transduce external signals. Considerable experimental evidence suggests that evolutionarily many features or components are highly conserved. The phenomenon of reversible protein phosphorylation, widely accepted as a conserved mechanism underlying a variety of cell metabolic and regulatory processes, also appears to play an integral role in signal transduction pathways.[4] For example, numerous peptide signals, including many hormones and growth factors, mediate their physiological activities by modulating the phosphorylation state of one or more proteins (enzymes) that comprise the signal transfer cascade. The importance and extensiveness of reversible protein phosphorylation in these transmembrane signaling systems only now is beginning to be appreciated.[5,6] In spite of the fact that the number of external signs is large, the mechanisms whereby transmembrane receptors generate intracellular signals may be few in number. A general overview of these mechanisms, with particular emphasis on the role of protein phosphorylation in early mammalian development, is provided in Chapter 3. Broader aspects of cellular protein phosphorylation, particularly as it relates to the discovery and characterization of receptor tyrosine kinases and serine kinase cascades, have been reviewed recently.[4,5,6]

Although classic experiments in amphibia showed that the fate of a developing cell could be totally altered by transplanting it to a different site, factors that govern the fate and rate of proliferation of embryonic cells only now are being studied at the cellular and molecular levels. Among peptide growth factors that are known to stimulate cellular proliferation, insulin is of unique interest. Whereas insulin classically is considered a hormone product of vertebrate pancreatic beta cells, functional insulin receptors have been characterized on a number of nontraditional target organs in numerous phyla. Additionally, both the insulin receptor and the ligand itself show extraordinary evolutionary conservation. Chapter 4 reviews the comparative aspects of insulin and the insulin receptor.

Closely related to insulin, both structurally and functionally, are the insulin-like growth factors (IGFs), IGF-I and IGF-II. These peptide hormones share a high degree of amino acid homology, as well as overlapping biological activities. Although the receptors for insulin and IGF-I are structurally similar, a distinct, high-affinity receptor exists for each peptide. Based on considerable evidence from *in vivo* studies, insulin and the IGFs also have distinct physiological roles. Insulin primarily regulates rapid, anabolic responses such as glucose uptake into muscle and fat cells, glycogen synthesis in hepatocytes, and fat synthesis in adipocytes. On the other hand, IGF-I, an important growth regulator, stimulates long-term cellular responses such as DNA synthesis. The physiological function of IGF-II continues to be debated. All three peptides are known to stimulate growth and differentiation in a number of embryonic systems *in vitro*. The chick provides a uniquely accessible embryonic system for experimental manipulation, and a great deal of information regarding the action of insulin and IGFs has been obtained from studies of avian embryos. These are reviewed and discussed in Chapter 5.

Despite intensive study for more than two decades, the reasons why most mammalian preimplantation embryos do not develop as well *in vitro* in simple media as they do *in vivo*, remain unclear. In the mouse, development of zygotes from most strains *in vitro* is blocked at the two-cell stage. Embryos grown *in vitro* have reduced cell numbers and exhibit lowered viability.[8] Nutritional requirements for other mammalian preimplantation embryos to develop in culture are not well defined, and many species are difficult to culture in media currently available.[9] Further, a number of species demonstrate superior embryo development *in vitro* in an organ cultured oviduct, or in complex media containing sera from various sources, rather than in a simple defined medium.[9]

During the past three decades, numerous studies have provided convincing evidence for the role of growth factors as mediators in cell-cell contact, cellular proliferation, and differentiation.[10] Most of these data have been obtained using established cell lines. However, circumstantial evidence for the existence of factors that influence proliferation exists for some mammalian embryos. Efforts to study the role of growth factors in early mammalian development have been hindered by the time and expense required to obtain sufficiently large numbers of mammalian embryos. Techniques such as indirect immunofluorescence, autoradiography, and high resolution electron microscopy circumvent the requirement for large amounts of tissue. These techniques have been used successfully for studies at the cellular level of growth factors and their receptors and have been used in particular to examine embryos for evidence of the expression of receptors that bind insulin and IGFs in the preimplantation mouse embryo. These, and other studies that discuss functional correlates, are reviewed in Chapter 6. Newer techniques, for example, the polymerase chain reaction,[11] have removed some of the previous constraints placed on molecular studies of preimplantation mammalian embryos. It is now possible to examine a relatively small number of embryos and gain information on the temporal expression of developmentally regulated genes. A recent report describes a modification of the polymerase chain reaction that permits detection of messenger RNAs coding for platelet-derived growth factor (PDGF) and transforming growth factors alpha and beta (TGF-α, TGF-β) in the preimplantation mouse embryo.[12] The presence of mRNA at these early stages raises fascinating questions with respect to autocrine production and functional correlations.

Teratocarcinomas are tumors that can be induced in mice by the transplantation of gonadal ridges or primordial germ cells. These tumors possess a stem cell population of highly malignant embryonal carcinoma (EC) cells that have the ability to differentiate into a wide range of somatic cell types. A number of investigators have suggested that EC cells correspond to the cells of the inner cell mass. These cells provide an alternative approach to the use of mammalian embryos and have provided important data regarding the role of growth factors in early mammalian development. These studies are reviewed in Chapter 7.

In addition to insulin and the IGFs, a number of other growth factors have been implicated in early mammalian development. Of particular interest are the members of the TGF-β family. These growth factors can initiate differentiation or transformation, depending on the target cell. Further, there is evidence to show that one member is involved in morphogenesis in *Drosophila*. The C-terminus of the decapentaplegic gene complex, which determines the dorsal structures of the embryo and the morphogenesis of the imaginal discs, demonstrates a high degree of homology to the C-terminal region of mammalian TGF-β. Thus, the TGF-β growth factors may be implicated in cellular hierarchy and pattern formation. Evidence to support this and other roles of TGF and other growth factors are described in Chapter 8.

The concept that oncogenes may play a role in the regulation of normal cell growth and differentiation is linked to the discovery of proto-oncogenes, the normal cellular homologues of viral transforming genes. This idea has been fueled by recent evidence revealing that several oncogenes encode products that share singular properties with key regulatory elements that control cell growth and differentiation. Oncogene products resembling peptide growth factors, hormones, or their receptors are prime suspects, since these multifunctional molecules are capable of generating mitogenic signals. Although the results of numerous experiments have documented the activity or activation of several oncogenes during development,[13,14] the relationship of these phenomona to gene activity governing normal cellular growth and development was not conclusive. This situation has changed recently with the identification of the normal counterpart of the *int-1* oncogene in *Drosophila*. This oncogene is responsible for the induction of mammary tumors in mice, while its normal counterpart has been shown to be highly homologous, if not identical, to the developmental gene *wingless*.[15,16] This exciting finding has stimulated considerable interest in the search for other oncogenes with normal cellular homologues involved in the regulation of embryonic growth and differentiation. Chapter 9 provides a review and critical discussion of the limited information available regarding this important new field.

The mechanisms regulating the coordinated growth and differentiation of the mammalian organism just now are beginning to be defined. An understanding of early mammalian development is fundamentally important, both from an academic as well as a practical viewpoint. At the simplest level, we need to expand our knowledge of the basic cellular and biochemical processes that underlie development. From a pragmatic viewpoint, *in vitro* fertilization and culture conditions for both humans and endangered species, or economically important species can be improved.

I. Y. Rosenblum
S. Heyner

REFERENCES

1. **Johnson, M. H.,** The molecular and cellular basis of preimplantation mouse development, *Biol. Rev.,* 56, 463, 1981.
2. **Johnson, M. H. and Maro, B.,** Time and space in the mouse early embryo: a cell biological approach to cell diversification, in *Experimental Approaches to Mammalian Embryonic Development,* Rossant, J. and Pederson, R. A., Eds., Cambridge University Press, New York, 1986.
3. **Braude, P., Bolton, V., and Moore, S.,** Human gene expression first occurs between the four- and eight-cell stages of preimplantation development, *Nature (London),* 332, 459, 1988.
4. **Shenolikar, S.,** Protein phosphorylation: hormones, drugs and bioregulation, *FASEB J.,* 2, 2753, 1988.
5. **Czech, M. P., Klarlund, J. K., Yagaloff, K. N., Bradford, A. P., and Lewis, R. E.,** Insulin receptor signalling, *J. Biol. Chem.,* 263, 11017, 1988.
6. **Yarden, Y. and Ullrich, A.,** Growth factor receptor tyrosine kinases, *Annu. Rev. Biochem.,* 57, 443, 1988.
7. **Harlow, G. M. and Quinn, P.,** Development of preimplantation mouse embryos *in vivo* and *in vitro, Aust. J. Biol. Sci.,* 35, 187, 1982.
8. **Bowman, P. and McLaren, A.,** Viability and growth of mouse embryos after *in vitro* culture and fusion, *J. Embryol. Exp. Morph.,* 23, 693, 1970.
9. **Bavister, B. D., Ed.,** *The Mammalian Preimplantation Embryo: Regulation of Growth and Differentiation In Vitro,* Plenum Press, New York, 1987.
10. **Gospodarowicz, D.,** Epidermal and nerve growth factors in mammalian development, *Annu. Rev. Physiol.,* 43, 251, 1981.
11. **Saiki, R. K., Gelfand, D. H., Stoffel, S., Scharf, S. J., Higuchi, R., Horn, G. T., Mullis, K. B., and Erlich, H. A.,** Primer-directed enzymatic amplification of DNA with a thermostable DNA polymerase, *Science,* 239, 487, 1988.
12. **Rappolee, D. A., Brenner, C. A., Schultz, R., Mark, D., and Werb, Z.,** Developmental expression of PDGF, TGF-α, and TGF-β genes in preimplantation mouse embryos, *Science,* 241, 1823, 1988.
13. **Tsonis, P. A.,** Oncogenes take a place in pattern formation, *Trends Biochem. Sci.,* 13, 4, 1988.
14. **Adamson, E. D.,** Oncogenes in development, *Development,* 99, 449, 1987.
15. **Baker, N. F.,** Molecular cloning of sequences from *wingless,* a segment polarity gene in *Drosophila*: the spatial distribution of a transcript in embryos, *EMBO J.,* 6, 1765, 1987.
16. **Rijsewijk, F., Schuermann, M. E., Wagenaar, E., Parren, P., Weigel, D., and Nusse, R.,** The *Drosophila* homolog of the mouse mammary oncogene *int-1* is identical to the segment polarity gene *wingless, Cell,* 50, 649, 1987.

THE EDITORS

I. Y. Rosenblum, Ph.D., is head of the Pharmacology and Toxicology Section, Department of Medical Affairs, Warner-Lambert Company, Morris Plains, New Jersey. He is Adjunct Professor of Toxicology at the Philadelphia College of Pharmacy and Science, Philadelphia, and Visiting Scientist in the Department of Obstetrics and Gynecology at the Albert Einstein Medical Center, Philadelphia.

Dr. Rosenblum graduated from the Pennsylvania State University in 1964 with a B.S. degree in Biochemistry and from the University of Wisconsin in 1969 with a Ph.D. in Physiological Chemistry. He was a postdoctoral fellow in Biochemistry at the University of Hawaii (1969 to 1970) and at the University of Arizona (1975 to 1976).

Dr. Rosenblum has presented papers at a variety of regional and national scientific meetings and has lectured at numerous colleges, universities, and research institutes. He has published more than 50 scientific papers and abstracts on topics which span the biochemical, pharmacological, and toxicological disciplines. Among other scientific accomplishments, he pioneered models of chemical-induced and spontaneous diabetes in nonhuman primates. His current major research interests have focused on developmental aspects of insulin and the insulin-like growth factors and their receptors in mammalian embryos.

Dr. Rosenblum is a member of the American Society for Pharmacology and Experimental Therapeutics, the Society of Toxicology, the American Association for the Advancement of Science, the American Society of Cell Biology, and the Society for Developmental Biology. He also has served the American Diabetes Association, the American Society of Primatologists, and the International Primatological Society.

Susan Heyner, Ph.D., is Director of the Obstetrics and Gynecology Research Laboratories, Albert Einstein Medical Center, Philadelphia and Professor, Department of Obstetrics and Gynecology and the Department of Anatomy, Temple University School of Medicine, Philadelphia.

Dr. Heyner graduated in 1957 from the University of Southampton with a B.Sc. in Zoology (Special Honours) and obtained the Ph.D. degree from the University of London in 1960.

Dr. Heyner has been a member of the National Institutes of Health Study Sections on Human Embryology and Development, and Reproductive Biology. She has been a member of the Board of Trustees for the Society for Developmental Biology. She is on the Editorial Boards of *Journal of Reproductive Immunology* and *Contraception*. She is a member of the American Association for the Advancement of Science, the American Society for Cell Biology, the Society for Developmental Biology, the Society for the Study of Reproduction, and is a Fellow of the College of Physicians in Philadelphia.

Dr. Heyner has presented more than 50 invited lectures nationally and internationally and has published more than 75 research papers, book chapters, and abstracts. Her research interests include the regulation of gene expression in early mammalian development, immunological aspects of early development, and *in vitro* fertilization.

CONTRIBUTORS

John D. Biggers, Ph.D., D. Sc.
Professor
Laboratory of Human Reproduction and
  Reproductive Biology
Department of Physiology
Harvard Medical School
Boston, Massachusetts

Scott Allen Chambers, Ph.D.
Patent Examiner
Biotechnology Unit
Patent and Trademark Office
Arlington, Virginia

Geoffrey Cooper, Ph.D.
Professor of Pathology
Dana-Farber Cancer Institute
Harvard Medical School
Boston, Massachusetts

Flora de Pablo, M.D., Ph.D.
Visiting Scientist
Diabetes Branch
National Institute of Diabetes, Digestive,
  and Kidney Diseases
National Institutes of Health
Bethesda, Maryland

Martin Farber, M.D.
Chairman
Department of Obstetrics and Gynecology
Albert Einstein Medical Center
Philadelphia, Pennsylvania

David K. Gardner, D.Phil.
Research Fellow
Department of Biology
University of York
York, England

Ann Anderson Kiessling, Ph.D.
Associate Professor
Department of Obstetrics, Gynecology,
  and Reproductive Biology
Harvard Medical School
Boston, Massachusetts

Henry J. Leese, Ph.D.
Lecturer in Biochemistry
Department of Biology
University of York
York, England

Britta A. Mattson, Ph.D.
Research Associate
Department of Anatomy and Cellular
  Biology
Tufts University Schools of Medicine
Boston, Massachusetts

Marit Nilsen-Hamilton, Ph.D.
Professor
Department of Biochemistry and
  Biophysics
Molecular, Cellular, and Developmental
  Biology Program
Iowa State University
Ames, Iowa

Michael Femi Obasaju, Ph.D., D.V.M.
Postdoctoral Fellow
Department of Obstetrics and Gynecology
Division of Reproductive Biology and
  Medicine
University of California
Davis, California

M. Lynn Pritchard, Ph.D.
Medical Communications Associate
Medical Communications
Glaxo Inc.
Research Triangle Park, North Carolina

Kenneth Ramos, Ph.D.
Assistant Professor
Department of Pharmacology and
  Therapeutics
Health Science Center
School of Medicine
Texas Tech University
Lubbock, Texas

Angie Rizzino, Ph.D.
Associate Professor
Eppley Institute for Cancer Research
University of Nebraska Medical Center
Omaha, Nebraska

Robert M. Smith
Senior Research Specialist
Department of Pathology and Laboratory
  Medicine
University of Pennsylvania School of
  Medicine
Philadelphia, Pennsylvania

Lynn Maxey Wiley, Ph.D.
Associate Professor
Department of Obstetrics and Gynecology
Division of Reproductive Biology and
 Medicine
University of California
Davis, California

To our students, past and present

*"At ebb tide I wrote a line upon the sand
and gave it all my heart and all my soul.*

*At flood tide I returned to read what I had inscribed
and found my ignorance upon the shore."*

Khalil Gibran

TABLE OF CONTENTS

Chapter 1

GENETICS OF EARLY MOUSE DEVELOPMENT

Lynn M. Wiley and Michael Femi Obasaju

TABLE OF CONTENTS

I. INTRODUCTION

A. OVERVIEW OF EARLY MOUSE DEVELOPMENT

The eutherian mammalian embryo begins development following fertilization within the ampulla of the oviduct. As it travels down the oviduct, the embryo undergoes three to four reduction cleavages, during which the blastomeres become progressively smaller with each cleavage. After a characteristic number of reduction cleavages, the embryo undergoes a series of morphogenetic events that transform it from a solid sphere of cells into a blastocyst. These events usually overlap with the time during which the embryo enters the uterus but may occur after uterine entry. In some species, including the mouse, the embryo is a morula when it enters the uterus.

The blastocyst is a cystic structure, whose wall, the trophectoderm, develops from the outer blastomeres of the morula into a polarized transporting epithelium that encloses and maintains the blastocoele.[1-3] The former inner blastomeres of the morula now comprise the inner cell mass (ICM), a cluster of cells that adhere to the inner surface of the trophectoderm. These first two cell types to evolve within the embryo synthesize cell-type-specific gene products that distinguish them from each other and from their shared blastomere progenitors. Neither trophectoderm nor ICM can resume the genetic repertoire of their progenitor blastomeres. Trophectoderm is the only tissue that can implement implantation and development of the placenta, while the ICM is the only tissue that can develop into the fetus.[4,5]

In the uterus, the blastocyst expands, escapes from the zona pellucida, and proceeds to implant and develop additional epithelial layers and embryonic cavities that comprise the extraembryonic membranes, the placenta, and the definitive embryo. As these events proceed, they are accompanied by several genetic correlates. Some of these correlates are associated with the formation of the primary germ layers, while others are associated with whether a given cell finds itself a component of the embryo proper or of an extraembryonic membrane. In most cases, the functional rationales for these genetic correlates are not yet appreciated, and the belief that such rationales exist inspires much of the current research in mammalian development.

B. MAJOR QUESTIONS REGARDING THE GENETIC ASPECTS OF EARLY DEVELOPMENT

It is the purpose of this chapter to examine some of the major research questions now posed regarding these genetic correlates with respect to what they imply about developmental control. The questions we will examine here include (1) when does the embryonic genome become functional? (2) do male and female pronuclei make equivalent contributions to the embryo? (3) are the male and female genetic contributions expressed differentially within different tissues? (4) when do embryonic cells and embryonic nuclei lose their pluripotentiality? (5) do male and female gametes contribute equivalently to the embryo? and (6) what regulates cellular proliferation in the embryo?

This chapter, for the most part, will be limited to information obtained from the mouse embryo, simply because most of what we know about developmental genetics has been derived from studies on the mouse. However, where known, information pertaining to other species shall be included. The period of development surveyed in this chapter will extend from fertilization to the formation of the primitive streak, since most of current knowledge applies to this time period.

II. ONSET OF EMBRYONIC GENE EXPRESSION

Gamete fusion provides the embryo with several potential sources of messenger RNA (mRNA), namely, mRNA from the sperm cytoplasm and from the oocyte cytoplasm and newly transcribed mRNA from the embryonic genome. In the mammal, there is no evidence for sperm contributing translatable mRNA to the embryo. The oocyte, on the other hand, does contribute

TABLE 1
Onset of Embryonic Gene Expression in Different Mammalian Species

Species	Embyonic stage for initial appearance of embyonic proteins (time postovulation)	Length of gestation (days after fertilization)	Ref.
Mouse	2-cell stage (22—32 h)	19—20	19,26,27
Sheep	8-cell stage (37—48 h)	145—155	21
Pig	4-cell stage (1—3 d)	112—115	20
Human	4-8 cell stage (43—50 h)	252—274	28

TABLE 2
Onset of Expression of Paternal Alleles in Preimplantation Mouse Embryos[32]

Embyonic stage	Gene production	Chromosome no.	Ref.
1 cell	—	—	—
2 cell	B_2-Microglobulin	2	26
	Hypoxanthine phosphoribosyl transferase (HPRT)	X	29
2—4 cell	β-Glucuronidase	7	27
6—8 cell	Non H-2 (H-3, H-6) alloantigens	2	30
8 cell	Glucophosphate isomerase (GPI)	8	31

mRNA to the embryo, some of which is polyadenylated and, presumably, available for translation. In most metazoan species, maternal mRNA that is stored in the oocyte prior to fertilization controls the majority of protein synthesis up to gastrulation when germ layer formation begins.[6-8] In mammalian embryos, however, proteins encoded by maternal mRNA virtually cease being synthesized prior to trophectoderm/ICM differentiation. In the mouse, maternal mRNA-encoded proteins are synthesized only into the two-cell stage, 25 to 28 h after fertilization,[9,10] after which the major portion becomes degraded.[11-13] There is speculation, however, that some translatable maternal RNAs or additional unidentified cytoplasmic elements[14,15] persist longer during cleavage, because maternally inherited effects on development have been observed well beyond the two-cell stage.[16-18]

Again, in contrast to most non-mammalian species, the first embryonic encoded mRNAs in the mouse embryo are detectable very early, some time during the first three cleavages, which is well before overt cell differentiation has occurred (Table 1). Interestingly, the onset of embryonic gene expression appears to correlate temporally with the spontaneous cleavage arrest many embryos undergo during development *in vitro*.[19-21] Cultured one-cell mouse embryos from most strains will cleave once and no further (two-cell block), and the first embryonic mRNAs appear before the first cleavage in the mouse[22-25] with the corresponding proteins becoming detectable shortly after the first cleavage. Some of the proteins that appear initially in the two-cell mouse embryo include B_2-microglobulin[26] and B-glucuronidase.[27] With each successive cleavage, additional embryo-encoded gene products begin to appear, but the order of their appearance does not seem to follow any obvious relationship to coincident developmental events (Table 2).

At first glance, mammalian embryonic mRNA may seem to have a shorter half-life than the mRNA found in non-mammalian embryos. It must be remembered, however, that eutherian mammalian development takes weeks to months to produce an individual that can ingest food originating from non-maternal sources. As an illustration, the period of time a mouse embryo takes to complete its first two cell cycles is equivalent to the same period of time it takes a frog embryo to aquire 37,000 cells and for the sea urchin embryo to develop into a multicellular, self-feeding, free-swimming pluteus.[33] Consequently, the temporal stability of maternal mRNA may not differ significantly among embryos from different metazoan phyla, and one reason why maternal mRNA plays a larger role in non-mammalian development than it does in mammalian development may result in part from the relative rates at which development occurs.

The relatively leisurely pace of mammalian development may account for another aspect in which it differs from development in non-mammalian species; the mammalian embryo has an obligatory requirement for early embryonic gene expression. This requirement can be revealed experimentally as early as the first cleavage in the mouse[34—36] or during the third cleavage in the human[28] by drugs that inhibit RNA synthesis. This requirement is revealed naturally by the existence of several embryonic lethal mutations that cause developmental arrest at characteristic times in homozygotes during preimplantation development.[16,18,37—40,42] Studies on these lethal mutations imply that there are maternally inherited, essential gene products whose stabilities set the lifespan of the affected homozygous embryo. It is unfortunate that the gene products involved have remained so far unidentified except in rare instances; furthermore, the function(s) of the gene product(s) has not been defined.[43] Consequently, it is not known why these mutations result in developmental arrest of homozygous embryos.

III. CONTRIBUTIONS OF THE MALE AND FEMALE GAMETES TO THE EMBRYO: MATERNAL/PATERNAL NON-EQUIVALENCY

Because of the morphological differences between the sperm and the oocyte, one would expect intuitively that there might be also heritable non-equivalent genetic contributions by the two gametes to the embryo. Two traditionally recognized non-equivalent contributions are sex determination by the sperm and sex-linked traits whose phenotypic expression depends upon the sex of the embryo. It has been more recently appreciated that following gamete fusion the male pronucleus, but not the female one, exchanges its protamines for histones.[41] However, there are more profound indications of non-equivalency that imply strongly that the genetic information encoded by the male gamete is interpreted differently than that encoded by the female gamete, even when both pronuclei contain chromosomes carrying matching alleles. Two general groups of observations support this implication: those where an embryo is the product of pronuclear transfers and those where an embryo is the product of natural fertilization.

A. EMBRYOS PRODUCED BY PRONUCLEAR TRANSFER
These experiments involved the construction of fertilized oocytes containing either two paternal pronuclei (androgenetic embryos) or two maternal pronuclei (gynogenetic embryos).[44,46,47] Androgenetic embryos develop relatively normal extraembryonic tissues (ecto-placental cone, trophoblast, and yolk sac), but the embryonic portion is very poorly developed. In contrast, gynogenetic embryos develop normal embryonic structures, while failing to develop normal extraembryonic structures. Neither androgenetic nor gynogenetic embryos develop to term, although gynogenetic embryos may attain the forelimb bud stage (approximately 25 somites).

B. EMBRYOS PRODUCED BY NATURAL FERTILIZATION
Additional examples of maternal/paternal non-equivalency involve specific chromosomes

rather than the entire genome.[17,38,39,48] In otherwise genetically balanced individuals, maternal disomy 11 (two maternal chromosome 11, no paternal chromosome 11) results in newborn mice that are smaller than their normal littermates. The opposite situation (paternal disomy 11) produces newborn mice that are larger than their normal littermates.[49] Maternal disomy 6 is lethal[49] as is maternal duplication/paternal deficiency of the proximal end of chromosome 17,[50, 51] again, in both cases, in individuals that have been shown to be genetically balanced.

It has been found, however, that maternal or paternal disomy of chromosomes 1, 4, 5, 9, 13, 14, and 15, respectively, produces (apparently) normal offspring.[49, 52] Consequently, not all of the genome may be interpreted differentially by the embryo with respect to the parent of origin, although all of the genome may still prove to be imprinted differentially.

The "ovum" mutation in the mouse is exhibited by the DDK inbred strain. Female mice of this strain exhibit low fertility when mated to males of other strains, whereas reciprocal crosses (DDK males mated to females from other strains) and intrastrain DDK matings (DDK males mated to DDK females) exhibit normal fertility. Affected embryos do not develop much beyond the blastocyst stage, although in some cases they do elicit a decidual reaction.[53—55] These observations are consistent with the presence of an autosomal gene that encodes an oocyte factor that interacts specifically with a gene of sperm origin and is incompatible with sperm of non-DDK origin.[48] Pronuclear transfer experiments using DDK zygotes have shown recently that there is, indeed, a cytoplasmic oocyte factor whose interaction with the male pronucleus modifies the oocyte cytoplasm so that other female pronuclei, regardless of their strain of origin, become developmentally deficient.[56]

With the T^{hp} deletion, we are dealing with the T-complex[42,43,57—60] region on chromosome 17. In T^{hp} homozygotes, there is a failure in development shortly after the blastocyst stage.[48]

When the T^{hp} deletion is in combination with the wild-type+ allele, the outcome depends upon the origin of the parental deletion. If it is of maternal origin, $T^{hp}/+$ embryos die late in development,[61, 62] and if it is of paternal origin in a $T^{hp}/+$ embryo, an adult with a hairpin tail is the outcome. To understand the cause of this non-equivalency, aggregation chimeras were constructed from +/+ and T^{hp} (maternal)/+ embryos, transferred to foster mothers, and allowed to develop to term. The resulting newborn were mosaic in many tissues and also contained germ cells containing the deletion. However, such females failed to transmit the deletion to their offspring,[63] showing that the defect was somehow related to oogenesis and/or the oocyte. Subsequent nuclear transfer experiments have shown that this defect is of nuclear origin and not cytoplasmically inherited.[95]

The occurrence of the DDK ovum mutation and the T^{hp} deletion illustrate the concept that interactions between the paternal and maternal pronuclei are as important to the outcome of fertilization as is the genetic information that they encode. This concept will be illustrated further in subsequent sections of this chapter.

X-chromosome inactivation in the female is a situation where the embryonic interpretation of an entire chromosome is associated with the parental origin of the chromosome. However, in contrast to the DDK-ovum mutation and the T^{hp} deletion, X-chromosome inactivation exhibits tissue specificity. The somatic cells of female eutherian mammals have only one X chromosome that is transcriptionally active. Whether the paternal or maternal X chromosome is active in a given cell is determined randomly during embryogenesis,[64—67] making the female eutherian mammal a natural genetic mosaic.

The only cell type in the adult female that contains two active X chromosomes is the oocyte, which aquires this condition as a primordial germ cell entering meiosis.[68] This condition prevails during oocyte maturation; when the oocyte becomes fertilized, it receives an inactive sperm-borne X chromosome, which is activated soon after fertilization. The cleaving female embryo retains two active X chromosomes in all of its cells until blastocyst formation and differentiation of the trophectoderm.[66]

When trophectodermal cells begin to express their phenotype as transporting epithelia, one of the X chromosomes becomes inactivated in each cell. However, in cells of the ICM, both X chromosomes remain active, and as successive differentiated cell types develop from the ICM, they, in turn, exhibit inactivation of one of their X chromosomes.[67,69] The temporal correlation of X chromosome inactivation with cellular differentiation has been demonstrated convincingly using isozyme markers. These studies also indicate that X-chromosome inactivation has occurred in all of the somatic cells of the female embryo by the time of primitive streak formation.[70—72]

In contrast to the situation in the female's somatic cells, X-chromosome inactivation in her extraembryonic cells is not random and exhibits maternal/paternal non-equivalency. In these extraembryonic tissues, i.e., trophectoderm and primitive endoderm derivatives from the ICM, it is the paternal X-chromosome that is inactivated.[73—75]

Currently, no definitive role has been identified for either the random X-chromosome inactivation observed in somatic cells, or the preferential inactivation of the paternal X chromosome in the extraembryonic cells. There is some speculation that X-chromosome inactivation in the human is associated with sex determination[76] and that in general it may be necessary for gene dosage compensation. Interestingly, in marsupials, preferential inactivation of the paternal X chromosome occurs in the female's somatic cells as well as in her extraembryonic cells.[77]

At this time, there is no mechanism(s) that has been identified conclusively for either the imprinting or for the inactivation (and maintenance of inactivation) of the X chromosome. However, there is some evidence that DNA modification is involved during inactivation,[78] perhaps in a manner involving DNA methylation.[79,80] However, because X-chromosome inactivation occurs in diploid parthenotes, it may not depend upon events associated with spermatogenesis and/or with fertilization (fusion of the sperm with the oocyte).[81] This observation suggests that the mechanism of imprinting is more likely to involve the maternal X chromosome rather than the paternal one, implying that such imprinting functions to protect the maternal chromosome from being inactivated in extraembryonic tissues.[82] If this is so, then such "protection" must become modified later in development to permit the random X-chromosome inactivation that is observed in female somatic cells.[82]

C. SPECULATION ON THE MOLECULAR BASIS OF MATERNAL/PATERNAL NON-EQUIVALENCY

It has been suggested that the male and female genomes are imprinted during gametogenesis to result in their differential interpretation by the embryo.[83,46] Studies with transgenic mice have shown that the paternally inherited copy of a specific locus was undermethylated in comparison with the maternally inherited copy.[84] This difference in DNA methylation could be associated with genetic imprinting,[84,85] although additional mechanisms may be involved.

It is not known whether imprinting of the X chromosome occurs by the same mechanism that is responsible for imprinting autosomes. It is also not known why embryonic interpretation of the imprinting of the X chromosomes in extraembryonic tissues differs from that in fetal somatic tissues.

IV. CLONING MAMMALS

Finally, the phenomenon of maternal/paternal non-equivalency is related to the topic of cloning mammals. Cloning, in turn, leads us to the fourth question posed in the Introduction; namely, when do embryonic nuclei and cells lose their pluripotentiality?

A. PLURIPOTENCY OF MAMMALIAN EMBRYONIC NUCLEI

To examine how long embryonic nuclei retain pluripotency, nuclear transfer experiments have been performed wherein nuclei from blastomeres of four- or eight-cell mouse embryos or

TABLE 3
Viability of Embryos Obtained by Transferring Cleavage Stage
Nuclei into Enucleated Oocytes or Zygotes

Species	Stage of donor nucleus	Stage of recipient cytoplasm	Development	Ref.
Mouse	2 cell	Zygote	Blastocyst	51
Sheep	8 cell	Unfertilized oocyte	Term	86
Cow	9—15 cell	Unfertilized oocyte	Term	87

from ICM cells have been introduced into enucleated zygotes. These reconstituted zygotes did not reach the blastocyst stage, and when the experiment was repeated with nuclei from two-cell stage embryos, only 19% of the recipient zygotes were able to form blastocysts.[51]

It is important to point out, however, that mouse embryos are precocious in comparison with most other mammalian species, with respect to when embryonic genomic activation and birth occur (Table 1). Developmental restriction may be similarly precocious. In other species, notably sheep and cows, nuclear transfers using nuclei from later cleavage-stage blastomeres placed into enucleated oocytes have led to the birth of live young (Table 3).[86,87]

Experiments utilizing pronuclear transfers have been carried out with mouse embryos and involved the production of haploid, androgenetic zygotes (removal of female pronucleus) or haploid, gynogenetic zygotes (removal of the male pronucleus). The androgenetic and gynogenetic haploid zygotes were then permitted to develop into four- and eight-cell embryos, respectively (androgenetic embryos will only develop to the four-cell stage while gynogenetic embryos can develop into blastocysts) and used as a source of donor nuclei.

Recipient haploid zygotes were prepared from which either the male or female pronucleus was removed, providing recipient "cytoplasm". Those with a resident male pronucleus received a gynogenetic 2-, 4-, 8-, or 16-cell-stage haploid nucleus, while those with a resident female pronucleus received an androgenetic haploid 4-cell-stage nucleus. These reconstituted "zygotes" now all contained a male and a female set of chromosomes and were diploid. However, their two sets of chromosomes were developmentally asynchronous. Reconstituted zygotes were cultured overnight, during which time they cleaved and were then transferred to foster mother mice. Development to term was achieved for both types of reconstituted zygote. This experiment shows that genetic material from an embryo as advanced as the 16-cell stage is still pluripotent when placed into the cytoplasm of the zygote. This observation implies that embryonic gene activation does not necessarily abolish pluripotentiality. In addition, some degree of developmental asynchrony between cytoplasm and nucleus may be permissible for normal development.

No one has yet reported the results obtained when zygotes lacking both resident pronuclei receive a gynogenetic and an androgenetic haploid nucleus from respective cleavage stage embryos. Aggregation chimeras have been constructed, however, between diploidized androgenetic and gynogenetic haploid stage four-cell-stage embryos.[90] Following transfer to foster mothers, none of these embryos developed to term. These results suggest that cell interactions within chimeras of this genotype cannot substitute for the interaction between the two sets of parental genes that are present normally within the same cytoplasm.

B. PLURIPOTENCY OF MAMMALIAN EMBRYONIC CELLS

The developmental potential of embryonic cells has been tested experimentally by three methods: (1) by culturing individual blastomeres from different stages of cleaving embryos[90,92] (2) by rearrangements of blastomeres (or whole embryos)[93] or (3) by blastomere injection experiments where the developmental fate of progeny blastomeres could be followed.[94] The collective evidence from these experiments is that pluripotency, defined by the ability of a blastomere to develop along either the trophectoderm- or ICM-lineage, is retained by all blastomeres within the mouse embryo until sometime between the fifth and sixth cleavages.

V. MALE AND FEMALE GAMETES CONTRIBUTE DIFFERENT HERITABLE CYTOPLASMIC COMPONENTS TO THE EMBRYO: MITOCHONDRIAL INHERITANCE

The mammalian zygote inherits at least five potential sources of heritable genetic material from its parent gametes, namely, the maternal pronucleus, the paternal pronucleus, vertically transmitted retroviruses, maternal mitochondrial DNA, and paternal mitochondrial DNA. We have already discussed the participation of both pronuclei in development and will examine the role of retroviruses shortly. Here, let us consider mitochondrial DNA.

Intuitively, the chances seem slight that any of the sperm's mitochondria become incorporated into the embryonic complement of mitochondria. There are several reasons for this. First, the number of sperm mitochondria is insignificant[96] compared to the numbers of oocyte mitochondria (approximately 92,500 per mouse oocyte).[97] Second, the sperm mitochondria retain their association with the sperm tail that becomes incorporated into the oocyte and are usually sequestered within the same blastomere during cleavage. Third, for geometric reasons, the likelihood of a given blastomere residing on the outside of the embryo is greater than the likelihood of it residing on the inside of the embryo; only the inner blastomeres become the ICM, and subsequently, the embryo.

However, because mitochondria carry identifiable alleles, it has been possible to obtain evidence that mitochondria are maternally inherited in mammals; studies on the fragment patterns of mitochondrial DNA (mtDNA) purified from the tissues of many strains of laboratory mice suggest strongly that the common laboratory strains of mice are all descended from a single female. This mouse might have lived perhaps as recently as 1920 or as long ago as 3200 B.C.[98]

VI. CHIMERAS: GENETIC MOSAICS THAT ILLUSTRATE THE EXISTENCE OF EMBRYONIC BIOGENIC FACTORS

Chimeras are composite individuals in which more than one fertilized egg contributes to the developing cell population.[99] Mammalian chimeras can be produced by aggregating cleavage stage blastomeres from two or more embryos (aggregation chimeras) or by injecting blastomeres into the blastocoel of an embryo of the same or a different strain (injection chimeras). Aggregation and injection chimeras have been used to examine questions regarding cell lineage, immunological self-recognition, sex determination, and gamete formation. Tumor cell-blastocyst injection chimeras have revealed some startling behavioral properties of tumor cells placed in an embryonic milieu, while aggregation chimeras provide sensitive dosimeters of environmental genotoxic conditions. Finally, both types of chimera have been used to determine whether or not a given embryonic lethal mutation is cell-type specific, affecting a specific cellular process.[99]

In the following sections, we will examine three applications of chimeras since they provide direct evidence that the very early embryo contains biologically active factors, some of which appear to be secreted into the environment. The chemical identity is not known for most of these factors, and none has been assigned, so far, a specific function or mode of action during the initial appearance in the embryo.

A. EMBRYONAL CARCINOMA STEM CELLS WITHIN CHIMERAS
Embryonal carcinoma (EC) (teratocarcinoma) is a highly malignant carcinoma of germ-cell

origin that can kill a mouse in 6 weeks. The stem cells of the tumor (EC cells) can differentiate into cell types of all three germ-layer derivations, and their presence enables the tumor to be transplanted to syngeneic hosts.[100] If the EC cells are transplanted to an adult tissue site, they continue to grow as a malignant tumor containing various differentiated cell types and a self-renewing population of EC cells.[101] When the tumors are transplanted to the mouse blastocyst, EC cells can colonize the inner cell mass, and their progeny cells can participate in normal embryogenesis and even form normal gametes.[102–106] Similar results have been obtained by combining EC cells with cleavage-stage embryos in aggregation chimeras,[107] suggesting that the blastocoel cavity per se is not essential for the EC cell to participate in normal development.

These observations suggest that the embryo has the capacity to superimpose normal growth behavior onto the malignant phenotype. The mechanism(s) responsible for this capacity remains unknown, nor has it been established whether cell-cell contact is required for the preimplantation embryo to regulate cellular proliferation and cytodifferentiation in EC cells. However, in the following section, a situation is described wherein an embryo within an aggregation chimera can affect cell proliferation within the partner embryo.

B. CHIMERAS AS BIODOSIMETERS

When a chimera constructed of blastomeres from two different strains of mice is permitted to develop to term, it will develop into an individual containing a cellular predominance of the strain having the proliferative advantage.[108,109] Strain-related influences on cellular proliferation have been shown to be evident in interstrain chimeras prior to implantation. In chimeras constructed of a fast-growing Balb/cbyj embryo paired with a relatively slow growing CBA/HT6 embryo, there is enhancement of cleavage rate of the CBA/HT6 partner after only 40 h in culture, to the morula stage of development.[110] The enhanced proliferative rate of the CBA/HT6 partner is not dependent on cell-cell contact, suggesting that there might be strain-dependent quantitative differences in secretion rates, or in the receptor activity for biologically active factors that may influence embryonic cellular proliferation. It has already been established that preimplantation embryos do secrete factors that are "biogenic" in other sytems.[111,112] However, whether these factors are indeed biogenic in the preimplantation embryo (or to cells of the surrounding maternal tissues) has not yet been established. Additional evidence for the secretion of factors that influence proliferation comes from studies showing that developmental rates and embryonic cell numbers are enhanced under crowded conditions.[113]

If one of the partner embryos of an intrastrain aggregation chimera is treated with very low doses of ionizing radiation prior to pairing it with an untreated same-age embryo, the irradiated one will proliferate more slowly than if it had been cultured alone or paired with another embryo receiving a similar dose of radiation.[114] The same observation has been made when blastomeres prelabeled with very low concentration of tritiated thymidine are aggregated together with unlabeled same-age blastomeres.[115] In both these cases, the biogenic effect appears to be mediated via direct cell-cell contact between the cells of the normal embryo and those of the treated embryo.

Taken together, these observations are consistent with the general case where genetic alterations such as point mutations[116,117] and aneuploidies[118–120] in embryonic cells generally confer a proliferative disadvantage, which can be revealed when they are paired with normal cells in chimeras. These observations also show that the use of appropriately constructed aggregation chimeras may be exploited to reveal exogenously induced, sublethal genetic alterations, which are expressed as an alteration in cellular proliferation rate.

C. RESCUE: LETHALITY CIRCUMVENTED BY DEVELOPMENT

Chimeras can be used to rescue embryos that carry lethal embryonic mutations that usually result in death. This is accomplished by combining affected embryos with normal partners, as aggregation chimeras. Viable chimeras have been produced using parthenogenetic diploid embryos paired with normal embryos;[121,122] the parthenote not only participated in organogenesis, but their progeny cells formed normal gametes.[121] Similarly, tetraploid mouse embryos[120] and embryos homozygous for the lethal yellow Agouti (A^Y) allele[123] have been rescued in chimeric aggregates. The latter case is particularly interesting because progeny mutant cells colonized only the inner cell mass derivatives.

However, other embryo lethal mutations, specifically, the t-alleles of the T-complex and oligosyndactyly (Os) were not rescued when combined with wild-type embryos in chimeras.[48] The use of aggregation chimeras demonstrates that in some situations (T-complex, Os) result in cell autonomous lethality; in others, (Agouti locus, polyploidy) may disrupt development at an organizational rather than a cellular level.

VII. RETROVIRUSES

The genetic repertoire encoded by RNA tumor viruses or retoviruses contains genetic material (viral oncogenes), homologous to endogenous cellular genes (cellular protooncogenes) in several vertebrate classes. Several of these cellular oncogenes have been found to encode growth factors, or their receptors[124—127] and Chapter 9, this volume. Specific examples of oncogene expression modulation have been described during mouse embryonic and fetal development[128,129] and some cellular oncogene expression is thought to be associated with genes that regulate entry into the cell cycle.[130]

Two morphologically distinguishable retroviruses, intracisternal A-type and cytoplasmic C-type, have been detected in mouse preimplantation embryos.[131—135] Evidence for their participation in development has been based partly on the presence of viral particles in embryos from all strains of laboratory and feral mice investigated so far[131] and partly on the observations of stage-dependent increases and decreases in the incidence of viral particles in the the cytoplasm during cleavage. For example, "small A" particles, a subset of the intracisternal A-type viruses, are found predominantly between the two- to eight- cell stage of development, while the other subset, the "large A" particles appear to increase in number from the eight-cell stage onwards.[134—137] C-type virus particles are more prevalent after implantation, although they have been reported in the cytoplasm of cleavage-stage mouse preimplantation embryos.[137] Retroviruses and/or their cellular counterparts represent a potential vehicle for the control of cellular proliferation and growth. Information should be available soon on their role in growth regulation during mammalian development.

VIII. INTRODUCTION OF FOREIGN MATERIAL INTO THE EMBRYO

One method for the developmental consequences of experimentally manipulated genetic material is provided by the construction of transgenic embryos. This is done by introducing foreign genetic material into the nuclear DNA and allows a more detailed analysis of the role of individual genes in development. In the mouse, relatively few mutants affecting early development have been identified, and fewer still have been cloned. Nevertheless, a considerable body of literature has emerged in some areas of developmental genetics; e.g., gene therapy for the correction of genetic defects and numerous studies of the consequence of exogenous DNA integration into the genome,[138,139] much of which has been derived from the study of transgenic animals.

For a number of obvious reasons, the mouse has been the favored species for the production of transgenic animals so far. Methods used to produce these animals include (1) injection of DNA into one of the pronuclei of the fertilized ovum,[138,140,141] which has the advantage that all the cells of the transgenic animal will carry the foreign gene; (2) infection of cleavage-, (or later), stage emrbyos with viral vectors,[142,143] in which not all of the blastomeres will necessarily become infected; and (3) injection of genetically altered stem cells of teratocarcinomas into morulae[107,144] or blastocysts,[123] which so far has not resulted in the development of animals capable of producing viable gametes.

Of the numerous transgenic mice that have been produced so far, several hold potential as models for studies of growth and development. Those that carry the human growth-hormone-releasing-factor gene (which causes increased growth),[145] the gonadotropin-releasing hormone gene (which restores reproductive function in the hypogonadal mouse),[146] and the human beta-globin gene (which is expressed only in erythropoietic tissues),[147] have been instrumental in providing evidence that foreign genes are not only functional but can also be regulated endogenously in a manner similar to native genes. Transgenic mice carrying a mutant gene resulting in limb deformities[148] can provide insight into the control of morphogenesis and pattern formation. In addition to being regarded as a research vehicle, transgenic animals have been viewed as a means of enhancing the disease resistance and food production potential of agriculturally important animals.[149]

IX. CONCLUSION

It is clear from the discussion on the onset of embryonic gene expression that the embryonic genetic repertoire is essential from the onset of mammalian development. Consequently, mutations, chromosomal imbalances, and other genetic anomalies will have a potential for perturbing developmental events from fertilization onwards.

From pronuclear transfer experiments, it appears that both a male and a female pronucleus are required for normal embryogenesis. It has also been shown that the absence of the paternal autosome set, per se, is sufficient for embryo lethality of gynogenetic embryos.[150] The collective evidence from the pronuclear experiments suggests that the male pronucleus is essential for the development of extraembryonic structures, while the female pronucleus is essential for the development of the embryo itself.

This conclusion is consistent with the clinical experience regarding the hydatidiform mole tumor. These tumors are of trophoblastic origin and develop from a fertilization event that produces a zygote containing a diploid set of sperm-borne chromosomes but lacking a maternal set.[151]

The studies with chromosomes 11 and 6 and the T^{hp} deletion on chromosome 17 show that embryonic interpretation of imprinting is somehow contingent upon the parental origin of the genetic material. Embryonic interpretation of imprinting of genetic material based on parental origin can also exhibit tissue specificity, as shown by studies comparing X chromosome inactivation in extraembryonic and somatic embryonic cells of the female embryo.

Regarding the X chromosome, descriptive studies on its inactivation have led to the definition of a cell lineage for assigning relationships between progenitors and differentiated extraembryonic and embryonic germ layers.[152] From such studies, the ICM has become identified as the stem cell population of the early embryo. Observations on the temporal correlation between X-chromosome inactivation and cell differentiation have also identified the origin of primordial germ cells as the epiblast (embryonic ectoderm, the same progenitor shared by epidermis and neural tissue)[153,154] rather than endoderm, as had been traditionally believed.

From the discussions on nuclear pluripotency in mammals, it appears as though some form of cloning may be technically possible in those species where the nuclei retain pluripotentiality for 3 to 4 cleavages; e.g., the cow.[87] In addition, these experiments have led to the question of why a haploid genome exhibits a greater apparent degree of pluripotency than a diploid genome. The implication here is that an interaction between the maternal and paternal set of chromosomes (or additional pronuclear components?) may be correlated with restriction in the ability to re-initiate development when the nuclei are placed in the cytoplasm of the zygote.

This discussion on the pluripotency of mammalian cells raises another question, namely, why the nucleus from the cleavage-stage blastomere is not pluripotent when placed within the zygote cytoplasm. The pronuclear transfer experiments have shown that nuclear-cytoplasmic asynchrony is tolerated by haploid nuclei placed within zygote cytoplasm. Consequently, the answer to this question may be that the interaction between the maternal and paternal set of chromosomes limits the extent to which nuclear-cytoplasmic asynchrony is compatible with developmental pluripotency.

The discussion of experiments on mitochondrial inheritance suggests that this may contribute to maternal/paternal non-equivalence in mammalian development. There is evidence supporting this suggestion in the form of a maternally transmitted antigen, Mta, which is a murine cell surface Class I-like antigen defined by cytotoxic lymphocytic reactivity. The expression of Mta requires cooperation between genetic elements in the Qa/T1a region of chromosome 17 with genetic elements of mitochondria. In other words, the genotype of the mitochondria appears to modulate the phenotypic expression of a cell surface molecule encoded by the genotype of the nucleus.[155]

Although the applications derived from the use of transgenics appear limitless, enthusiasm for their application is tempered by the instances where transgenes have acted as mutagens, altering or destroying specific gene functions.[156] A case in point is the integration of the collagen I gene in the mouse genome, which has been shown to cause embryo-lethal mutations.[157] Findings of this nature are important for they serve to identify mutant genes of potential developmental importance which could be cloned and studied specifically for insights into the mechanisms governing early embryonic development, morphogenesis, and pattern formation.

Finally, the ability to detect sperm-borne genetic alterations as changes in cellular proliferation rates provides another research tool for investigating the genetic regulation of growth rates in genetically nonidentical embryonic cell populations. It would be interesting to determine whether these alterations involve alterations in the activities and levels of embryonic biogenic factors and/or their receptors.

REFERENCES

1. **Lewis, W. A. and Wright, E. S.,** On the early development of the mouse egg, *Contrib. Embryol. Carnegie Inst.,* 148, 115, 1935.
2. **Calarco, P. G. and Brown, E. H.,** An ultrastructural and cytological study of preimplantation development of the mouse, *J. Exp. Zool.,* 171, 253, 1969.
3. **DiZio, S. M. and Tasca, R. J.,** Sodium-dependent amino acid transport in preimplantation mouse embryos. III. Na+-K+-ATPase-linked mechanisms in blastocysts, *Dev. Biol.,* 59, 198, 1977.
4. **Gardner, R. L.,** Analyses of determination and differentiation in the early mammalian embryos using intra- and inter-specific chimaeras, in *The Developmental Biology of Reproduction,* Market, C. L. and Papaconstantinou, J., Eds., Academic Press, Orlando, FL, 1975, 207.
5. **Gardner, R. L. and Rossant, J.,** Determination during embryogenesis, in *Embryogenesis in Mammals,* CIBA Foundation Symposium 40, North/Holland, Amsterdam, 1976, 5.
6. **Davidson, E. H.,** in *Gene Activity in Early Development,* Academic Press, Orlando, FL, 1976.
7. **Woodland, H. R., Flynn, J. M., and Wyllie, A. J.,** Utilization of stored mRNA in *Xenopus* embryos and its replacement by newly synthesized transcripts: histone H1 synthesis using interspecies hybrid, *Cell.,* 18, 165, 1979.
8. **Wells, D. E., Showman, R. M., Klein, W. H., and Raff, R. A.,** A delayed recruitment of maternal histone H3 MRNA in sea urchin embryos, *Nature (London),* 292, 477, 1981.
9. **Braude, P., Pelham, H., Flach, G., and Lobatto, R.,** Post-transcriptional control in the early embryo, *Nature (London),* 282, 102, 1979.
10. **Howlett, S. K. and Bolton, V. N.,** Sequence and regulation of morphological and molecular events during the first cell cycle of mouse embryogenesis, *J. Embryol. Exp. Morphol.,* 87, 175, 1985.
11. **Piko, L. and Clegg, K., B.,** Quantitative changes in total RNA, total poly (A), and ribosomes in early mouse embryos, *Dev. Biol.,* 89, 362, 1982.

12. **Flach, J. B., Johnson, M. H., Braude, P. R., Taylor, A. S., and Bolton, V. N.,** The transition from maternal to embryonic control in the 2-cell mouse embryo, *EMBO J.,* 1, 681, 1982.

13. **Bolton, V. N., Oades, P. J., and Johnson, M. H.,** The relationship between cleavage, DNA replication, and gene expression in the mouse 2-cell embryo, *J. Embryol. Exp. Morphol.,* 79, 139, 1984.

14. **Fischer, L. K., Hausmann, B., and Chapman, V. M.,** A new H-2 linked Class I gene whose expression depends on a maternally inherited factor, *Nature (London),* 306, 383, 1983.

15. **Smith, R., Huston, M. M., Jenkins, R. N., Huston, D. P., and Rich, R. R.,** Mitochondria control expression of a murine cell surface antigen, *Nature (London),* 306, 599, 1983.

16. **McLaren, A.,** The impact of pre-fertilization events on post-fertilization development in mammals, in *Maternal Effects in Development,* Newth, O. R. and Balls, M., Eds., Cambridge University Press, London, 287, 1979.

17. **Magnuson, T.,** Genetic abnormalities and early mammalian development, in *Development in Mammals,* Vol. 5, Johnson, M. H., Ed., Elsevier Science Publishers, Amsterdam, 209, 1983.

18. **Yee, D., Golden, W., Debrot, S., and Magnuson, T.,** Short-term rescue by RNA injection of a mitotic arrest mutation that affects the preimplantation mouse embryo, *Dev. Biol.,* 122, 256, 1987.

19. **Goddard, M. J. and Pratt, H. P. M.,** Control of events during early cleavage of the mouse embryo: an analysis of the "2-cell block", *J. Embryol. Exp. Morphol.,* 73, 111, 1983.

20. **Hunter, R. H. F.,** Chronological and cytological details of fertilization and early embryonic development in the domestic pig, *Sus scrofa, Anat. Rec.,* 178, 169, 1974.

21. **Crosby, I. M., Gandolfi, F., and Moor, R. M.,** Control of protein synthesis during early cleavage of sheep embryos, *J. Reprod. Fertil.,* 82, 769, 1988.

22. **Moore, G. P. M.,** The RNA polymerase activity of the preimplantation mouse embryo, *J. Embryol. Exp. Morphol.,* 34, 291, 1975.

23. **Young, R. J., Sweeney, K., and Bedford, J. M.,** Uridine and guanosine incorporation by the mouse one-cell embryo, *J. Embryol. Exp. Morphol.,* 44, 133, 1978.

24. **Clegg, K. B. and Piko, L.,** RNA synthesis and cytoplasmic polyadenylation in the one-cell mouse embryo, *Nature (London),* 295, 342, 1982.

25. **Clegg, K. B. and Piko, L.,** Poly (A) length, cytoplasmic adenylation and synthesis of poly (A) + RNA in early mouse embryos, *Dev. Biol.,* 95, 331, 1983.

26. **Sawicki, J. A., Magnuson, T., and Epstein, C. J.,** Evidence for expression of the paternal genome in the two-cell embryo, *Nature (London),* 294, 450, 1981.

27. **Wudl, L. and Chapman, V.,** The expression of b-glucuronidase during preimplantation development of mouse embryos, *Dev. Biol.,* 48, 104, 1976.

28. **Braude, P., Bolton, V., and Moore,** Human Gene expression first occurs between the four- and eight-cell stages of preimplantation development, *Nature (London),* 332, 459, 1988.

29. **Kratzer, P. G.,** Expression of maternally and embryologically derived hypoxanthine phosphoribosyl transferase (HPRT) activity in mouse eggs and early embryo, *Genetics,* 104, 685, 1983.

30. **Muggleton-Harris, A. L. and Johnson, M. H.,** The nature and distribution of serologically detectable alloantigens on the preimplantation mouse embryo, *J. Embryol. Exp. Morphol.,* 35, 59, 1976.

31. **Brinster, R. L.,** Parental glucose phosphate isomerase activity in three-day mouse embryos, *Biochem Genet.,* 9, 187, 1973.

32. **Dyban, A. P. and Baranov, V. S.,** Influence of genes on early mammalian development, in *Cytogenetics of Mammalian Embryonic Development,* Oxford Science Publications, Clarendon Press, Oxford, England, 1987, 242.

33. **Gilbert, S. F.,** *Developmental Biology,* Sinauer Associates, Sunderland, MA 1985.

34. **Golbus, M. S., Calarco, P. G., and Epstein, C. J.,** The effects of inhibitors of RNA synthesis (α-amanitin and actinomycin D) on preimplantation mouse embryogenesis, *J. Exp. Zool.,* 186, 207, 1973.

35. **Warner, C. J. and Versteegh, L. R.,** *In vivo* and *in vitro* effect of α-amanitin on preimplantation embryo RNA polymerase, *Nature (London),* 248, 678, 1974.

36. **Epstein, C. J.,** Gene expression and macromolecular synthesis during preimplantation embryonic development, *Biol. Reprod.,* 12, 82, 1975.

37. **Chapman, V. M, West, J. D., and Adler, D. A.,** Genetics of early mammalian embryogenesis, in *Concepts in Mammalian Embryogenesis,* Sherman, M. I., Ed., MIT Press, Cambridge, 1977, 95.

38. **Magnuson, T.,** Mutations and chromosomal abnormalities: how are they useful for studying genetic control of early mammalian development?, in *Experimental Approaches to Mammalian Embryonic Development,* Rossant, J. and Pedersen, R., Eds., Cambridge University Press, London, 1986, 437.

39. **Magnuson, T. and Epstein, C. J.,** Genetic control of very early mammalian development, *Biol. Rev.,* 56, 369, 1981.

40. **McLaren, A.,** Genetics of the early mouse embryo, *Annu. Rev. Genet.,* 10, 361, 1976.

41. **Wolgemuth, D. J.,** Synthetic activities of the mammalian early embryo: molecular and genetic alterations following fertilization, in *Mechanisms and Control of Animal Fertilization,* Hartmann, J. F., Ed., Academic Press, Orlando, FL, 1983, 415.

42. **Sherman, M. I. and Wudl., L. R.,** T-complex mutations and their effects, in *Concepts in Mammalian Embryogenesis,* MIT Press, Cambridge, MA, 1977, 136.

43. **Silver, L. M.,** Genetic organization of the mouse t complex, *Cell,* 27, 239, 1981.

44. **McGrath, J. and Solter, D.,** Completion of mouse embryogenesis requires both the maternal and paternal genomes, *Cell,* 37, 179, 1984.

45. **Kaufman, M. H., Barton, S. C., and Surani, M. A. H.,** Normal postimplantation development of mouse parthenogenetic embryos to the forelimb bud stage, *Nature (London),* 265, 53, 1977.

46. **Surani, M. A. H., Barton, S. C., and Norris, M. L.,** Development of reconstituted mouse eggs suggests imprinting of the genome during gametogenesis, *Nature (London),* 308, 548, 1984.

47. **Barton, S. C., Surani, M. A. H., and Norris, M. L.,** Role of paternal and maternal genomes in mouse development, *Nature (London),* 311, 374, 1984.

48. **Magnuson, T. and Epstein, C. J.,** Genetic expression during early mouse development, in *The Mammalian Preimplantation Embryo: Regulation of Growth and Differentiation in Vitro,* Bavister, B. D., Ed., Plenum Press, New York, 1987, 133.

49. **Cattanach, B. M. and Kirk, M.,** Differential activity of maternally and paternally derived chromosome regions in mice, *Nature (London),* 315, 496, 1985.

50. **Lyon, M. F. and Glenister, P. H.,** Factors affecting the observed number of young resulting from adjacent-2 disjunction in mice carrying a translocation, *Genet. Res.,* 29, 83, 1977.

51. **McGrath, J. and Solter, D.,** Inability of mouse blastomere nuclei transferred to enucleated zygotes to support development *in vitro, Science,* 226, 1317, 1984.

52. **Lyon, M. F., Ward, H. C., and Simpson, G. M.,** A genetic method for measuring non-disjunction in mice with Robertsonian translocations, *Genet. Res.,* 26, 283, 1976.

53. **Wakasugi, N.,** Studies on fertility of DDK mice: reciprocal crosses between DDK and C57BL/6J strains and experimental transplantation of the ovary, *J. Reprod. Fertil.,* 33, 283, 1973.

54. **Wakasugi, N.,** A genetically determined incompatiblity system between spermatozoa and eggs leading to embryonic death in mice, *J. Reprod. Fertil.,* 41, 85, 1974.

55. **Wakasugi, N., Tomita, T., and Kondo, K.,** Differences of fertility in reciprocal crosses between inbred strains of mice: DDK, KK, and NC, *J. Reprod. Fertil.,* 13, 41, 1967.

56. **Renard, J. P. and Babinet, C. H.,** Identification of a paternal developmental effect on the cytoplasm of one-cell-stage mouse embryos, *Proc. Natl. Acad. Sci. U.S.A.,* 83, 6883, 1986.

57. **Dobrovolskaia-Zavadskaia, N.,** Sur la mortification spontanee de la queue chez le souris nouveau-nee et sur l'existence d'un charactere (facteur) hereditaire "non-viable", *C. R. Seances Soc. Biol.,* Paris, 97, 114, 1927.

58. **Bennett, D.,** The t-locus of the mouse, *Cell,* 6, 441, 1975.

59. **Klein, J. and Hammerberg. C.,** The control of differentiation by the t complex, *Immunol. Rev.,* 33, 70, 1977.

60. **Wudl, L. R. Sherman, M. I., and Hillman, N.,** Nature of lethality of t mutations in embryos, *Nature, (London),* 270, 137, 1977.

61. **Johnson, D. R.,** Hairpin-tail: a case of post-reductional gene action in the mouse egg?, *Genetics,* 76, 795, 1974.

62. **Johnson, D. R.,** Further observations on the hairpin-tail (T^{hp}) mutation in the mouse, *Genet. Res.,* 24, 207, 1975.

63. **Bennett, D.,** Rescue of a lethal T/t locus genotype by chimaerism with normal embryos, *Nature (London),* 272, 539, 1978.

64. **Lyon, M.,** Chromosomal and subchromosomal inactivation, Annu. Rev. *Genet.,* 2, 31, 1968.

65. **Lyon, M. F.,** Mechanisms and evolutionary origins of variable X-chromosome activity in mammals, *Proc. R. Soc. London,* B187, 243, 1974.

66. **Martin, G. R.,** X-Chromosome inactivation in mammals, *Cell,* 29, 721, 1982.

67. **Monk, M.,** Biochemical studies on mammalian X-chromosome activity, in *Development in Mammals,* Johnson, M. H., Ed., Elsevier/North Holland, New York, 1978, 189.

68. **Monk, M. and McLaren, A.,** X-chromosome activity in foetal germa cells of the mouse, *J. Embryol. Exp. Morphol.,* 63, 75, 1981.

69. **Monk, M. and Kathuoria, H.,** Dosage compensation for an X-linked gene in preimplantation mouse embryos, *Nature, (London),* 270, 599, 1977.

70. **Monk, M. and Harper, M.,** Sequential X-chromosome inactivation coupled with cellular differentiation in early mouse embryos, *Nature (London),* 281, 311, 1979.

71. **Johnson, M. M.,** X-chromosome inactivation and the control of gene expression, *Nature (London),* 296, 493, 1982.

72. **Rastan, S.,** Timing of X-chromosome inactivation in post implantation mouse embryos, *J. Embryol. Exp. Morphol.,* 71, 11, 1982.

73. **Takagi, N. and Sasaki, M.,** Preferential inactivation of the paternally derived X-chromosome in the extra embryonic membrane of the mouse, *Nature (London),* 256, 640, 1975.

74. **Wake, N., Takaki, N., and Sasaki, M.,** Non-randon inactivation of X-chromosome in the rat yolk sac, *Nature (London),* 262, 580, 1976.

75. **Harper, M., Fosten, M., and Monk, M.,** Preferential paternal X inactivation in extraembryonic tissues of early mouse embryos, *J. Embryol. Exp. Morphol.,* 67, 127, 1982.

76. **Chandra, H. S.,** Is human X-chromosome inactivation a sex-determining device?, *Proc. Natl. Acad. Sci. U.S.A.,* 82, 6947, 1985.

77. **Cooper, D. W.,** in *Isozymes. III. Developmental Biology,* Markert, L. C., Ed., Academic Press, Orlando, FL, 1975, 559.

78. **Chapman, V. M., Kratzer, P. C., Siracusa, L. D., Quarantillo, B. A., Evans, R., and Liskay, R. M.,** Evidence for DNA modification in the maintenance of X-chromosome inactivation of adult mouse tissues, *Proc. Natl. Acad. Sci. U.S.A.,* 79, 5357, 1982.

79. **Mohandas, T., Sparks, R. S., and Shapiro, L. J.,** Reactivation of an inactive human X-chromosome: evidence for X inactivation by DNA methylation, *Science,* 211, 393, 1981.

80. **Yen, P. H., Patel, P., Chinault, A. C., Mohandas, T., and Shapiro, L. J.,** Differential methylation of hypoxanthine phosphoribosyltransferase genes on active and inactive human X chromosomes, *Proc. Natl. Acad. Sci U.S.A.,* 81, 1759, 1984.

81. **Kaufman, M. H., Guc-Cubrilo, M., and Lyon, M. F.,** X-chromosome inactivation in dipolid parthenogenetic mouse embryos, *Nature (London),* 271, 547, 1978.

82. **Lyon, M. F. and Rastan, S.,** Paternal source of chromosome imprinting and its relevance for X chromosome inactivation, *Differentiation,* 26, 63, 1984.

83. **Chandra, H. S. and Brown, S. W.,** Chromosome imprinting and the mammalian X chromosome, *Nature (London),* 253, 165, 1975.

84. **Reik, W., Collick, A., Norris, M. L., Barton, S. C., and Surani, M. A.,** Genomic imprinting determines methylation of parental alleles in transgenic mice, *Nature (London),* 328, 248, 1987.

85. **Sapienza, C., Peterson, A. C., Rossant, J. and Balling, R.,** Degree of methylation of transgenes is dependent of gamete of origin, *Nature (London),* 328, 251, 1987.

86. **Willadsen, S. M.,** Nuclear transplantation in sheep embryos, *Nature* (London), 320, 63, 1986.

87. **Prather, R. S., Barnes, F. L., Sims, M. M., Robl, J. M., Eyestone, W. H., and First, N. L.,** Nuclear transplantation in the bovine embryo: assessement of donor nuclei and recipient oocyte, *Biol. Reprod.,* 37, 859, 1987.

88. **Surani, M. A. H., Barton, S. C., and Norris, M. L.,** Nuclear transplantation in the mouse: heritable differences between parental genomes after activation of the embryonic genome, *Cell,* 45, 127, 1986.

89. **Barra, J. and Renard, J. -P.,** Diploid mouse embryos constructed at the late 2-cell stage from haploid parthenotes and androgenotes can develop to term, *Development,* 102, 773, 1988.

90. **Surani, M. A. H., Barton, S. C., and Norris, M. L.,** Influence of parental chromosomes on spatial specificity in androgenetic ↔ parthenogenetic chimaeras in the mouse, *Nature (London),* 326, 395, 1987.

91. **Tarkowski, A. K.,** Experimental studies on regulations in the development of isolated blastomeres of mouse eggs, *Acta Theriol.,* 3, 191, 1959.

92. **Tarkowski, A. K. and Wroblewska, J.,** Development of blastomeres of mouse eggs isolated at the 4- and 8-cell stage, *J. Embryol. Exp. Morphol.,* 18, 155, 1967.

93. **Hillman, N., Sherman, M. I., and Graham, C.,** The effect of spatial arrangement on cell determination during mouse developement, *J. Embryol. Exp. Morphol.,* 28, 263, 1972.

94. **Pedersen, R. A., Wu, K., and Balakier, H.,** Origin of the inner cell mass in mouse embryos: cell lineage analysis by microinjection, *Dev. Biol.,* 117, 581, 1986.

95. **McGrath, J. and Solter, D.,** Maternal T^{hp} lethality in the mouse is a nuclear, not cytoplasmic, defect, *Nature* (London), 308, 550, 1984.

96. **Szollosi, D. G.,** The fate of sperm middle-piece mitochondria in the rat egg, *J. Exp. Zool.,* 159, 367, 1965.

97. **Piko, L. and Matsumoto, L.,** Number of mitochondria and some properties of mitochondrial DNA in the mouse egg, *Dev. Biol.,* 49, 1, 1976.

98. **Ferris, S. D., Sage, R. D., and Wilson, A. C.,** Evidence from mtDNA sequences that common laboratory strains of inbred mice are descended from a single female, *Nature (London),* 295, 163, 1982.

99. **Le Douarin, N., and McLaren, A.,** *Chimeras in Developmental Biology,* Academic Press, London, 1984.

100. **Martin, G. R. ,** Teratocarcinomas and mammalian embryogenesis, *Science,* 209, 768, 1980.

101. **Damjanov, I. and Solter, D.,** Experimental teratoma, in *Current Topics in Pathol.,* 59, 69, 1974.

102. **Brinster, R. L.,** The effect of cells transferred into the mouse blastocyst on subsequent development, *J. Exp. Med.,* 140, 1049, 1974.

103. **Mintz, B. and Illmensee, K.,** Normal genetically mosaic mice produced from malignant teratocarcinoma cells, *Proc. Natl. Acad. Sci. U.S.A.,* 72, 3585, 1975.

104. **Mintz, B., Illmensee, K., and Gearhart, J. D.,** Developmental and experimental potentialities of mouse teratocarcinoma cells from embryoid body cores, in *Teratomas and Differentiation,* Sherman, M. I. and D. Solter, Eds., Academic Press, Orlando, FL, 1975, 59.

105. **Stewart, C. L. and Mintz, B.,** Successive generations of mice provided from an established culture line of euploid teratocarcinoma cells, *Proc. Natl. Acad. Sci. U.S.A.,* 78, 6314, 1981.

106. **Bradley, A., Evans, M. Kaufman, M. H., and Robertson, E.,** Formation of germ line chimaeras from embryo derived teratocarcinoma cell lines, *Nature (London),* 309, 255, 1984.

107. **Stewart, C. L.,** Formation of viable chimaeras by aggregation between teratocarcinomas and preimplantation mouse embryos, *J. Embryol. Exp. Morphol.,* 67, 167, 1982.

108. **Mintz, B. and Palm, J.,** Gene control of haematopoiesis. I. Erythrocyte mosaicism and permanent immunological tolerance in allophenic mice, *J. Exp. Med.,* 129, 1013, 1969.

109. **West, J. D.,** Red blood cell selection in chimaeric mice, *Exp. Hematol.,* 5, 394, 1977.

110. **Obasaju, M. F., Wiley, L. M., and Overstreet, J. W.,** Cell proliferation in mouse embryo aggregation chimeras, submitted.

111. **Fishel, S. B. and Surani, M. A. H.,** Evidence for the synthesis and release of a glycoprotein by mouse blastocysts, *J. Reprod. Fertil.,* 59, 181, 1980.

112. **Choroszewska, A. S. and Mankowske, E.,** Ability of mouse embryos to degranulate mast cells *in vitro, J. Reprod. Fertil.,* 75, 95, 1985.

113. **Wiley, L. M., Yamami, S., and Van Muyden, D.,** Effect of postassium concentration, type of protein supplement, and embryo density on mouse preimplantation development *in vitro, Fertil. Steril.,* 45, 111, 1986.

114. **Obasaju, M. F., Wiley, L. M., Oudiz, D. J., Miller, L., Samuels, S. J., Chang, R, J., and Overstreet, J. W.,** An assay using embryo aggregation chimeras for the detection of non-lethal changes in X-irradiated mouse preimplantation embryos, *Radiat. Res.,* 113, 289, 1988.

115. **Kelly, S. J. and Rossant, J.,** The effect of short-term labelling in [^{3}H] thymidine on the viability of mouse blastomeres: alone and in combination with unlabeled blastomeres, *J. Embryol. Exp. Morphol.,* 35, 95, 1976.

116. **Mintz, B.,** Formation of genetically mosaic mouse embryos and early development of "lethal" (t^{12}/t^{12})-normal-mosaics, *J. Exp. Zool.,* 61, 273, 1964.

117. **Spiegleman, M.,** Fine structure of cells in embryos chimeric for mutant genes at the T/t locus, in *Genetic Mosaics and Chimaeras in Mammals,* Russell, L. B., Ed., Plenum Press, New York, 1978, 59.

118. **Epstein, C. J., Smith, S. A., Zamora, T., Sawicki, J. A., Magnuson, T. R., and Cox, D. R.,** Production of viable adult trisomy A ↔ diploid mouse chimeras, *Proc. Natl. Acad. Sci. U.S.A.,* 79, 4376, 1982.

119. **Magnuson, T., Smith, S., and Epstein, C. J.,** The development of monosomy in 19 mouse embryos, *J. Embryol. Exp. Morphol.,* 69, 223, 1982.

120. **Lu, T. and Markert, C. L.,** Manufacture of diploid-tetraploid chimeric mice, *Proc. Natl. Acad. Sci. U.S.A.,* 77, 6012, 1980.

121. **Stevens, L. C., Varnum, D. S., and Eicher, E. M.,** Viable chimeras produced from normal and parthenogenetic mouse embryos, *Nature (London),* 269, 517, 1977.

122. **Stevens, L. C.,** Totipotent cells of parthenogenetic origin in a chimaeric mouse, *Nature (London),* 276, 266, 1978.

123. **Papaioannou, V. E. and Gardner, R. L.,** Investigation of the lethal yellow A^y/A^y embryo using mouse chimeras, *J. Embryol. Exp. Morphol.,* 52, 1, 1979.

124. **Heldin, C. H. and Westermark, B.,** Growth factors: mechanism of action and relation to oncogenes, *Cell,* 37, 9, 1984.

125. **Bishop, J. M.,** Viral oncogenes, *Cell,* 42, 23, 1985.

126. **Klein, G. and Klein, E.,** Evolution of tumours and the impact of molecular oncology, *Nature (London),* 315, 190, 1985.

127. **Muller, R.,** Proto-oncogenes and differentiation, *Trends Biochem. Sci.,* 11, 129, 1986.

128. **Slamon, D. J. and Cline, M. J.,** Expression of cellular oncogenes during embryonic and fetal development of the mouse, *Proc. Natl. Acad. Sci. U.S.A.,* 81, 7141, 1984.

129. **Adamson, E. D.,** Oncogenes in development, *Development,* 99, 449, 1987.

130. **Ferrari, S. and Baserga, R.,** Oncogenes and cell cycle genes, *Bioessays,* 7, 9, 1987.

131. **Calarco, P. and Szollosi, D.,** Intracisternal A particles in ova and preimplantation stages of the mouse, *Nature (London), New Biol.,* 243, 124, 91, 1973.

132. **Biczysko, W., Pienkowski, M., Solter, D., and Koprowski, H.,** Virus particles in early mouse embryos, *J. Natl. Cancer Inst.,* 51, 1041, 1973.

133. **Piko, L.,** Immunocytochemical detection of a murine leukaemia virus-related nuclear antigen in mouse oocytes and early embryos, *Cell,* 12, 697, 1977.

134. **Huang, T. T. I., Jr. and Calarco, P. C.,** Evidence for the cell surface expression of intracisternal A particle-associated antigens during early mouse development, *Dev. Biol.,* 82, 388, 1981.

135. **Wival, N. A. and Smith, G. H.,** Distribution of intracisternal A-particles in a variety of normal and neoplastic mouse tissues, *Int. J. Cancer,* 7, 167, 1971.

136. **Calarco, P.,** Intracisternal A particle formation and in preimplantation mouse embryos, *Biol. Reprod.,* 12, 448, 1975.
137. **Chase, D. and Piko, L.,** Expression of A- and C-type particles in early mouse embryos, *J. Natl. Cancer Inst.,* 51, 1971, 1973.
138. **Gordon, J. W. and Ruddle, F. H.,** DNA-mediated genetic transformation of mouse embryos and bone marrow — a review, *Gene,* 33, 121, 1985.
139. **Wagner, E. F., Ruther, U., and Stewart, C. L.,** Gene transfer into mouse stem cells, in *Biotechnology: Potentials and Limitations, Dahlem Workshop Report, Life Sciences Report,* 35, Silver, S., Ed., Springer-Verlag, Berlin, 1986, 185.
140. **Palmiter, R. D. and Brinster, R. L.,** Transgenic mice, *Cell,* 14, 343, 1985.
141. **Gordon, J. W. and Ruddle, F. H.,** Gene transfer into mouse embryos: production of transgenic mice by pronuclear injection, *Methods Enzymol.,* 101, 411, 1983.
142. **Jaenisch, R.,** Germ line intergration and mendelian transmission of the exogenous Moloney leukaemia virus, *Proc. Natl. Acad. Sci. U.S.A.,* 73, 1260, 1976.
143. **Jaenisch, R., Janner, D., Nobis, P., Simon, L., Lohler, J., Harbers, K., and Grotkopp, D.,** Chromosomal position and activation of retroviral genomes inserted into the germ line of mice, *Cell,* 24, 519, 1981.
144. **Fuji, J. T. and Martin, G. R.,** Developmental potential of teratocarcinoma stem cells in utero following aggregation with cleavage stage embryos, *J. Embryol. Exp. Morphol.,* 74, 79, 1983.
145. **Hammer, R. E., Brinster, R. L., Rosenfeld, M. G., Evans, R. M., and Mayo, K. E.,** Expression of human growth hormone-releasing factor in transgenic mice results in increased somatic growth, *Nature (London),* 315, 413, 1985.
146. **Mason, A. J., Pitts, S. L., Nikolics, K., Szonyi, E., Wilcox, J. N., Seegurg, P. H., and Stewart, T. A.,** The hypogonadal mouse: reproductive functions restored by gene therapy, *Science,* 234, 1372, 1986.
147. **Magram, J., Chada, K., and Constantini, F.,** Developmental regulation of a cloned adult β-globin gene in transgenic mice, *Nature (London),* 315, 338, 1985.
148. **Woychik, R. P., Stewart, T. A., Davis, L. G., D'Evstachio, P., and Leder, P.,** An inherited limb deformity created by insertional mutagenesis in a transgenic mouse, *Nature (London),* 318, 6041, 1985.
149. **Hammer, R. E., Pursel, V. G., Rexroad, C. E., Jr., Wall, R. J., Bolt, D. J., Ebert, K. M., Palmiter, R. D., and Brinster, R. L.,** Production of transgenic rabbits sheep and pigs by microinjection, *Nature (London),* 315, 680, 1985.
150. **Lushbaugh, C. C. and Ricks, R. C.,** Some cytokinetic and histopathologic considerations of irradiated male and female gonadal tissues, *Radiat. Ther. Oncol.,* 6, 228, 1972.
150a. **Mann, J. R. and Lovell-Badge, R. H.,** The development of XO gynogenetic mouse embryos, *Development,* 99, 411, 1987.
151. **Edwards, R. G.,** Growth of the foetus after implantation until the first missed period, in *Conception in the Human Female,* Academic Press, London, 1980, 892.
152. **Monk, M.,** A stem-line model for cellular and chromosomal differentiation in early mouse development, *Differentiation,* 19, 71, 1981.
153. **McMahon, A., Fosten, M., and Monk, M.,** Random X-chromosome inactivation in female primordial germ cells in the mouse, *J. Embryol. Exp. Morphol.,* 64, 251, 1981.
154. **Gardner, R. L., Lyon, M. F., Evans, E. P., and Burtenshaw, M. D.,** Clonal analysis of X-chromosome inactivation and the origin of the germ line in the mouse embryo, *J. Embryol. Exp. Morphol.,* 88, 349, 1985.
155. **Huston, M. M., Smith, R., III, Hull, R., Huston, D. P., and Rich, R. R.,** Mitochondrial modulation of maternally transmitted antigen: analysis of cell hybrids, *Proc. Natl. Acad. Sci. U.S.A.,* 82, 3286, 1985.
156. **Westphal, H.,** Transgenic mice, *Bioessays,* 6, 73, 1987.
157. **Lohler, J. R., Timpl, R., and Jaenisch, R.,** Embryonic lethal mutation in mouse collagen I gene causes rupture of blood vessels and is associated with erythropoietic and mesenchymal cell death, *Cell,* 38, 597, 1984.

Chapter 2

CONTROL OF CARBOHYDRATE METABOLISM IN PREIMPLANTATION MAMMALIAN EMBRYOS

J. D. Biggers, D. K. Gardner, and H. J. Leese

TABLE OF CONTENTS

I. INTRODUCTION

This review focuses on the mouse preimplantation embryo, in which the requirements for growth and development are better defined than in other mammalian species (see Bavister[1] for chapters on the culture of different mammalian species). During the first 3 d following fertilization, the early mouse embryo travels down the oviduct. Implantation occurs on days 4 to 5, after the embryo has spent a further 2 d "free-living" in the uterus. Thus the embryo is exposed to a changing environment provided by the fluids formed in the oviduct and uterus during the first 5 d after fertilization (Figure 1). The composition of oviductal fluids has been reviewed recently by Leese.[2] Little information is available on the composition of the fluids of the uterus during preimplantation development.

Unusual findings obtained during the formulation of chemically defined media that could substitute for the natural fluids in culture experiments led to an interest in the metabolism of preimplantation mammalian embryos. Perhaps the most surprising finding that has emerged from these *in vitro* nutritional studies is that Krebs Ringer bicarbonate supplemented with glucose does not support development of the mouse preimplantation embryo until the eight-cell stage.[4] However, when calcium lactate was substituted for calcium chloride, the medium was found to support development from the two-cell stage.[5] Subsequent nutritional studies, stimulated by this serendipitous finding, led to the discovery that the energy (or carbon) sources change during oogenesis and the initial stages of development.[6-10] The results are summarized in Figure 2.

Only pyruvate or oxaloacetate can support the first and subsequent cleavage divisions. The second and subsequent cleavage divisions can also be supported by lactate. It is not until the eight-cell stage that glucose and several other carbon sources can support development. Later work suggests that it is during oogenesis that the oocyte is limited to a dependence on pyruvate.[11] Another important observation made in the development of methods for culturing the mouse preimplantation embryo is that phosphate cannot replace bicarbonate as a buffer.[12]

These findings, in combination with the consistent observation that development of the mouse embryo *in vitro* lags behind that *in vivo,* have stimulated considerable research into the metabolism of preimplantation mouse embryos. Clearly, the oviductal environment must contribute factors that influence development, and yet these factors remain unidentified. In this context, peptide hormones have received increasing attention as mediators of proliferation and differentiation, although their presence in the reproductive tract remains to be established. Most studies of preimplantation metabolism have focused on the demonstration of the existence of classical metabolic pathways, and the concentration of their constituent enzymes and substrates.[13—17] The description of metabolic pathways, however, is only a prerequisite for explaining the apparently unique physiological processes suggested by the *in vitro* nutritional studies. Ultimate interest centers on the control of flux through a complex network of metabolic pathways.[18,19] It is this aspect of the metabolism of preimplantation embryos that will be emphasized in this review.

II. METHODS FOR THE ANALYSIS OF ENZYMES AND METABOLITES IN SINGLE PREIMPLANTATION MAMMALIAN EMBRYOS

Technical limitations to physiological studies of the control mechanisms of metabolism in the initial stages of mammalian development have been (1) the paucity of embryos that can be made available at any given stage and (2) the insufficient sensitivities of the methods required for the analysis of single or small groups (<5) of embryos. As a result, the majority of studies have used batches of synchronized embryos at the appropriate stages. The heterogeneity which exists between embyros is assumed to be distributed randomly between treatment groups when

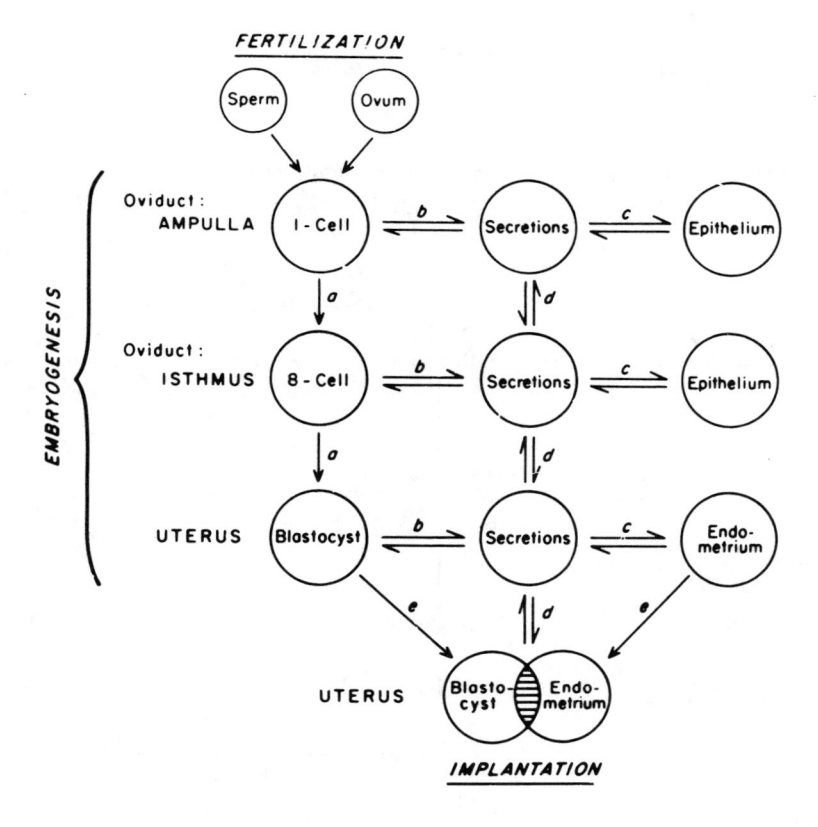

FIGURE 1. A theoretical model showing the changes between various compartments involved in the establishment of pregnancy. (From Biggers, J. D., *Cellular and Molecular Aspects of Implantation,* Glasser, S. R. and Bullock, D. W., Eds., Plenum Publishing, New York, 1981, 39. With permission.)

comparisons are made between treatments. The experimental strategy has the defect that fast changes may be masked when measurements are made on grouped embryos.

Both invasive and non-invasive methods of analysis of single embryos are now becoming available. These methods permit the study of heterogeneity between embryos. They also provide an ideal experimental situation which allows the design of comparative experiments with few embryos. This capability is particularly important since it makes possible the study of species where the production of large batches of synchronous embryos is impractical, such as some large domestic species and primates.

The ultramicrochemical methods that can be used for the study of single mammalian ova and embryos may be grouped under the following four headings:

1. Fluorometric methods
2. Bioluminescence methods based on the luciferase enzyme
3. Radiochemical methods
4. Cytophotometric methods

A. FLUOROMETRIC ASSAYS

Fluorometric assays are usually based on the generation or consumption of the reduced pyridine nucleotides NADH or NADPH. For example, the activity of the enzyme lactic acid dehydrogenease (LDH, E.C. 1.1.1.27) may be measured by monitoring the rate of NADH oxidation in the following reaction:

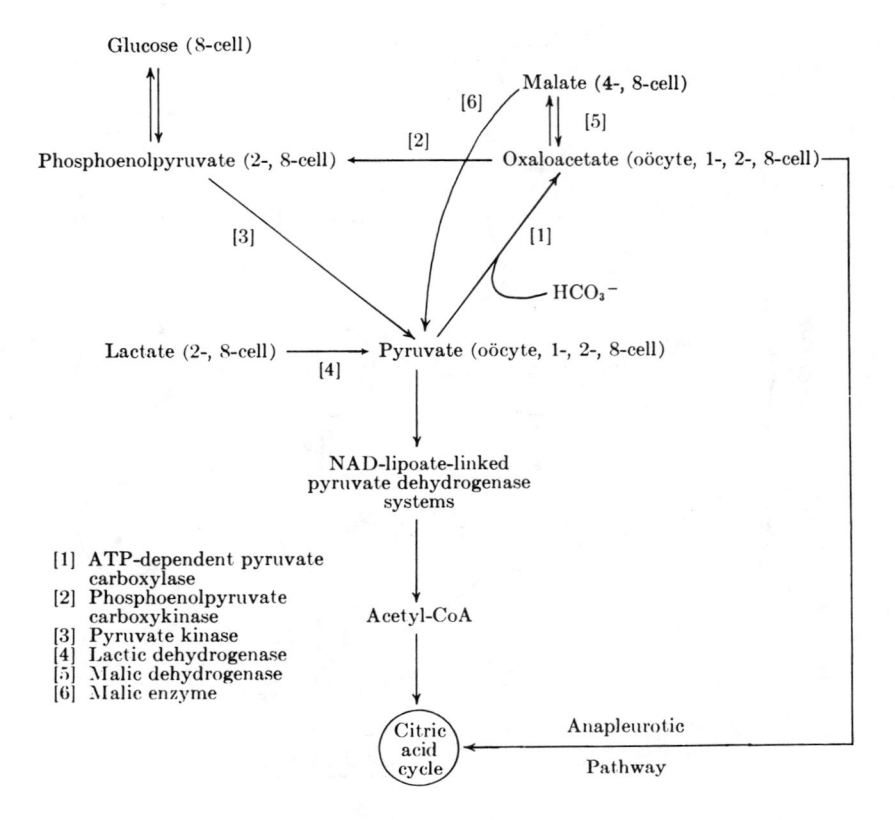

FIGURE 2. The compounds able to act as sources of energy to mouse oocytes and early cleavage stages and their metabolic relations. The stages supported by each compound are shown in parentheses. (From Biggers et al., *Proc. Natl. Acad. Sci. U.S.A.*, 58, 560, 1967. With permission.)

$$\text{pyruvate} + \text{NADH} + \text{H}^+ \leftrightarrow \text{lactate} + \text{NAD}^+$$

By varying conditions the same reaction may be used to measure the concentrations of pyruvate, lactate, NAD^+, and NADH concentrations. Compendia of assays based on such reactions have been provided by Lowry and Passoneau[20] and Bergmeyer and Gawehn.[21]

There are two main problems which have to be overcome in order to adapt these methods for the analysis of enzymes and metabolites in single oocytes or single preimplantation embryos. First, the oocytes or embryos have to be extracted in small, nanoliter volumes of medium so as not to dilute their content of enzymes or metabolites excessively. Second, sensitive methods are required to detect small quantities of NAD(P)H. Oliver H. Lowry et al.[22,23] overcame these problems by freeze-drying single embryos on microscope slides prior to extraction and amplifying the NAD(P)H formed in the subsequent assays by the technique of enzymatic cycling. For example, the NAD^+ formed in a given reaction may be amplified by the following reactions:

$$\text{NAD}^+ + \text{lactate} \leftrightarrow \text{NADH} + \text{H}^+ + \text{pyruvate} \tag{1}$$

$$\text{NADH} + \alpha\text{-ketoglutarate} + \text{NH}_4^+ \leftrightarrow \text{NAD}^+ + \text{glutamate} \tag{2}$$

These reactions result in the production of 6,000 to 8,000 times more glutamate than the original NAD^+. The glutamate may then be measured separately by reversing reaction 2.

Leese et al.[24] adopted a different approach based on the ultramicrofluorometric techniques devised by Mroz and Lechene.[25] Because of the possibility that the concentrations of ATP, ADP and AMP regulate the activities of the glycolytic pathway and the TCA cycle,[15] a method was devised which measured all three compounds simultaneously on a single mouse embryo. These methods are a miniaturization of the conventional techniques of enzymatic analysis, but with a sensitivity that eliminates the need for enzyme cycling. Instead of being done in cuvettes, the reactions are carried out in nano- and picoliter-sized droplets of medium on siliconized microscope slides under a layer of mineral oil. The NAD(P)H is quantified by a fluorescence microscope with photomultiplier and photometer attachments. Micropipettes in the nanoliter and picoliter range have to be constructed using a microforge. Such micropipettes may also be used to dispense the small volumes of medium required for the extraction of metabolites from single oocytes and embryos.[24] Because the enzymes in these small droplets may become denatured at the oil/aqueous interface, the concentration of the coupling enzymes used in metabolite assays are increased above those used in conventional assays. When enzymes, as opposed to metabolites, need to be determined, the reactions may be carried out under the coverslip of a hemocytometer slide,[26,27] thus avoiding the use of oil.

Ultramicrofluorometric assays can also be developed for the analysis of single oocytes or embryos using other fluorochromes. Thus Wudl and Chapman[28] measured the activities of beta-glucuronidase in different stages of mouse preimplantation development. The method depended on the hydrolysis of 4-methylumbelliferyl-β-D-glucuronic acid to give the fluorescent product 4-methylumbelliferone.

B. BIOLUMINESCENCE METHODS

Bioluminescence methods have been used to measure the amounts of ATP, ADP, and AMP in mouse preimplantation mouse embryos. The method is based on the following reaction:

$$\text{ATP} + \text{D-luciferin} + O_2 \xrightarrow[\text{Mg}^{2+}]{\text{firefly luciferase}} \text{oxyluciferin} + PP_i + \text{AMP} + CO_2 + \text{light}$$

The intensity of the light produced is proportional to the ATP concentration and may be quantified using a luminometer. Although luciferase is specific for ATP, the adenine nucleotides ADP and AMP may also be measured by this method after enzymatic conversion into ATP. These methods have been used by Quinn and Wales[29] and Spielmann et al.[30] to measure the adenylate nucleotide concentrations in preimplantation mouse embryos.

C. RADIOCHEMICAL METHODS

Wales[31] and Wales et al.[32] have examined the metabolism of single mouse and human embryos by incubating them with U-^{14}C glucose and DL (1-^{14}C) lactic acid at very high specific activities. The method is an adaptation of the radiochemical techniques of Brinster[33,34] and Wales and Whittingham.[35] The embryos are incubated in approximately 500 nl of medium in an appropriate chamber which allows the CO_2 evolved to be collected. The method may be limited in its application to the study of single embryos by the need to use substrates at very high specific activities. This limits the amount of unlabeled substrate which may be added and hence the total concentration of substrate in the incubation medium. There is also the risk of possible isotope effects inducing deleterious changes in the embryos. For example, Wiebold and Anderson[36] have shown that tritiated amino acids can impair the development of early mouse embryos.

D. CYTOPHOTOMETRIC METHODS

De Schepper et al.[37] have described a quantitative method for measuring the activity of enzymes in single preovulatory mouse oocytes. The oocytes are incorporated into a poly-

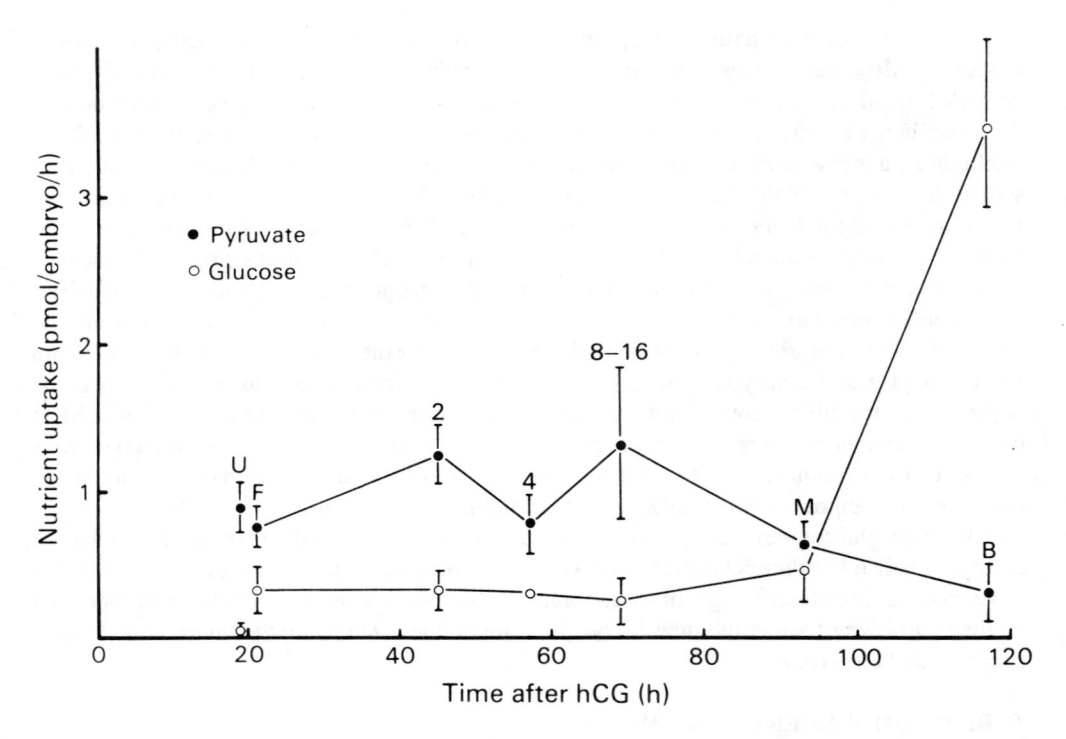

FIGURE 3. The uptake of pyruvate and glucose by (U) unfertilized and (F) fertilized mouse ova, and the (2) 2-cell, (4) 4-cell, (8-16) 8-16-cell, (M) morula and (B) blastocyst stages of development. Values are mean ± SEM of at least 6 determinations except for the 4-cell values, which are the means of 3 (pyruvate) and 2 (glucose) determinations. (From Leese, H. J. and Barton, A. M., *J. Reprod. Fertil.*, 72, 9, 1984. With permission.)

acrylamide matrix to preserve their morphology, prior to appropriate enzyme cytochemical staining. The electrons from the NADH or NAD(P)H formed are transferred via 1-methoxyphenazine methosulfate to tetranitroblue tetrazolium, the absorbance of which is measured with a scanning and integrating cytophotometer. The values for enzyme activity correlated well with those obtained using standard biochemical assays performed on 20 to 50 oocytes. An analogous method has been published by Van Noorden et al.[38]

III. UPTAKE AND UTILIZATION OF GLUCOSE AND PYRUVATE BY THE PREIMPLANTATION MOUSE EMBRYO

The fact that glucose cannot support the development of mouse preimplantation embryos prior to the eight-cell stage may be explained in part by the observation that glucose uptake by the embryos is low during this embryonic period.[39] Using a non-invasive ultramicrofluorescence method, and groups of one to five freshly recovered mouse embryos, Leese and Barton measured the uptake of both glucose and pyruvate (Figure 3). Prior to the eight-cell stage, the uptake of pyruvate is high and the uptake of glucose is low. Just before the eight-cell stage, the uptake of pyruvate falls and that of glucose increases rapidly. Similar changes have also been observed in one- and two-cell mouse embryos cultured *in vitro*.[40] The crossover point of the uptakes of glucose and pyruvate occurs around the time of compaction, at which time the embryo acquires the ability to control the composition of its extracellular fluid.[41] It is at this time that energy-dependent vectorial transport processes develop which form the blastocoel fluid.

Two hypotheses can be proposed to explain why pyruvate is the preferred substrate up to the eight-cell stage and why glucose is the preferred substrate after this stage of development. The

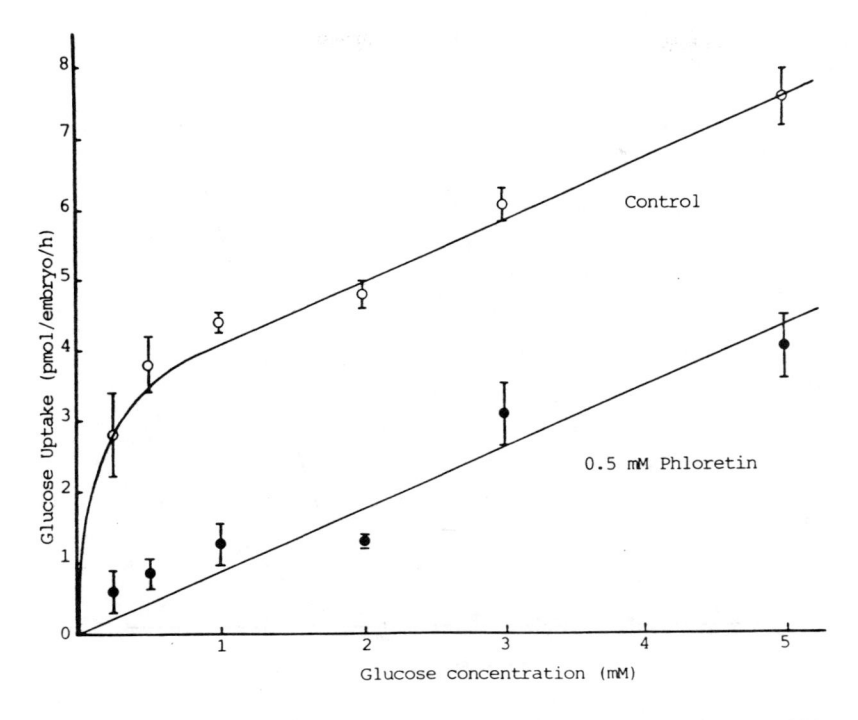

FIGURE 4. Glucose uptake by single mouse blastocysts in the presence and absence of 0.5 m*M* phloretin. Values are mean ± SEM of 6 embryos. (From Gardner D. K. and Leese, H. J., *Development,* 104, 423, 1988. With permission from Company of Biologists, Ltd.)

first hypothesis, called the transport hypothesis, is that the uptakes of the two compounds are controlled at the plasma membrane. The fluxes at each stage of development will be determined by the existence or efficiency of the mechanisms that transport pyruvate and glucose into the embryonic cells. The second hypothesis, called the metabolic hypothesis, is that the uptakes are controlled by the existence or efficiency of the metabolic fluxes of the two substrates after they have entered the embryo.

A. TRANSPORT HYPOTHESIS

The classical approach used to elucidate the mechanism of sugar transport is to determine uptake at a range of glucose concentrations in the incubation medium. If the transport is passive, there will be a linear relationship, but if it is facilitated, the rate of uptake will reach a maximum. Gardner and Leese[42] measured the rate of glucose transport by single mouse blastocysts as a function of glucose concentration (Figure 4). The results were best explained in terms of a combination of facilitated and passive diffusion, one being superimposed on the other. Further evidence for the presence of a glucose carrier was provided by competitive inhibition studies using sugar analogues and by incubating single blastocysts in the presence of phloretin and cytochalasin B, two classical inhibitors of glucose-facilitated diffusion. When phloretin was used, the relationship between the rate of glucose uptake and glucose concentration was linear, presumably representing the passive sugar transport component (Figure 4). When this passive component was subtracted from the overall rate of uptake, a conventional double reciprocal analysis could be performed on the data to give values for the J_{max} and K_t for the carrier-mediated component of 3.53 pmol per embryo per h and 0.14 m*M*, respectively. A proviso with this type of kinetic analysis relates to the use of the terms J_{max} and K_t in an intact cellular system. On strict kinetic grounds, J_{max} and K_t are used to define the passage of molecules across a membrane layer

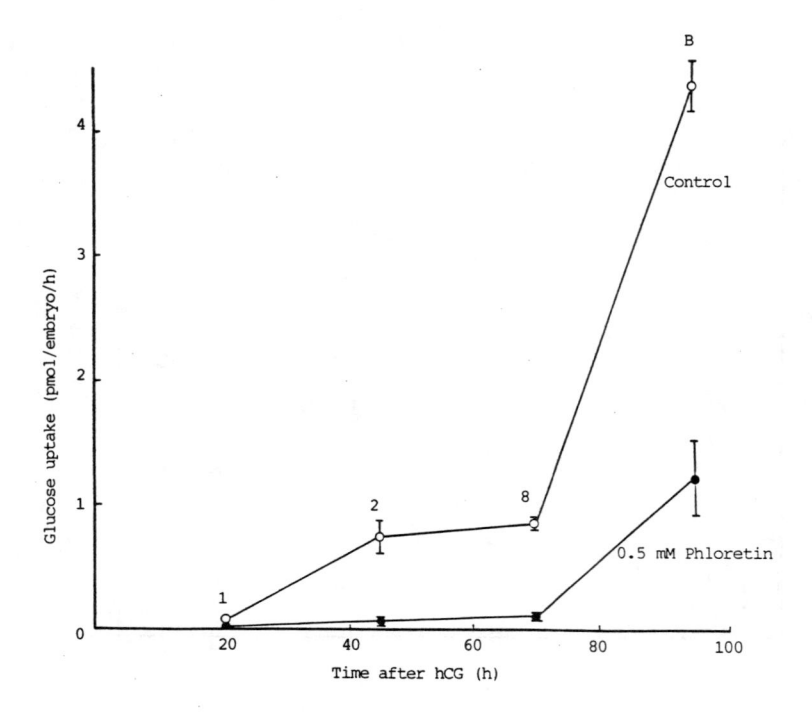

FIGURE 5. The uptake of 1 m*M* glucose by single mouse embryos in the presence and absence of 0.5 m*M* phloretin. (1) 1-cell, (2) 2-cell, (8) 8-cell, (B) blastocyst. Values are mean ± SEM of at least 6 embryos. (From Gardner, D. K. and Leese, H. J., *Development*, 104, 423, 1988. With permission from Company of Biologists, Ltd.)

alone and are analogous to the terms K_m and V_{max} used to define the kinetic properties of an isolated enzyme. In the present system, glucose may be metabolized within the embryo after crossing the plasma membrane, lowering its intracellular concentration and acting as a driving force for further glucose entry. The term should more correctly be considered as an "apparent K_t" which is affected by processes by which glucose disappears from the incubation medium (i.e., a combination of glucose membrane transport and intracellular metabolism). Gardner and Leese[42] also determined the extent to which the phloretin-inhibitable component of glucose uptake was present at the earlier stages of preimplantation development. The results (Figure 5) showed that the facilitated glucose entry mechanism was present in the two- and eight-cell stages, though its activity was less than that in the blastocyst. Although this suggests a role for membrane transport in limiting glucose utilization, it would be necessary to determine the kinetic parameters of the carrier at the earlier developmental stages.

α-Cyano-4-hydroxycinnamate is a specific inhibitor of pyruvate transport in erythrocytes and liver mitochondria.[43] In the presence of this inhibitor, Leese and Barton[39] found that pyruvate uptake by unfertilized mouse oocytes was greatly reduced, suggesting the presence of a carrier-mediated entry mechanism for pyruvate. Gardner and Leese[42] extended this observation by showing that the entry of pyruvate into preimplantation mouse embryos was carrier-mediated at each developmental stage up to the blastocyst (Figure 6). It is unlikely, therefore, that the characteristic decline in pyruvate uptake after compaction is due to a limitation in pyruvate transport capacity.

B. METABOLIC HYPOTHESIS

The observations of Leese and Barton[39] and Gardner and Leese[40] demonstrate that glucose and pyruvate are taken up significantly by the preimplantation mouse embryo at all stages of development. Studies with radiolabeled glucose show that prior to the eight-cell stage, when

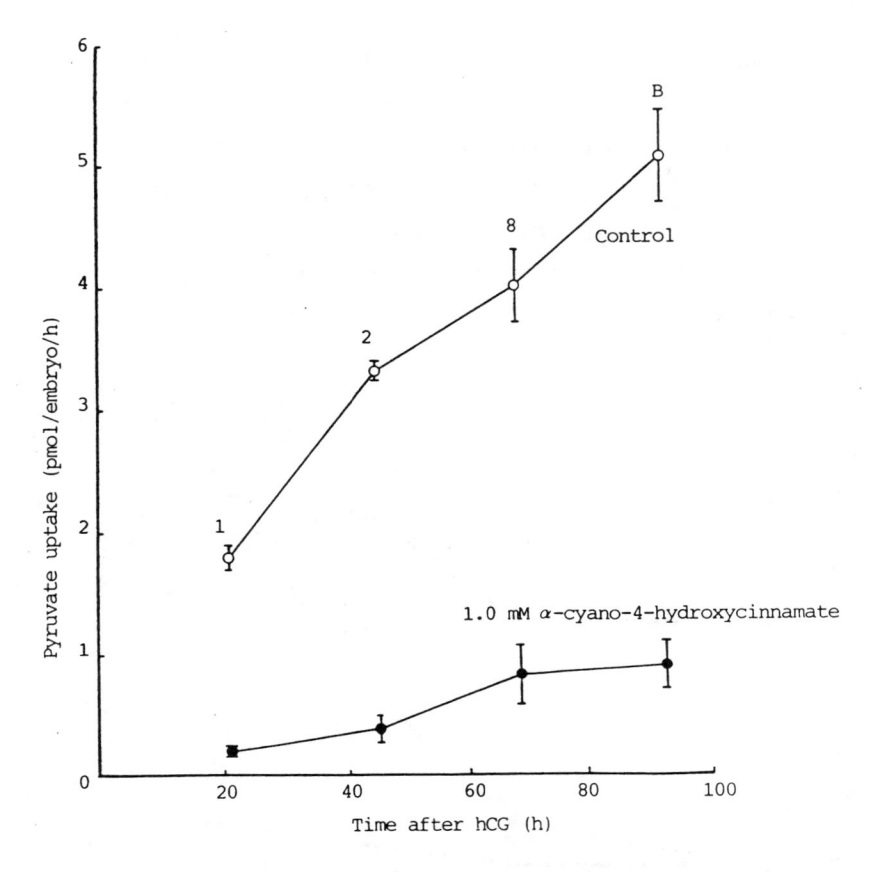

FIGURE 6. The uptake of 0.5 mM pyruvate by single mouse embryos in the presence and absence of 1 mM a-cyano-4-hydroxycinnamate: (1) 1-cell, (2) 2-cell, (8) 8-cell, (B) blastocyst. Values are mean ± SEM of 6 embryos. (From Gardner, D. K. and Leese, H. J., *Development,* 104, 423, 1988. With permission from Company of Biologists, Ltd.)

glucose utilization is low, some glucose is metabolized to lactate[31,44] or carbon dioxide[45] (Figure 7). Ten to fifteen times more of the carbon from pyruvate is metabolized to carbon dioxide than into other molecules. In contrast, half to equal amounts of the carbon from glucose is metabolized to carbon dioxide than into other molecules. Little glucose is used for the synthesis of glycogen until after the two-cell stage despite a high content of glycogen synthetase.[46,48] A recent study has shown that at the two-cell stage, 16% of the glucose is used through the pentose phosphate pathway,[49] while at the time of blastocyst formation, the value falls to 10%. Tracer studies also show that pyruvate readily enters the TCA cycle and is oxidized at all stages of development[33,50] These observations indicate that the major metabolic pathways which are fed by the substrates — the glycolytic pathway, the pentose phosphate pathway and the TCA cycle — are in place at all stages of preimplantation development of the mouse.

It seems probable that the fate of exogenous glucose is determined by the factors that control the fluxes through the metabolic pathways at each stage of development. A metabolic flux in its simplest form is a sequence of enzymatically catalyzed reactions, the product of one reaction being the substrate of the next:

$$A \rightarrow B \rightarrow C \rightarrow D \rightarrow P$$
$$E_1 \quad E_2 \quad E_3$$

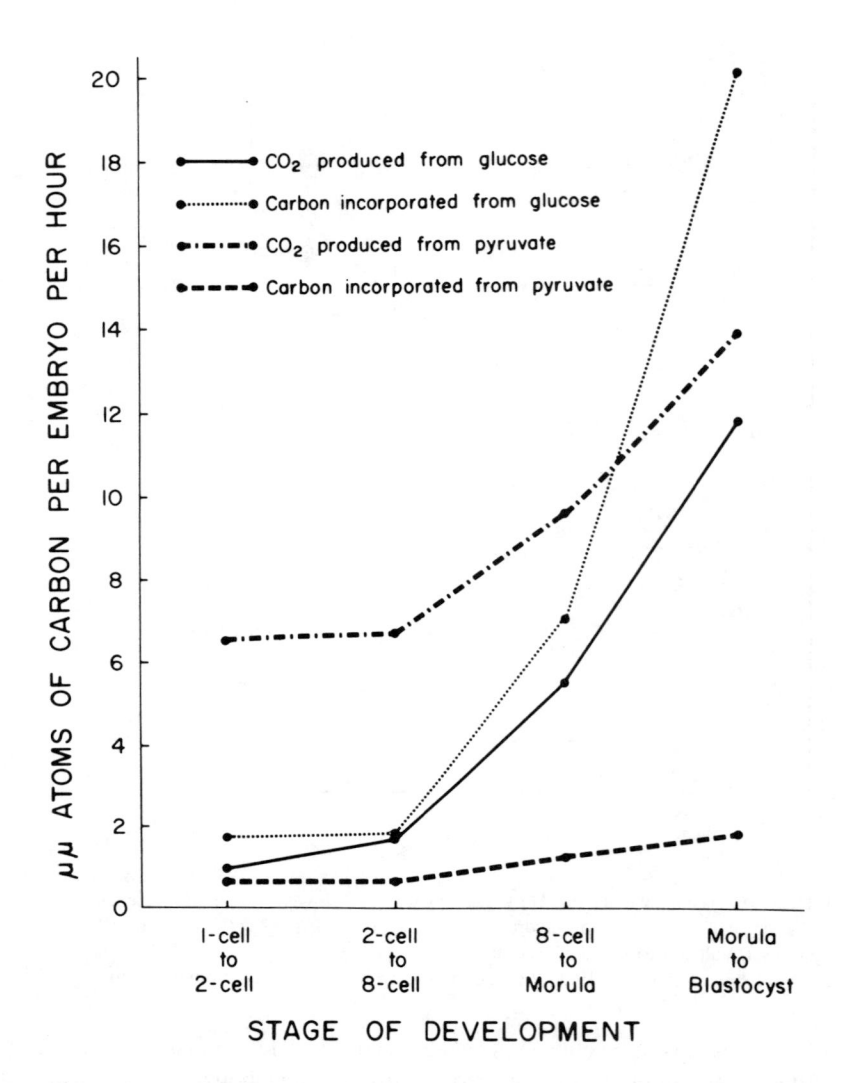

FIGURE 7. Utilization of glucose and pyruvate by the preimplantation mouse embryo. (From Brinster, R. L., *Handbook of Physiology,* Greep, R. O. and Astwood, E. B., Eds., American Physiological Society, Washington, D.C., 1973, 165. With permission.)

A method for locating control points in such a sequence is the starvation/refeeding experimental strategy. In this method, the effects of starvation of the first substrate (A) on the amount of particular substrates further down the chain are observed initially. The first substrate is then refed and the effect on substrates further down the chain observed. This experimental strategy was used by Barbehenn et al.[22,23] to locate control points in the glycolytic cycle of single preimplantation mouse embryos at various stages of development.

Figure 8 shows the fructose-1,6-P_2 levels in mouse embryos at the two-cell, eight-cell, morula, and blastocyst stages after 60 min starvation in an energy-substrate-free medium, followed by realimentation with glucose alone and glucose together with pyruvate. The fructose-1,6-P_2 levels did not fall during starvation and sometimes increased. Realimentation with glucose alone had no effect on the levels of fructose-1,6-P_2 in two-cell embryos but progressively greater effects in eight-cell embryos, morulae, and blastocysts. As development proceeds, there is greater conversion of glucose-6-P to fructose bisphosphate. When the

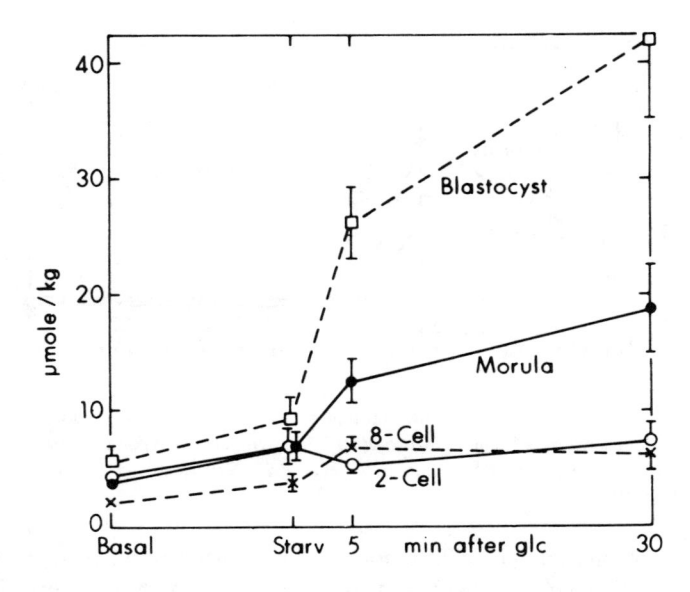

FIGURE 8. Fructose-1,6-P$_2$ levels before and after 60 min of starvation and after realimentation in a medium containing 5.5 mM glucose. Each point represents 7 to 12 embryos in most cases. There were only 3 embryos for the starvation point on the blastocyst curve. (From Barbehenn, E. K., et al., *Proc. Natl. Acad. Sci. U.S.A.*, 71, 1056, 1974. With permission.)

embryos are realimenated with both glucose and pyruvate, the level of fructose-1,6-P$_2$ falls to zero in two-cell and eight-cell embryos and morulae, but increases in blastocysts (Figure 9). Up to the eight-cell stage, pyruvate inhibits completely the conversion of glucose-6-P to fructose-1,6-P$_2$.

Similar starvation and realimentation studies on single-mouse embryos by Barbehenn et al.[22] produced evidence to show that at least four enzymes between the two-cell and morula stages of development were potentially rate limiting. These are hexokinase, 6-phosphofructokinase (PFK), glycogen phosphorylase, and an enzyme in the citric acid cycle between citrate and malate. In a subsequent paper, Barbehenn et al.[23] showed that this enzyme is isocitrate dehydrogenase.

It needs to be recognized that the enzyme data reported is for the maximum catalytic activity *in vitro* and may not necessarily reflect the properties of enzymes within intact cells. For example, enzymes such as those discussed above have traditionally been assigned to the "soluble" compartment of the cell, whereas they may well associate with particulate elements. It is notable that Swezey and Epel[51] have provided evidence suggesting that the majority of the enzyme glucose 6-phosphate dehydrogenase in unfertilized *Strongylocentrotus purpuratus* eggs is "bound", which lowers its catalytic activity, whereas following fertilization, it becomes "free".

IV. CONTROL OF METABOLISM BY ALLOSTERIC EFFECTORS

A great amount of work has been done on the adenine nucleotides, ATP, ADP, and AMP. The amounts of each have been measured on single unfertilized mouse oocytes and preimplantation embryos up to the late blastocyst stage using ultramicrofluormetric[24] and luciferase-based[30] assays. The levels of ATP and ADP have also been measured in groups of delayed and activated mouse blastocysts using the Lowry enzyme cycling method.[52,53] The ATP/ADP ratio can be calculated for each stage of development. The ATP content of single unfertilized mouse ova is

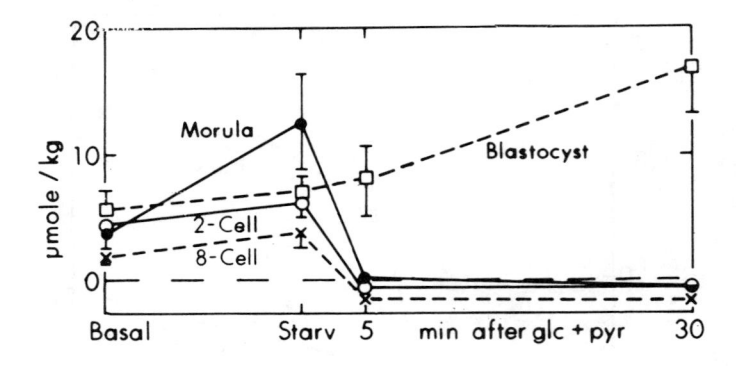

FIGURE 9. Fructose-1,6-P$_2$ after 60 min of starvation and realimentation in medium containing 5.5 mM glucose and 0.5 mM pyruvate. Each point represents 5 to 9 embryos. (From Barbehenn, E. K., et al. *Proc. Natl. Acad. Sci. U.S.A.*, 71, 1056, 1974. With permission.)

approximately 1 pmol and declines to 0.5 pmol per embryo in the morula/early blastocyst. There is a corresponding fall in the ATP/ADP ratio. The significance of these observations lies in the sensitivity to adenine nucleotides of PFK, which is allosterically inhibited by high ATP concentrations and a high ATP/ADP ratio, and deinhibited by ADP. Since Barbehenn et al. have shown that PFK is potentially rate limiting for glycolysis in the early mouse embryo, it is possible that the concentration of these adenine nucleotides may play a role in limiting glucose utilization in the early preimplantation stages.[15] This idea is reinforced by the observation that the marked drop in ATP content and ATP/ADP ratio, and a rise in ADP content, is coincident with the activation in glucose metabolism at the eight-cell/morula stage. Furthermore, Nieder and Weitlauf[52] observed a marked decline in the ATP/ADP ratio upon activation of the delayed mouse blastocyst from 3.08 to 0.81, which coincided with an increase in lactate formation via glycolysis.

Taken together, these data do not allow us to state unequivocally whether the transport or metabolic hypothesis is the more correct to explain why pyruvate rather than glucose is the preferred substrate up to the eight-cell stage. It is possible that the transport of glucose across the blastomere cell membrane is rate limiting under some circumstances, perhaps at locally high glucose concentrations in the oviduct or uterus, and that glucose utilization, via the enzymes and allosteric effectors discussed above, is limiting in other cases. In order to resolve these questions, it would be necessary to measure the intracellular glucose concentration, but this is likely to prove difficult.

V. CONCLUSION

More than 30 years have passed since Whitten[5] obtained the first evidence that the preimplantation mouse embryo had unusual metabolic characteristics. The biochemical analysis of these characteristics was slow to evolve due to the paucity of biological material for analysis and the lack of sensitivity in the available methods of quantitative analysis. New sensitive methods of analysis have now made the study of single embryos possible so that kinetic analyses can be done. This is a major advance, since it opens up the possibility of analyzing the fluxes through the metabolic pathways at different stages of development. Information concerning the control of these fluxes is essential for elucidating the genetic control of metabolism in the preimplantation stages of mammalian development. The development of these techniques is equally important for comparative physiology. It has been known for some years that the metabolism of mouse and rabbit preimplantation stages differ.[13] We now have the techniques for determining whether the preimplantation mouse embryo is a representative animal model for the study of metabolism in other early mammalian embryos.

ACKNOWLEDGMENTS

The preparation of this paper has been supported in part by grants (HD21581 and HD21988) to J. D. Biggers from the National Institutes of Health. D. K. Gardner and H. J. Leese acknowledge financial support from the U.K. Science and Engineering Research Council.

REFERENCES

1. **Bavister, B. D., Ed.,** *The Mammalian Preimplantation Embryo,* Plenum Press, New York, 1987.
2. **Leese, H. J.,** The formation and function of oviduct fluid, *J. Reprod. Fertil.,* 82, 843, 1988.
3. **Biggers, J. D.,** Introduction: cell biology of the developing egg, in *Cellular and Molecular Aspects of Implantation,* Glasser, S. R. and Bullock, D. W., Eds., Plenum Publishing, New York, 1981, 39.
4. **Whitten, W. K.,** Culture of tubal mouse ova., *Nature (London),* 177, 96, 1956.
5. **Whitten, W. K.,** Culture of tubal ova, *Nature (London),* 179, 1081, 1957.
6. **Brinster, R. L.,** Studies on the development of mouse embryos *in vitro.* II. The effect of energy source, *J. Exp. Zool.,* 158, 59, 1965.
7. **Brinster, R. L.,** Studies on the development of mouse embryos *in vitro* IV. Interaction of energy sources, *J. Reprod. Fertil.,* 10, 227, 1965.
8. **Brinster, R. L. and Thomson, J. L.,** Development of eight-cell mouse embryos *in vitro*, *Exp. Cell Res.,* 42, 308, 1966.
9. **Biggers, J. D., Whittingham, D. G. and Donahue, R. P.,** The pattern of energy metabolism in the mouse oocyte and zygote, *Proc. Natl. Acad. Sci. U.S.A.,* 58, 560, 1967.
10. **Cross, P. C. and Brinster, R. L.,** The sensitivity of one-cell mouse embryos to pyruvate and lactate, *Exp. Cell Res.,* 77, 57, 1973.
11. **Brinster, R. L. and Harstad, H.,** Energy metabolism in primordial germ cells of the mouse, *Exp. Cell Res.,* 109, 111, 1977.
12. **Quinn, P. and Wales, R. G.,** Growth and metabolism of preimplantation mouse embryos cultured in phosphate-buffered medium, *J. Reprod. Fertil.,* 35, 289, 1973.
13. **Biggers, J. D. and Stern, S.,** Metabolism of the preimplantation mammalian embryo, in *Advances in Reproductive Physiology,* Vol. 6, Bishop, M. W. H., Ed., Paul Elek Scientific Books, London, 1973, 1.
14. **Wales, R. G.,** Maturation of the mammalian embryo: biochemical aspects, *Biol. Reprod.,* 12, 66, 1975.
15. **Biggers, J. D.,** Bioenergetic aspects of fertilization and embryonic development of the mouse, in *Biological and Clinical Aspects of Reproduction,* Ebling, F. J. G. and Henderson, I. W., Eds., Excerpta Medica, Amsterdam, 1976, 128.
16. **Biggers, J. D. and Borland, R. M.,** Physiological asepcts of growth and development of the preimplantation mammalian embryo, *Annu. Review of Physiol.,* 38, 95, 1976.
17. **Kaye, P. L.,** Metabolic aspects of the physiology of the preimplantation embryo, in *Experimental Approaches to Mammalian Embryonic Development,* Rossant, J. and Pedersen, R. A., Eds., Cambridge University Press, 1986, 267.
18. **Crabtree, B. and Newsholme, E. A.,** A systematic approach to describing and analyzing metabolic control systems, *Trends Biochem. Sci.,* 12, 4, 1987.
19. **Kacser, H. and Porteous, J. W.,** Control of metabolism: what do we have to measure?, *Trends Biochem. Sci.,* 12, 5, 1987.
20. **Lowry, O. H. and Passoneau, J. V.,** *A Flexible System of Enzymatic Analysis,* Academic Press, New York, 1972.
21. **Bergmeyer, H. U. and Gawehn, K., Eds.,** *Methods of Enzymatic Analysis.,* 2nd English ed., Academic Press, Orlando, FL, 1974.
22. **Barbehenn, E. K., Wales, R. G., and Lowry, O.H.,** The explanation for the blockade of glycolysis in early mouse embryos, *Proc. Natl. Acad. Sci., U.S.A.,* 71, 1056, 1974.

23. **Barbehenn, E. K., Wales, R. G., and Lowry, O. H.,** Measurement of metabolites in single preimplantation embryos: a new means to study metabolic control in early embryos, *J. Embryol. Exp. Morphol.,* 43, 29, 1978.

24. **Leese, H. J., Biggers, J. D., Mroz, E. A., and Lechene, C.,** Nucleotides in a single mammalian ovum or preimplantation embryo, *Anal. Biochem.,* 140, 443, 1984.

25. **Mroz, E. A. and Lechene, C.,** Fluorescence analysis of picoliter samples, *Anal. Biochem.,* 102, 90, 1980.

26. **Hooper, M. A. K. and Leese, H. J.,** Measurement of enzymes in single embryos, *Human Reprod.,* 1, 25, 1986.

27. **Leese, H. J.,** Non-invasive methods for assessing embryos, *Human Reprod.,* 2, 435, 1987.

28. **Wudl, L., and Chapman, V.,** The expression of beta-glucuronidase during preimplantation development of mouse embryos, *Dev. Biol.,* 48, 104, 1976.

29. **Quinn, P. and Wales, R. G.,** The relationships between the ATP content of preimplantation mouse embryos and their development *in vitro* during culture, *J. Reprod. Fertil.,* 35, 301, 1973.

30. **Spielmann, H., Jacob-Mueller, U., Schulz P., and Schimmel, A.,** Changes of adenyl ribonucleotide content during preimplantation development of mouse embryos *in vivo* and *in vitro, J. Reprod. Fertil.,* 71, 467, 1984.

31. **Wales, R. G.,** Measurement of metabolic turnover in single mouse embryos, *J. Reprod. Fertil.,* 76, 717, 1986.

32. **Wales, R. G., Whittingham, D. G., Hardy, K., and Craft, I. L.,** Metabolism of glucose by human embryos, *J. Reprod. Fertil.,* 79, 289, 1987.

33. **Brinster, R. L.,** Carbon dioxide production from lactate and pyruvate by the preimplantation mouse embryo, *Exp. Cell Res.,* 47, 634, 1967.

34. **Brinster, R. L.,** Carbon dioxide production from glucose by the preimplantation mouse embryo, *Exp. Cell Res.,* 47, 271, 1967.

35. **Wales, R. G. and Whittingham, D. G.,** A comparison of the uptake and utilization of lactate and pyruvate in one- and two-cell mouse embryos, *Biochim. Biophys. Acta,* 148, 703, 1967.

36. **Wiebold, J. L. and Anderson, G. B.,** Lethality of a tritiated amino acid in early mouse embryos, *J. Embryol. Exp. Morphol.,* 88, 209, 1985.

37. **De Schepper, G. G., Van Noorden, C. J. F., and Koperdraad, F.,** A cytochemical method for measuring enzyme activity in individual preovulatory mouse oocytes, *J. Reprod. Fertil.,* 74, 709, 1985.

38. **Van Noorden, C. J. F., Tas, J., and Vogels, I. M. C.,** Cytophotometry of glucose-6-phosphate dehydrogenase activity in individual cells, *Histochem. J.,* 15, 583, 1983.

39. **Leese, H. J. and Barton, A. M.,** Pyruvate and glucose uptake by mouse ova and preimplantation embryos, *J. Reprod. Fertil.,* 72, 9, 1984.

40. **Gardner, D. K. and Leese, H. J.,** Non-invasive measurement of nutrient uptake by single cultured preimplantation mouse embryos, *Human Reprod. 1,* 25, 1986.

41. **Biggers, J. D., Bell, J. E., and Benos, D. J.,** The mammalian blastocyst: transport functions in a developing epithelium., *Am. J. Physiol.,* 255; *Cell Physiol.,* 24, C419, 1988.

42. **Gardner, D. K. and Leese, H. J.,** The role of glucose and pyruvate transport in regulating nutrient utilization by preimplantation mouse embryos, *Development,* 104, 423, 1988.

43. **Halestrap, A. P. and Denton, R. M.,** Specific inhibition of pyruvate transport in rat liver mitochondria and human erythrocytes by a-cyano-4-hydroxycinnamate, *Biochem. J.,* 138, 313, 1974.

44. **Wales, R. G.,** Accumulation of carboxylic acids from glucose by the preimplantation mouse embryo, *Aust. J. Biol. Sci.,* 22, 701, 1969.

45. **Brinster, R. L.,** Nutrition and metabolism of the ovum, zygote and blastocyst, in *Handbook of Physiology,* Section 7, Vol. II, Part 2, Greep, R. O. and Astwood, E. B., Eds., American Physiological Society, Washington, D.C., 1973, 165

46. **Brinster, R. L.,** Incorporation of carbon from glucose and pyruvate into the preimplantation mouse embryo, *Exp. Cell Res.,* 58, 153, 1969.

47. **Edirisinghe, W. R., Wales, R. G., and Pike, I. L.,** Synthesis and degradation of labelled glycogen pools in preimplantation mouse embryos during short periods of *in vitro* culture, *Aust. J. Biol. Sci.,* 37, 137, 1984.

48. **Stern, S.,** The activity of glycogen synthetase in the cleaving mouse embryo, in Proc. 3rd Annual Meeting, Society for the Study of Reproduction, Columbus, Ohio, 1970, 3.

49. **O'Fallon, J. and Wright, R. W., Jr.,** Quantitative determination of the pentose phosphate pathway in preimplantation mouse embryos, *Biol. Reprod.,* 34, 58, 1986.

50. **Wales, R. G. and Whittingham, D. G.,** Further studies on the accumulation of energy substrates by two-cell mouse embryos, *Aust. J. Biol. Sci.,* 27, 519, 1974.

51. **Swezey, R. R. and Epel, D.,** Regulation of glucose-6-phosphate dehydrogenase activity in sea urchin eggs by reversible association with cell structural elements, *J. Cell Biol.,* 103, 1509, 1986.

52. **Nieder G. L. and Weitlauf, H. M.,** Regulation of glycolysis in the mouse blastocyst during delayed implantation, *J. Exp. Zool.,* 231, 121, 1984.

53. **Weitlauf, H. M. and Nieder, G. L.,** Metabolism in preimplantation mouse embryos, *J. Biosocial. Sci.,* 6 (Suppl. 2), 33, 1984.

Chapter 3

HORMONAL SIGNALING MECHANISMS: THE ROLE OF PROTEIN PHOSPHORYLATION IN EARLY DEVELOPMENT

I. Y. Rosenblum, M. Farber, K. Ramos, and M. L. Pritchard

TABLE OF CONTENTS

I. INTRODUCTION

The evolution of a fertilized egg to form a complex organism comprised of different cell types is regulated by multiple biochemical mechanisms at all stages of development. Fundamental to the growth and development of all multicellular organisms is the operation of communication networks that convey information both intra- and intercellularly. Considerable evidence supports the existence of at least four information-transfer systems found in most eukaryotic cells.[1] The cornerstone of each system is the receptor which selectively receives and transfers an extracellular signal to the appropriate intracellular target. Receptors for non-lipophilic molecules, such as peptide hormones, growth factors, and some neurotransmitters, are allosteric transmembrane proteins that consist of functional extracellular peptide domains. Transmembrane signaling is triggered when extracellular signal molecules bind selectively to cell-surface receptor sites. Ligand binding activates a unique reaction, or chain of reactions, that culminates in the modification of key biochemical pathways.

The mechanisms underlying the pathways by which the fertilized egg becomes a multicellular organism still remain to be elucidated. In this chapter, we will review the role of phosphorylation-dephosphorylation reactions as putative regulatory mechanisms in early mammalian development. Our approach will focus on studies that provide evidence to support the notion that receptor-mediated phosphorylation of biological substrates represents an important cell regulatory mechanism during early embryogenesis.

In one transmembrane signaling system, the external signal, a neurotransmitter, binds to receptor proteins that function as ion channels in the cell membrane. Neurotransmitter binding triggers conformational changes that "open" the channels and allow the movement of ions into the cell. On the other hand, the transmembrane receptors for several peptide hormones and growth factors are enzymes (kinases) that catalyze a phosphate transfer reaction. Ligand binding to these receptors is associated with a rapid enhancement of kinase activity, resulting in the stimulation of one or more protein phosphorylation reactions. These reactions will be discussed in greater detail in subsequent sections of this chapter.

Another class of transmembrane receptors contains those that are not associated with ionic fluxes or intrinsic enzymatic activity. Instead, these receptors are coupled to enzymes via the mediation of a guanine nucleotide-binding protein, or "G-protein". G-protein-coupled receptors include those that respond selectively to photons (e.g., photoreceptors of the rods and cones) and certain neurotransmitters (e.g., epinephrine) and the hormone-sensitive adenylate cyclase signaling proteins. The major receptor-mediated mechanisms of transmembrane signaling are illustrated in Figure 1.

The importance of phosphorylation reactions in cell regulation was recognized first in the 1950s when the mechanism of catecholamine-induced liver glycogenolysis was elucidated. The discovery of cyclic AMP (cAMP) as the second messenger in this transmembrane signaling cascade paved the way for our current understanding of how information may be communicated intra- and intercellularly. The actions of a large number of hormones appear to be mediated via cAMP by means of kinase-catalyzed phosphorylation of protein substrates that underlie the cascade of reactions responsible for signal transduction. Among the numerous cellular trafficking systems demonstrated to be regulated by phosphorylation-dephosphorylation reactions, the cAMP-initiated cascades are the most thoroughly characterized to date. The mechanisms responsible for the ubiquitous and pleiotropic nature of cAMP as a regulatory molecule have been reviewed recently.[2]

Phosphorylation can be defined as the covalent modification of target substrates by phosphate transfer. The process is catalyzed by various kinases and phosphatases and has been recognized as a central mechanism by which a variety of external stimuli regulate cellular functions. Phosphorylation modulates the enzymatic activity of target proteins and has been shown to be a mediator of hormonal transduction in cells of both normal and diseased tissues.

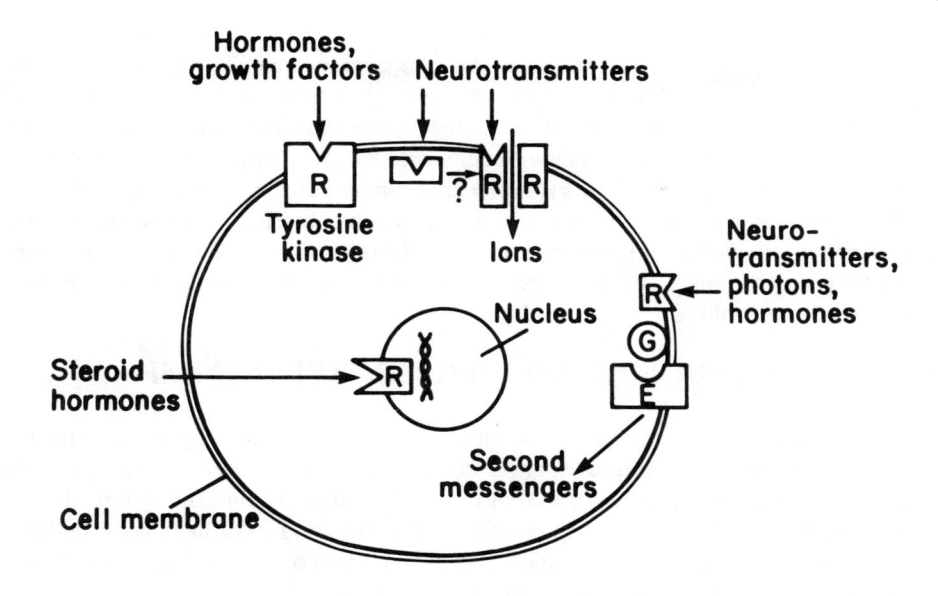

FIGURE 1. Receptor-mediated mechanisms of transmembrane signaling.

A wealth of information exists to suggest that phosphorylation of target proteins is an intermediate step in the physiological actions of various growth hormones and growth factors. Until a decade ago, hormone-induced protein phosphorylation was documented to occur only on serine and threonine residues. In 1979, another class of protein kinases was discovered which specifically phosphorylated tyrosine residues in target proteins.[3] The critical role of specific amino acid residue phosphorylation was firmly established following the demonstration that several growth factor receptors, including epidermal growth factor (EGF), platelet derived growth factor (PDGF), insulin-like growth factor I (IGF-I), and insulin, exhibit intrinsic tyrosine kinase activity.[4-9] The enzyme activities of these receptors appear to be regulated by both autophosphorylation and phosphorylation catalyzed by other kinases. For example, the EGF receptor is believed to be phosphorylated on threonine residues by protein kinase C.[10] The observation that phorbol esters, potent stimulants of protein kinase C, are able to induce increased phosphorylation of the EGF receptor in A431 cells supports this hypothesis.[11] The possibility that protein phosphorylation on tyrosine residues may be important in the regulation of cell proliferation has been supported by the recent finding that cells transformed by acutely oncogenic retroviruses have elevated levels of phosphotyrosine. It is known now that at least seven retroviral oncogenes possess tyrosine-specific protein kinase activity.[12] In addition, the receptors for EGF, PDGF, and insulin have amino acid sequences highly homologous to oncogene products.[13,14] These data suggest that phosphorylation is a critical component of the process of signal transduction required for the expression of a number of biological effects. The role of phosphorylation of cell surface receptors in the regulation of signal transduction pathways has been reviewed recently.[15]

A major hurdle to the biochemical study of mammalian preimplantation embryos has been the paucity of methods sufficiently sensitive to quantitate changes in biochemical parameters. The mouse zygote, for example, has a protein content of 23 ng, and this decreases by approximately 20% during preimplantation development.[16] Some of these problems have been minimized by the application of high-resolution gel electrophoresis and other techniques. As a result, it has been possible to assess the role of phosphorylation during early development in a more systematic fashion.

In order to demonstrate that phosphorylation plays a role in a given biological process, the four criteria of Krebs and Beavo[17] must be met. First, there must be a demonstration *in vitro* that the enzyme can be phosphorylated at a reasonable physiologic rate in a reaction catalyzed by a protein kinase. The phosphorylated form of the enzyme must serve as a substrate for an appropriate phosphoprotein phosphatase to regenerate the dephosphorylated form of the enzyme. Second, the properties of the protein must change appropriately with the degree of phosphorylation. Third, the phosphorylation/dephosphorylation must be associated with functional modification in an intact cell preparation. Last, cellular levels of the kinase or phosphatase must correlate with the level of phosphorylation of the target protein. The latter two criteria are the most difficult to satisfy and are especially challenging when working with the preimplantation mammalian embryo.

II. TYROSINE-SPECIFIC PROTEIN KINASES

A high degree of tyrosine phosphorylation in proteins of developing sea urchins and *Drosophila* has been documented.[18,19] These observations are consistent with previous studies which have shown that receptors for EGF, PDGF, IGF-I, and insulin exhibit intrinsic tyrosine kinase activity.[4-9] Furthermore, receptors that bind insulin and IGFs have been shown to be expressed on the surface of early chick and mouse embryos,[20-23] although the functional significance of these observations remains to be elucidated.

The similarity between these receptor-associated protein tyrosine kinases suggests that enzymatic activity is involved at an early step in the mechanism of transmembrane signaling. In the case of the insulin receptor, the insulin-binding and protein kinase activities have been co-purified to homogeneity from Triton X-100 solubilized extracts of human placental membranes.[24,25] Insulin binding to the cell-surface receptor site (α-subunit, 130 kDa) stimulates autophosphorylation of cytoplasmic tyrosine residues of the receptor protein (β-subunit, 95 kDa). The properties of the insulin receptor kinase have been shown to remain consistent throughout purification of the crude receptor protein.[26] Structural and functional aspects of the insulin receptor are described in greater detail in Chapters 4 to 6.

An important feature of the tyrosine-specific autophosphorylation reaction is that it serves to activate the protein kinase function of the receptor towards exogenous substrates in cell-free systems, as well as several endogenous substrates in cell lines *in vitro*.[27] Exogenous protein substrates such as histone and casein have been found to be phosphorylated *in vitro* by lectin-purified or immunoprecipitated insulin receptors. At least two endogenous substrates, in addition to the receptor β-subunit itself, have been reported, using preparations that contain catalytically active insulin receptor kinase.[27] The proteins were shown to be phosphorylated on tyrosine residues and appeared to be associated with the plasma membrane. The identity and function of these putative endogenous substrates have yet to be established.

That the protein tyrosine kinase activity of the receptor is necessary for hormone action has been demonstrated recently in studies using monoclonal antibodies to the protein kinase domain, and site-directed mutagenesis of the ATP binding site of the kinase domain.[24] The results of these studies have shown that insulin action depends upon receptor-associated tyrosine kinase activity. Cells with receptors deficient in kinase activity exhibited depressed responses to insulin, including glycogen synthesis, thymidine incorporation into DNA, S6 kinase activation, and non-receptor tyrosine phosphorylation.[24] Studies with chimeric receptor proteins have provided direct evidence that activation of the receptor tyrosine kinase is necessary for hormone signal transduction. In one study, cells were transfected with a DNA construct that encoded the ligand-binding domain of the insulin receptor fused to the transmembrane and kinase domains of the EGF receptor.[28] Interaction of the expressed protein with insulin resulted in the activation of the protein tyrosine kinase activity of the chimeric receptor. In another study, substitution of the transmembrane and kinase domains of v-*ros* for the homologous β-subunit domain of the insulin receptor yielded a chimeric receptor whose kinase was stimulated by insulin, but failed

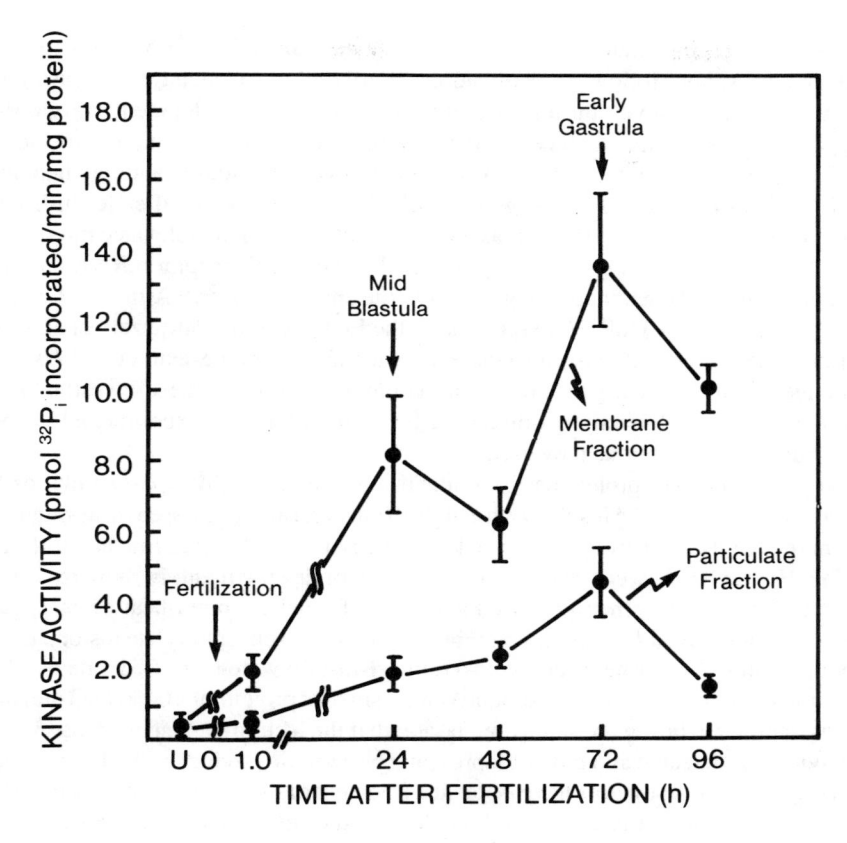

FIGURE 2. Tyrosine-specific protein kinase activity before and after fertilization of sea urchin eggs. (From Dasgupta, J. D. and Garbers, D. L., *J. Biol. Chem.*, 258, 6174, 1983. With permission.)

to promote deoxyglucose uptake.[29] These data emphasize the integral role of the receptor tyrosine kinase in the mechanism of transmembrane signaling.

The sea urchin provides a useful system for the manipulation of development. This is partly due to the availability of large quantities of eggs that can be fertilized and the rapid onset of cell division. Fertilization in the sea urchin leads to a two- to five-fold increase in tyrosine kinase activity within the first hour.[18,30] The specific activity of the tyrosine kinase is increased almost 20-fold over that of unfertilized eggs by the gastrula stage, as shown in Figure 2. A tyrosine kinase from the sea urchin *Strongylocentrotus purpuratas* phosphorylates a peptide similar to the pp60[v-src] autophosphorylation site with a K_m of 8.9 mM. Upon incubation of membrane fractions containing the tyrosine kinase with γ-[^{32}P]-ATP, nine sea urchin proteins were labeled with detectable amounts of phosphate on tyrosine residues. Phosphotyrosine could not be detected in membrane protein fractions from unfertilized eggs that had been incubated with radiolabeled ATP.[18] The increase in enzyme activity was shown to be independent of sperm components, RNA, DNA, or protein synthesis. The increase in kinase activity was present in fertilized eggs treated with puromycin or emetine (inhibitors of protein synthesis) and in enucleated unfertilized eggs that had been treated previously with 4.2 mM butyric acid.[31] Butyric acid has been shown to be a parthenogenetic activator of unfertilized eggs.[32] Similar results have been documented in a different species of sea urchin, *Lytechinus variegatus*.[19]

Although the precise identity of the tyrosine kinase protein in sea urchin embryos is not known, the tyrosine kinase activity that increases upon fertilization in *Lytechinus variegatus* does not belong to a c-*src*-related protein kinase, since the c-*src*-related protein kinase actually decreases in the first hour following fertilization.[33] Further, Levy et al. have demonstrated that c-*src* is expressed at high levels during later stages of embryonic development in vertebrates.[34]

As pointed out previously, the receptor for EGF possesses intrinsic protein tyrosine kinase activity.[4] Lev et al. have shown that the expression of a gene homologous to the human EGF receptor in unfertilized *Drosophila* eggs is low.[35] Two to five hours after fertilization, the expression of the EGF receptor gene increases dramatically and remains elevated throughout the embryonic stages of the *Drosophila* life cycle. The level of EGF receptor mRNA appeared to be lower during the larval and pupal stages than in the embryo proper.[35] Although Lev and his colleagues found poly (A+) mRNA expression of the EGF receptor, Petruzzelli et al. were not able to find EGF-dependent tyrosine kinase activity in membrane extracts of *Drosophila* embryos eluted from a wheat germ agglutinin column.[36] However, these investigators have pointed out that the receptor activity might have been present in their experimental system but undetected under the conditions they used.

An insulin-like receptor protein has been identified recently in *Drosophila* embryos.[36] The protein was capable of binding insulin with a high affinity and also possessed insulin-dependent protein kinase activity. The protein tyrosine kinase activity was detected first at 6 to 12 h after fertilization. Higher levels were detected by 12 to 18 h of development. In these experiments, protein kinase activity was measured by the ability of eluates from wheat germ agglutinin affinity chromatography of *Drosophila* membrane fractions to phosphorylate histone H2b with and without insulin. Two proteins cross-linked to ^{125}I-insulin were identified, one of 135 kDa (similar to the mammalian insulin α-subunit) and a smaller protein of 100 kDa. The 100 kDa protein was expressed during the same time period that the insulin-dependent protein tyrosine kinase activity was detected and did not represent a proteolytic product of the larger species.[36] More recently, these investigators have isolated a fragment of the *Drosophila* genome which is 11 kb in length, exhibits a high degree of homology to the gene encoding the β-subunit of the human insulin receptor, and is most prominent in 8 to 12 h embryos.[37] A sequence of the β-subunit of the human receptor, which is conserved in *Drosophila,* was used subsequently to generate an antibody. This antibody was reported to immunoprecipitate a 95 kDa phosphoprotein.[37] The latter protein appeared to be phosphorylated on tyrosine residues, since anti-phosphotyrosine antibodies precipitated the protein only when insulin was included in the kinase reaction, and the phosphate bond was stabilized under alkaline conditions. These studies represent the first direct evidence for the presence of a β-subunit of the insulin receptor in *Drosophila*.

III. PHOSPHATIDYLINOSITOL

The interaction of several hormones and neurotransmitters with their respective cell surface receptors results in the activation of guanine nucleotide regulatory (G) proteins. This family of proteins has been shown to couple receptor occupancy to second messenger systems involved in the regulation of cellular processes. In the case of β-adrenergic and muscarinic receptors, receptor occupancy by the agonists promotes the binding of GTP to oligomeric stimulatory (Gs) or inhibitory (Gi) G proteins. The binding of GTP results in the dissociation of the G protein into individual subunits which, in turn, modulate the activity of target proteins. The catalytic subunit of adenylate cyclase is regulated via modulation of the activities of Gs/Gi components. In addition, a wide variety of cell surface receptors are thought to be coupled to G proteins which regulate the rate of hydrolysis of polyphosphoinositides by activation of phospholipase C. The hydrolysis of phosphatidylinositol 4,5-bisphosphate [PtdIns(4,5)P$_2$], generates diacylglycerol (DAG) and inositol 1,4,5-triphosphate [Ins(1,4,5)P$_3$]. DAG activates protein kinase C, while Ins(1,4,5)P$_3$ binds to the smooth endoplasmic reticulum to cause the release of Ca^{++} into the cytosol. Elevation of intracellular calcium concentrations in turn mediates the activation of

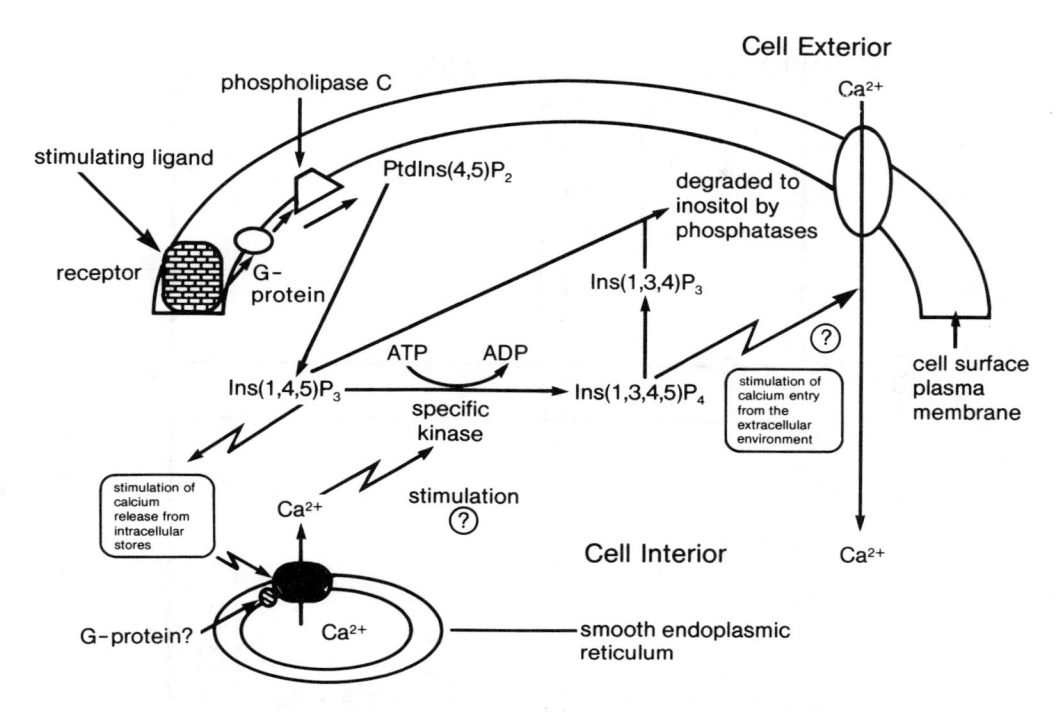

FIGURE 3. Proposed scheme for the involvement of phospholipids in the mobilization of Ca++. (From Houslay, M. D., *Trends Biochem. Sci.,* 12, 1, 1987. With permission.)

Ca++-sensitive processes.[38-40] The sequence of biochemical events associated with the hydrolysis of PtdIns(4,5)P$_2$ is presented in Figure 3.

The polyphosphoinositide system has been implicated in the regulation of cortical vesicle secretion in the sea urchin.[41] This represents the first association of a second messenger with a physiological response in an early embryo. Immediately following fertilization, the level of PtdIns(4,5)P$_2$ has been shown to decrease to 50% of the amount present in unfertilized eggs. This decrease correlated with a five-fold increase in Ins(1,4,5)P$_3$ 10 min after insemination.[42] In the sea urchin, *Lytechinus variegatus,* the injection of eggs with EGTA to maintain the intracellular Ca++ below 0.1 mM was shown to preclude formation of the fertilization envelope. The stimulation of exocytosis by microinjection of eggs with GTP-γ-S, a non-hydrolyzable analogue of GTP, was also prevented by EGTA. These observations raise the possibility that binding of sperm to sea urchin eggs is associated with the activation of a G protein and the subsequent hydrolysis of PtdIns(4,5)P$_2$.[43]

The oocyte is another example of a system in which cell signaling regulated by calcium-mobilizing receptors appears to operate via phosphoinositides. Recent work has suggested that both Ins(1,4,5)P$_3$ and inositol 1,3,4,5,-tetrakisphosphate [Ins(1,3,4,5)P$_4$], may act as "second messengers" to mediate an increase in intracellular calcium in sea urchin eggs. Ins(1,4,5)P$_3$ has been shown to be rapidly phosphorylated in brain and other tissue by a specific kinase to form Ins(1,3,4,5)P$_4$ which is rapidly dephosphorylated to Ins(1,3,4)P$_3$.[44-46] Although Ins(1,3,4,5)P$_4$ is not able to mobilize calcium from the smooth endoplasmic reticulum, it is thought to mediate the influx of calcium across the membrane.[47] Ins(1,3,4,5)P$_4$ has been implicated in the formation of the fertilization envelope when coinjected with Ins(2,4,5)P$_3$ (Figure 4). However, neither compound could raise the fertilization membrane alone.[47] Since Ins(1,4,5)P$_3$ is the only phosphoinositol that serves as a precursor to Ins(1,3,4,5)P$_4$, these observations support the notion that the generation of PtdIns(4,5)P$_2$ byproducts plays an important role in early development.

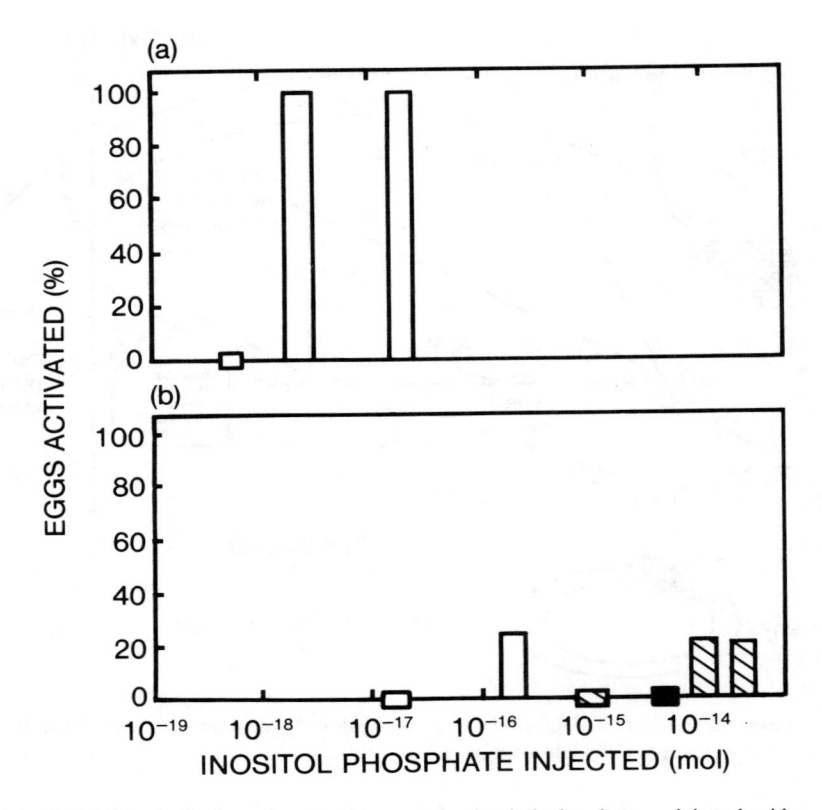

FIGURE 4. Activation of sea urchin eggs by inositol phosphates coinjected with Ins(2,4,5)P$_3$.[47] All eggs received 5×10^{-16} mol Ins(2,4,5)P$_3$ in addition to the following treatments Doses that produced no activation are shown by histograms that straddle the axis. (a) Open box, Ins(1,3,4,5)P$_4$; (b) open box, Ins(w,x,y,z)P$_4$; solid box, Ins(1,3,4)P$_3$; striped box, Ins(2,4,5)P$_3$. (Reprinted with permission from *Biochem. J.*, 240, 919, ©1986, The Biochemical Society, London.)

In *Xenopus* oocytes, the intracellular injection of Ins(1,3,4,5)P$_4$, and Ins(1,4,5)P$_3$ to a lesser degree, evoke an inward oscillatory chloride current.[48] In some but not all oocytes, activation of the chloride channel by Ins(1,3,4,5)P$_4$ has been shown to require a priming injection of Ins(1,4,5)P$_3$. The latter responses were not dependent on a source of extracellular calcium, but instead, were attributed to the release of intracellular calcium from stores within the oocyte.[48] Not only was the Ins(1,4,5)P$_3$ more potent in terms of inducing intracellular calcium release, but it was thought to be required for the activation of voltage-sensitive calcium channels in the cell membrane. Since the opening of chloride channels in oocyte membranes has been shown to be a calcium-dependent phenomenon,[48] these data suggest the existence of separate yet related messenger functions for the tri- and tetrakis-phosphoinositides. A specific phosphoinositol effect on cell membrane calcium channels cannot be excluded at the present time.

Several investigators have suggested that the phosphatidylinositol system may serve as a second messenger for tyrosine-specific protein kinases. The tyrosine kinases associated with v-*src* and v-*ros* have been found to phosphorylate phosphatidylinositol *in vitro*.[49,50] Similar results have been reported for the insulin receptor kinase.[51] It must be pointed out, however, that the tyrosine kinase activity and phosphatidylinositol kinase activity of v-*src* have been separated recently by means of selective immunoprecipitation.[52] The residual phosphatidylinositol kinase activity associated with purified insulin receptor preparations is significantly lower than that of the less-pure preparation. Although further investigation is needed to clarify this concept, collectively, these studies raise the possibility that oncogene products and growth factors may interact with components of the phosphatidylinositol pathway to mediate some of their biological effects.

IV. cAMP-DEPENDENT PROTEIN KINASE

The number of physiological processes mediated by cAMP is so large that studying them in the developing organism is extremely difficult. Oogenesis occupies a central role in development, and processes underlying oocyte maturation have received considerable attention. In most mammals, oocytes are arrested at the late diplotene stage of meiosis. In response to cyclical release of gonadotrophins, some oocytes resume meiosis and complete the first meiotic division, a necessary prelude to ovulation and fertilization. This process can be followed *in vitro* and is therefore more accessible to investigation than studies in the intact organism. It has been proposed that cAMP is involved in the maintenance of meiotic arrest, and that levels of cAMP in the oocyte are related to resumption of meiosis. Evidence in support of these concepts comes from experiments in which membrane-permeable cAMP analogues, such as dibutyryl cAMP and 8-bromo-cAMP, but not the corresponding cGMP analogues, and phosphodiesterase (PDE) inhibitors, have been shown to reversibly inhibit oocyte maturation *in vitro*.[53,54] Further support for the concept that a decline in cAMP levels is required for meiotic resumption comes from experiments in which forskolin, a potent reversible activator of adenylate cyclase produced dose-dependent increases in cAMP and resulted in a failure of both mouse and hamster oocytes to complete meiosis.[55-57]

In addition to its role in meiotic maturation, cAMP has been implicated in the control of the first division of the mouse embryo. Membrane permeable analogues of cAMP, PDE inhibitors, biologically active phorbol esters, and a synthetic diacylglycerol have been shown to inhibit the first cleavage of one-cell mouse embryos.[58] These effects suggest a putative role for protein phosphorylation in regulating the first division cycle of the mouse embryo. Even though DNA synthesis does not appear to be necessary for the first division, synthesis of transcription-requiring proteins in the embryo appears to be an essential component.

Cyclic AMP also has been implicated in early development, as demonstrated recently by Manejwala and co-workers.[59] Adenylate cyclase was shown to increase between the morula and blastocyst stages in the mouse embryo. The increase in adenylate cyclase activity at these stages of development appears to be dependent on transcription, since incubation of embryos in medium containing α-amanitin blocks the rise in activatable adenylate cyclase activity. In these studies, adenylate cyclase activity was determined by measuring the extent of cAMP formation in response to stimulation by activators of adenylate cyclase such as forskolin or cholera toxin. When early cavitating embryos were treated with cholera toxin, forskolin, or N_6-monobutyryl-cAMP, an increase in the rate of fluid accumulation was observed. Cyclic GMP analogues could not substitute for cAMP.[59]

During early stages in the development of *Drosophila* mutants, evidence has been obtained to show that cAMP is maintained at low levels during early embryogenesis. On the other hand, mutants that express a form of adenylate cyclase with lower activity lay eggs that fail to hatch. These findings suggest that maternal control of cAMP levels is vital for early development in *Drosophila*,[60] although the effect of low levels of cAMP on kinase and phosphatase activity at this stage is unknown.

Cyclic AMP-dependent protein kinase (cAMP-PK) is the major intracellular receptor for cAMP in most cell systems. Thus, the activity of cAMP-PK is affected by alterations in cellular cAMP levels. *In vitro,* small changes in cAMP have profound effects on the extent of cAMP-PK activation.[61] Two major classes of cAMP-PK isozymes, designated Type I and II, have been identified. These isozymes represent homologous proteins with a similar mechanism of activation.[17] In *Drosophila* embryos, the cAMP-PK is physically and kinetically similar to the Type II found in beef heart. These enzymes share several properties, such as a tetrameric structure (two regulatory and two catalytic subunits) and the formation of similar fragments when the catalytic subunit is subjected to proteolytic digestion. High intracellular levels of cAMP have also been associated with the inhibition of mouse embryogenesis. This effect is consistent with the inhibitory effects of analogues of cAMP and inhibitors of phosphodiesterase in the one-cell mouse embryo.[58] The latter inhibitory effects appear to be cell cycle dependent and only can be observed if inhibitors are added during the G1 phase. In contrast, protein kinase C activators are effective even if they are added late in the G2/M phase. The differences between the profile of cAMP-PK and protein kinase C-induced inhibition of cell division raise the possibility that the molecular mechanisms by which each enzyme system modulates cellular proliferation during early development may be intrinsically different.

In spite of the fact that tremendous progress has been made recently in identifying specific sites of action of cAMP, the mechanisms underlying the cAMP-mediated regulation of cellular proliferation and differentiation remain unknown. Although alterations in cellular cAMP levels are likely to modulate the activity of cAMP-PK and alter the extent of substrate phosphorylation, the possibility that cAMP exerts its effects by mechanisms unrelated to phosphorylation should not be ignored. This is particularly important in light of the controversy regarding the effects of cAMP as a positive or negative regulator of cellular proliferation. The opposing actions of cAMP under different conditions may also be associated with selective isozyme activation. Previous studies have suggested that the Type I isozyme is associated with cellular growth, whereas Type II appears to be involved in tissue differentiation.[62]

V. OTHER KINASES

Casein kinase II, an enzyme present in high levels in rapidly dividing tissues, is expressed in mouse embryos.[63] This cAMP- and calcium-independent enzyme uses either ATP or GTP as the preferred phosphate donor, phosphorylates proteins on serine or threonine residues, and its enzymatic activity is selectively inhibited by heparin. In the mouse embryo, the activity of casein kinase increases three- to four-fold at Day 12 of gestation. This increase in enzyme activity is associated with a proportional increase in the phosphorylation of a 110 kDa protein. Two lines of evidence suggest that the 110 kDa protein is a substrate for casein kinase II. First, the phosphorylation of the 110 kDa protein decreases dramatically with the inclusion of 1 mg/ml heparin in the enzyme assay. Second, the 110 kDa protein of the mouse embryo has an identical peptide profile to the known 110 kDa casein kinase II substrate from proliferating mouse tumor cells. The roles of casein kinase II and its protein substrate(s) have so far remained elusive.

VI. FUTURE DIRECTIONS

The role of proteins and/or second messenger systems that mediate phosphorylation/dephosphorylation reactions in embryogenesis are not understood fully, although evidence suggests that the phosphorylation of target substrates represents an important regulatory mechanism in early development. The discovery that various growth factor receptors and transforming retroviruses possess a tyrosine kinase activity has enhanced our understanding of the molecular basis of growth and development.

The plot thickens with the recent findings, based on amino acid sequence homologies and immunological cross-reactivities, that the endogenous substrates for both EGF- and pp60[src]-kinases may be a phospholipase A2-inhibitory protein, called "lipocortin".[64] These reports suggest a close relationship between the family of calcium/phospholipid/actin-binding proteins and protein tyrosine kinases. Perhaps these findings provide the long awaited bridge unifying previously unsuspected common pathways of signal transmission for several hormones, growth factors, and retroviral proteins.

REFERENCES

1. **Huganin, R. L. and Greengard, P.,** Regulation of receptor function by protein phosphorylation, *Trends Pharmacol. Sci.,* 8, 472, 1987.
2. **Roesler, W. J., Vandenbark, G. R., and Hansen, R. W.,** Cyclic AMP and the induction of eukaryotic gene transcription, *J. Biol. Chem.,* 263, 9063, 1988.
3. **Eckhart, W., Hutchinson, M. A., and Hunter, T.,** An activity phosphorylates tyrosine in polyoma T antigen immunoprecipitates, *Cell,* 18, 925, 1979.
4. **Ushiro, H. and Cohen, S.,** Identification of phosphotyrosine as a product of epidermal growth factor-activated protein kinase in A431 cell membranes, *J. Biol. Chem.,* 255, 8363, 1980.
5. **Ek, B., Westermark, B., Wasteson, A., and Heldin, C.-H.,** Stimulation of a tyrosine-specific phosphorylation by platelet-derived growth factor, *Nature (London),* 295, 419, 1982.
6. **Nishimura, J., Huang, J. S., and Deuel, T. F.,** Platelet-derived growth factor stimulates tyrosine-specific protein kinase activity in Swiss mouse 3T3 membranes, *Proc. Natl. Acad. Sci. U.S.A.,* 79, 4304, 1982.
7. **Jacobs, S., Kull, F. C., Earp, H. S., Svoboda, M. E., Van Wyck, J. J., and Cuatracasas, P.,** Somatomedin-C stimulates the phosphorylation of the beta-subunit of its own receptor, *J. Biol. Chem.,* 258, 9581, 1983.
8. **Rubin, J. B., Shia, M. A., and Pilch, P. F.,** Stimulation of tyrosine-specific phosphorylation in vitro by insulin-like growth factor I, *Nature (London),* 305, 438, 1983.
9. **Kasuga, M., Fujita-Yamaguchi, Y., Blithe, D. L., and Kahn, C. R.,** Tyrosine-specific protein kinase activity is associated with the purified insulin receptor, *Proc. Natl. Acad. Sci. U.S.A.,* 80, 2137, 1983.
10. **Cochet, C., Gill, G. N., Meisenhelder, J., Cooper, J. A., and Hunter, T.,** C-kinase phosphorylates the epidermal growth factor receptor and reduces its epidermal growth factor-stimulated tyrosine protein kinase activity, *J. Biol. Chem.,* 259, 2553, 1984.
11. **Iwashita, S., and Fox, C. F.,** Epidermal growth factor and potent phorbol esters induce epidermal growth factor receptor phosphorylation in a similar but distinctive manner in human epidermoid carcinoma A431 cells, *J. Biol. Chem.,* 259, 2559, 1984.
12. **Hunter, T. and Cooper, J. A.,** Protein-tyrosine kinases, *Annu. Rev. Biochem.,* 54, 897, 1985.
13. **Downward, J., Yarden, Y., Mayes, E., Scrace, G., Totty, N., Stockwell, P., Ullrich, A., Schlessinger, J., and Waterfield, M. D.,** Close similarity of epidermal growth factor receptor and v-*erb*-B oncogene protein sequences, *Nature (London),* 307, 521, 1984.
14. **Ullrich, A., Bell, J. R., Chen, E. Y., Herrera, R., Petruzzelli, L. M., Dull, T. J., Gray, A., Coussens, L., Liao, Y. C., Tsubokawa, M., Mason, A., Seeburg, P. H., Grunfeld, C., Rosen, O. M., and Ramachandran, J.,** Human insulin receptor and its relationship to the tyrosine kinase family of oncogenes, *Nature (London),* 313, 756, 1985.
15. **Sibley, D. R., Benovic, J. L., Caron, M. G., and Lefkowitz, R. L.,** Phosphorylation of cell surface receptors: a mechanism for regulating signal transduction pathways, *Endocrine Rev.,* 9, 38,1988
16. **Biggers, J. D. and Borland, R. M.,** Physiological aspects of the growth and development of the preimplantation mammalian embryo, *Annu. Rev. Physiol.,* 38, 95, 1976.
17. **Krebs, E. G. and Beavo, J. A.,** Phosphorylation-dephosphorylation of enzymes, *Annu. Rev. Biochem.,* 48, 923, 1979.
18. **Dasgupta, J. D. and Garbers, D. L.,** Tyrosine protein kinase activity during embryogenesis, *J. Biol. Chem.,* 258, 6174, 1983.
19. **Kinsey, W. H.,** Regulation of tyrosine-specific kinase at fertilization, *Dev. Biol.,* 105, 137, 1984.
20. **Hendricks, S. A., De Pablo, F., and Roth, J.,** Early development and tissue-specific patterns of insulin binding in chick embryo, *Endocrinology,* 115, 1315, 1984.
21. **Bassas, L., De Pablo, F., Lesniak, M. A., and Roth, J.,** Ontogeny of receptors for insulin-like peptides in chick embryo tissues: early dominance of insulin-like growth factor over insulin receptors in brain, *Endocrinology,* 117, 2321, 1985.
22. **Rosenblum, I. Y., Mattson, B. A., and Heyner, S.,** Stage-specific insulin binding in mouse preimplantation embryos, *Dev. Biol.,* 116, 261, 1986.
23. **Mattson, B. A., Rosenblum, I. Y., Smith, R. M., and Heyner, S.,** Autoradiographic evidence for insulin and insulin-like growth factor binding to early mouse embryos, *Diabetes,* 37, 585,1988.
24. **Rosen, O. M.,** After insulin binds, *Science,* 237, 1452, 1985.
25. **Petruzzelli, L., Herrera, R., and Rosen, O. M.,** Insulin receptor is an insulin-dependent tyrosine protein kinase: copurification of insulin-binding activity and protein kinase activity to homogeneity from human placenta, *Proc. Natl. Acad. Sci. U.S.A.,* 81, 3327, 1984.
26. **Avruch, J., Nemenoff, R. A., Pierce, M. W., Kwok, Y. C., and Blackshear, P. J.,** Protein phosphorylation as a mode of insulin action, in *Molecular Basis of Insulin Action,* Czech, M. P., Ed., Plenum Press, New York, 1985, 263.
27. **Caro, J. F., Shafer, J. A., Taylor, S. I., Raju, S. M., Perrotti, N., and Sinha, M. K.,** Insulin stimulated protein phosphorylation in human liver plasma membranes: detection of endogenous or plasma membrane associated substrates for insulin receptor kinase, *Biochem. Biophys. Res. Commun.,* 149, 1008, 1987.

28. **Riedel, H., Dull, T. J., Schlessinger, J., and Ullrich, A.,** A chimaeric receptor allows insulin to stimulate tyrosine kinase activity of epidermal growth factor receptor, *Nature (London),* 324, 68, 1986.

29. **Ellis, L., Morgan, D. O., Jong, S.-M., Wang, L.-H., Roth, R. A., and Rutter, W. J.,** Heterologous transmembrane signaling by a human insulin receptor-v-*ros* hybrid in Chinese hamster ovary cells, *Proc. Natl. Acad. Sci. U.S.A.,* 84, 5101, 1987.

30. **Ribot, H. D., Eisman, E. A., and Kinsey, W. H.,** Fertilization results in increased tyrosine phosphorylation of egg proteins, *J. Biol. Chem.,* 259, 5333, 1984.

31. **Satoh, N. and Garbers, D. L.,** Protein tyrosine kinase activity of the eggs of the sea urchin *Strongylocentrotus purpuratus:* the regulation of its increase after fertilization, *Dev. Biol.,* 111, 515, 1985.

32. **Dirkson, E. R.,** The presence of centrioles in artificially activated sea urchin eggs, *J. Biophys. Biochem. Cytol.,* 11, 244, 1961.

33. **Kamel, C., Veno, P. A., and Kinsey, W. H.,** Quantitation of a src-like tyrosine protein kinase during fertilization of the sea urchin egg, *Biochem. Biophys. Res. Commun.,* 138, 349, 1986.

34. **Levy, B. T., Sorge, L. K., Meymandi, A., and Maness, P. F.,** pp60^{c-src} is in chick and human embryonic tissues, *Dev. Biol.,* 104, 9, 1984.

35. **Lev, Z., Shilo, B. Z., and Kimchie, Z.,** Developmental changes in expression of the *Drosophila melanogaster* epidermal growth factor receptor gene, *Dev. Biol.,* 110, 499, 1985.

36. **Petruzzelli, L., Herrera, R., Garcia-Arenas, R., and Rosen, O. M.,** Acquisition of insulin-dependent protein tyrosine kinase activity during *Drosophila* embryogenesis, *J. Biol. Chem.,* 260, 16072, 1985.

37. **Petruzzelli, L., Herrera, R., Garcia-Arenas, R., Fernandez, R., Birnbaum, M. J., and Rosen, O. M.,** Isolation of a *Drosophila* genomic sequence homologous to the kinase domain of the human insulin receptor and detection of the phosphorylated *Drosophila* receptor with an anti-peptide antibody, *Proc. Natl. Acad. Sci. U.S.A.,* 83, 4710, 1986.

38. **Williamson, J. R.,** Role of inositol lipid breakdown in the generation of intracellular signals, *Hypertension,* 8, 110, 1986.

39. **Houslay, M. D.,** Egg activation unscrambles a potential role for IP$_4$, *Trends Biochem. Sci.,* 12, 1, 1987.

40. **Cockcroft, S.,** Polyphosphoinositide phosphodiesterase: regulation by a novel guanine nucleotide binding protein, Gp, *Trends Biochem. Sci.,* 12, 75, 1987.

41. **Turner, P. R., Sheetz, M. P., and Jaffe, L. A.,** Fertilization increases the polyphosphoinositide content of sea urchin eggs, *Nature (London),* 414, 1984.

42. **Kamel, L. C., Bailey, J., Schoenbaum, L., and Kinsey, W.,** Phosphatidylinositol metabolism during fertilization in the sea urchin egg, *Lipids,* 20, 350, 1985.

43. **Turner, P. R., Jaffe, L. A., and Fein, A.,** Regulation of cortical vesicle exocytosis in sea urchin eggs by inositol 1,4,5-trisphosphate and GTP-binding protein, *J. Cell Biol.,* 102, 70, 1986.

44. **Irvine, R. F., Letcher, A. J., Heslop, P., and Berridge, M. J.,** The inositol tris/tetrakisphosphate pathway — demonstration of Ins(1,4,5)P$_3$ kinase activity in animal tissues, *Nature (London),* 320, 631, 1986.

45. **Palmer, S., Hawkins, P. T., Michell, R. H., and Kirk, C. J.,** The labelling of polyphosphoinositides with [^{32}P]P$_i$ and the accumulation of inositol phosphates in vasopressin stimulated hepatocytes, *Biochem. J.,* 238, 491, 1986.

46. **Batty, I. R., Nahorski, S. R., and Levine, R. F.,** Rapid formation of inositol(1,3,4,4)tetrakisphosphate following muscarinic stimulation of rat cerebral cortical slices, *Biochem. J.,* 321, 211, 1985.

47. **Irvine, R. F. and Moor, R. M.,** Microinjection of inositol-1,3,4,5-tetrakisphosphate activates sea urchin eggs by a mechanism dependent on external Ca^{++}, *Biochem. J.,* 240, 917, 1986.

48. **Parker, I. and Miledi, R.,** Injection of inositol 1,3,4,5-tetrakisphosphate into *Xenopus* oocytes generates a chloride current, dependent upon intracellular calcium, *Proc. R. Soc. London Ser. B,* 232, 59, 1987.

49. **Sugimoto, Y., Whitman, M., Cantley, L. C., and Erikson, R. L.,** Evidence that the Rous sarcoma transforming gene product phosphorylates phosphatidylinositol and diacylglycerol, *Proc. Natl. Acad. Sci. U.S.A.,* 81, 2117, 1984.

50. **Macara, I., Marinetti, G. V., and Balduzzi, P. C.,** Transforming protein of avian sarcoma virus UR2 is associated with phosphatidylinositol kinase activity: possible role in tumorigenesis, *Proc. Natl. Acad. Sci. U.S.A.,* 81, 2728, 1984.

51. **Sale, G. Y., Fujita-Yamaguchi, Y., and Kahn, C. R.,** Characterization of a phosphatidylinositol kinase activity associated with the insulin receptor, *Eur. J. Biochem.,* 155, 345, 1986.

52. **MacDonald, M. L., Kuenzel, E. A., Glomset, J. A., and Krebs, E. G.,** Evidence from two transformed cell lines that the phosphorylations of peptide tyrosine and phosphatidylinositol are catalyzed by different proteins, *Proc. Natl. Acad. Sci. U.S.A.,* 82, 3993, 1985.

53. **Wassarman, P. M., Josefowicz, W. J., and Letourneau, G. E.,** Meiotic maturation of mouse oocytes *in vitro:* inhibition of maturation at specific stages of nuclear progression, *J. Cell Sci.,* 22, 431, 1976.

54. **Rice, C. and McGaughey, R. W.,** Effect of testosterone and dibutyryl cAMP on the spontaneous maturation of pig oocytes, *J. Reprod. Fertil.,* 62, 245, 1981.

55. **Schultz, R. M., Montgomery, R. R., and Belanoff, J. R.,** Regulation of mouse oocyte maturation: implication of a decrease in oocyte cAMP and protein dephosphorylation in commitment to resume meiosis, *Dev. Biol.,* 97, 264, 1983.

56. **Sato, E. and Koide, S. S.,** Forskolin and mouse oocyte maturation *in vitro, J. Exp. Zool.,* 230, 125, 1984.

57. **Racowsky, C.,** Effect of forskolin on the spontaneous maturation and cyclic AMP content of hamster oocyte-cumulus complexes, *J. Exp. Zool.,* 234, 87, 1985.

58. **Poueymirou, W. T. and Schultz, R. M.,** Differential effects of activators of cAMP-dependent protein kinase and protein kinase C on cleavage of one-cell mouse embryos and protein synthesis and phosphorylation in one- and two-cell embryos, *Dev. Biol.,* 121, 489, 1987.

59. **Manejwala, F., Kaji, E., and Schultz, R. M.,** Development of activatable adenylate cyclase in the preimplantation mouse embryo and a role for cyclic AMP in blastocoel formation, *Cell,* 46, 95, 1986.

60. **Bellen, H. J., Gregory, B. K., Olsson, C. L., and Kiger, J. A.,** Two *Drosophila* learning mutants, Dunce and Rutabaga, provide evidence for a maternal role for cAMP on embryogenesis, *Dev. Biol.,* 121, 432, 1987.

61. **Beebe, S. J. and Corbin, J. D.,** Cyclic nucleotide-dependent protein kinases, *The Enzymes,* 15, 43, 1986.

62. **Jungman, R. A. and Russell, D. H.,** Cyclic AMP, cyclic AMP-dependent protein kinase, and the regulation of gene expression, *Life Sci.,* 20, 1787, 1977.

63. **Schneider, H. R., Reichert, G. H., and Issinger, O. G.,** Enhanced casein kinase II activity during mouse embryogenesis, *Eur. J. Biochem.,* 161, 733, 1986.

64. **Hollenberg, M. D., Valentine-Braun, K. A., and Northup, J. K.,** Protein tyrosine kinase substrates: rosetta stones or simply structural elements, *Trends Pharmacol. Sci.,* 9, 63, 1988.

Chapter 4

COMPARATIVE ASPECTS OF INSULIN AND THE INSULIN RECEPTOR*

Britta A. Mattson, Scott A. Chambers, and Flora de Pablo

TABLE OF CONTENTS

---

* Abbreviations for terms used in this chapter are listed with their meanings at the end of this chapter, before the references.

I. INTRODUCTION

Traditionally, insulin has been viewed as a peptide hormone synthesized in the pancreatic beta cells of vertebrates. Upon secretion into the bloodstream and binding to classic target cells throughout the body, the physiological effects of insulin on carbohydrate metabolism and protein synthesis are elicited. However, as early as 1923, numerous reports have suggested that insulin is neither a unique product of pancreatic beta cells nor of vertebrate organisms. Functional insulin receptors have been characterized on a number of nonclassic target cells in both invertebrates and vertebrates. Moreover, recent studies have suggested a role for insulin in mediating fetal growth as well as oocyte maturation and embryonic development prior to the appearance of a functional pancreas. In this chapter, we review studies that characterize the highly conserved nature of insulin and insulin receptors throughout evolution. Although not all of these reports focus on the developmental period of the organisms, they support a role for insulin and insulin receptors in nontraditional growth and differentiation mechanisms. The first half of this chapter details insulin and insulin-related peptide biosynthesis, structure, and detection throughout evolution, whereas insulin receptor structure and functions are discussed subsequently.

II. INSULIN AND INSULIN-RELATED PEPTIDES

A. INSULIN BIOSYNTHESIS AND STRUCTURE

Insulin is a major anabolic peptide hormone synthesized and stored in the beta cells of the islets of Langerhans. Preproinsulin, a 109 amino acid single chain precursor in humans, is synthesized in the polyribosomes of the rough endoplasmic reticulum of the beta cells and is subsequently cleaved to an 86-amino acid single chain proinsulin. Proinsulin is converted to insulin in the Golgi complex by proteolytic cleavage during which four basic amino acids and the C-peptide are removed. The resultant insulin molecule consists of an A chain of 21 amino acids linked to a B chain of 30 amino acids by two disulfide bonds (Figure 1). The length of the C-peptide varies widely among species.

X-ray crystallographic analysis has shown that insulin can exist as a monomer, dimer, or hexamer, but the biologically active form of the hormone is thought to be the monomer.[1] Two molecules of Zn^{+2} are coordinated in the hexamer, which may have evolved to provide a granule storage form that is thermodynamically stable and more resistant to enzymatic degradation than the unassociated form. The insulin monomer has a well-defined globular, three-dimensional structure with a hydrophobic core and two predominately hydrophobic surfaces.

The amino acid sequences of insulins from a variety of vertebrates have been reported.[1] The rate of substitution for most insulins has been on the order of 1×10^{-9} per locus per year. The sequence data available indicate that the disulfide bridges and the hydrophobic core of the monomer are invariant, thus suggesting that natural selection has operated largely to preserve the tertiary and quaternary structures of the hormone. The conserved sequence proximal to the carboxyl terminus of the B chain, including the invariable Phe-Phe residues at B24 and B25, is believed to be the binding site for the insulin receptor. The most highly variable amino acid residues are A8 to A10 and B29 and B30, although subsitutions are not restricted to these locations. Porcine insulin is the most similar to human insulin, differing only by the substitution of an alanine residue at the carboxyl terminus of the B chain.[2] In contrast to the highly conserved nature of insulin, the C-peptide has undergone many modifications,[3,4] consistent with the belief that functionally less important molecules or parts of molecules undergo more mutant substitutions in evolution.[5]

Upon secretion into the bloodstream and binding to target cells throughout the body, the physiological effects of insulin include regulation of glucose, amino acid, and ion uptake; stimulation of RNA, protein, and lipid synthesis; and promotion of DNA synthesis and cell

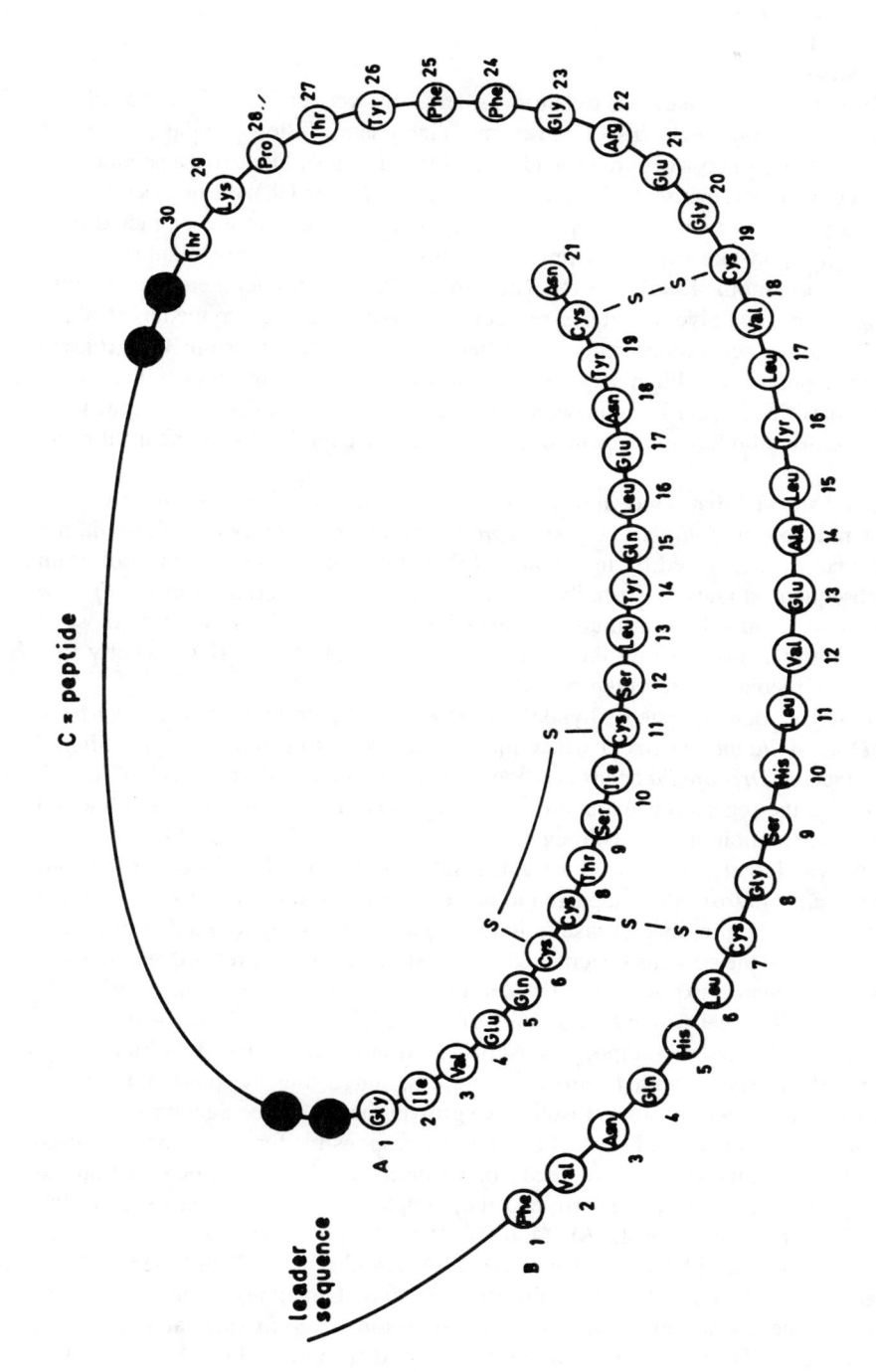

FIGURE 1. Preproinsulin is converted into proinsulin by enzymatic cleavage of the 23 amino acid leader sequence. Proinsulin undergoes a series of site-specific peptide cleavages which remove the C-peptide. The mature insulin molecule consists of a 21-amino acid A chain and a 30-amino acid B chain crosslinked by disulfide bonds. The solid circles represent dipeptides split off during the formation of the active hormone.

growth. Classically, insulin has been viewed as a vertebrate hormone synthesized by differentiated beta cells; nonetheless, multiple reports have suggested that insulin is neither a unique product of beta pancreatic cells nor of vertebrate organisms.

B. INVERTEBRATA
1. Insecta and Annelida

In insects, the main polysaccharide is trehalose rather than glycogen, and it is degraded into glucose during insect starvation.[6,7] Insect and mammalian tissues follow similar patterns of glucose metabolism from glucose to pyruvic acid,[8] therefore it is highly likely that a peptide with insulin-related activity exists in insects. Several reports have suggested that the neuroendocrine system of insects produces an insulin-related peptide. In 1976, Tager et al.[9] extracted both insulin-like and glucagon-like peptides from the corpus cardiacum-corpus allantum of the lepidopteran hornworm moth *Manduca sexta*. The partially purified immunoreactive peptides produced hypoglycemia and glycogenolysis, respectively, when injected into the larvae of the same species. The amino acid composition of purified insulin-like activity from *M. sexta* was found to differ from porcine insulin in only three residues.[10] In an immunohistochemical study, El-Salhy et al.[11] described specific neurosecretory cells in the brain of *M. sexta* that were immunoreactive with antibodies prepared to several mammalian peptides, including insulin and glucagon.

In 1978, Duve[12] showed that removal of the median neurosecretory cells (MNC) from the brain of the dipteran blowfly *Calliphora erythrocephala* resulted in both hypertrehalosemia and hyperglycemia that were reversed by injection of MNC homogenates or acid-ethanol brain extracts. In subsequent studies, a partially purified insulin-like material from *Calliphora vomitoria* heads was found to have immunological and biological activities similar to those of bovine insulin.[13] Further, this insulin-like material was localized immunofluorescently to a restricted number of median neurosecretory cells of *C. vomitoria*.[14]

Meneses and Ortiz[15] demonstrated a hypoglycemic effect of a crude protein extract from whole fruitfly *Drosophila melanogaster* when injected into mice. LeRoith et al.[16] gel filtered acid-ethanol extracts of *Drosophila* heads and bodies and recovered a discrete peak of insulin immunoreactivity in the region for mammalian insulins. In an insulin bioassay, this fraction stimulated glucose oxidation as well as lipogenesis in isolated rat adipocytes. Moreover, the biologic activity was largely neutralized by anti-insulin antibodies. Insulin concentrations measured in the heads of *Drosophila* appeared to be very similar to those previously reported in *Calliphora* heads.[13] The presence of insulin in *Drosophila* bodies suggested that insulin in insects is not restricted to the nervous system. In the same study, an insulin-related material was recovered from acid-ethanol extracts of both skin and internal structures of the earthworm *Annelida oligocheta*. The gel-filtered materials stimulated glucose oxidation and glucose incorporation into lipids in isolated adipocytes. Anti-insulin antibodies reduced the bioactivity, although not as effectively as with *Drosophila* extracts, suggesting the possibility of the presence of growth factors analogous to insulin-like growth factors.[13] (see Section III.A)

Prothoracicotropic hormone (PTTH) is a brain peptide responsible for activation of prothoracic glands to produce the molting hormone ecdysone in insects. By enzymatic degradation and microsequencing, Nagasawa et al.[17] determined the complete amino acid sequence of 4K-PTTH-II, one of the three forms of the M_r 4400 PTTH of the silkworm *Bombyx mori*. Like vertebrate insulin, it consisted of two nonidentical A and B peptide chains. The sequence of 4K-PTTH-II was shown to have considerable similarity (40%) with human insulin as well as with IGF-I. In particular, the distribution of six cysteine residues involved in intra- and interchain disulfide bonds in 4K-PTTH-II was identical to the cysteine distribution of insulin. These data suggest that the genes for the insulin-family peptides have arisen from a common ancestral gene and that these genes have been highly conserved during evolution from insects to mammals.

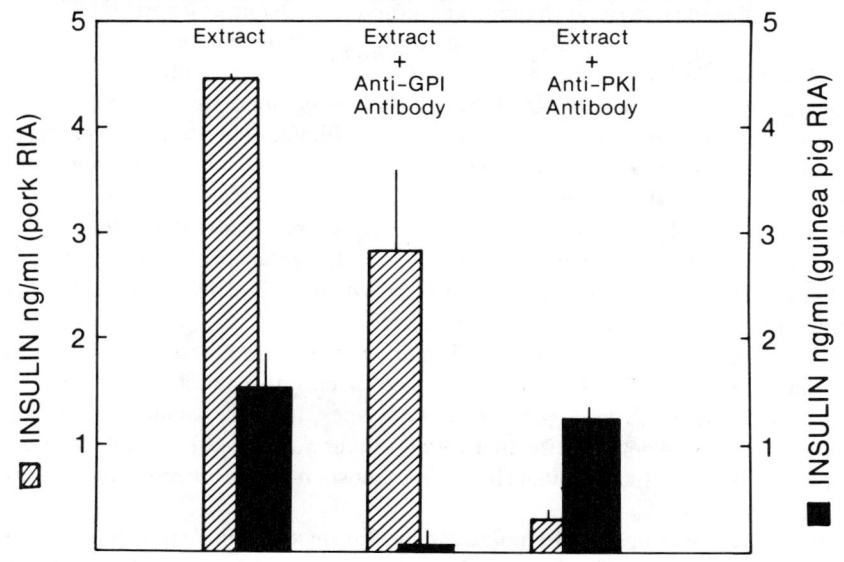

IMMUNO DEPLETION OF PORK AND GUINEA PIG–LIKE INSULINS IN TETRAHYMENA

FIGURE 2. *Tetrahymena pyriformis* were homogenized and an acid ethanol extract was gel filtered on Sephadex, G-50. The pooled fractions in the insulin region were tested in a pork radioimmunoassay (hatched bars) and in the guinea pig insulin radioimmunoassay (solid bars). The results before immunodepletion are depicted under "Extract". Equivalent aliquots of the extract were incubated overnight with either anti-guinea-pig-insulin (anti GPI) antibody or anti-pork-insulin (anti PKI) antibody which had been purified by Protein A affinity chromatography. Each mixture (extract plus antibody) was chromatographed on the same Protein-A column. An aliquot of the concentrated "flow through" of each sample was again tested in both immunoassays. Although recoveries of total activity were less than 100% there was specific immunodepletion of each type of immunoreactivity with the respective antibody.

2. Fungi and Protozoa

The detection of insulin in annelid skin suggested a broader distribution of insulin rather than a restriction to a neuroendocrine or gastrointestinal tissue derivative. This suggestion is further supported by the finding of insulin-related material that resembles rat/pork-type insulins in the unicellular undifferentiated eukaryotic fungi *Neurospora crassa* and *Aspergillus fumigatus* and the ciliated protozoan *Tetrahymena pyriformis* as well as in the conditioned medium in which the protozoa had been grown.[18] Previous to this study, an insulin-related material had not been detected in any organism that did not have either gut or neural elements. It is possible that the insulin-related material in these organisms may have no function, but the authors speculate that this material may function as an intra- or intercellular messenger. Interestingly, the insulin-related material from *T. pyriformis* has been shown to cross-react in two very different insulin radioimmunoassays, one for pork insulin and the other for guinea pig insulin.[19] Gel-filtered extracts of *T. pyriformis* were shown to be immunodepleted of the two separate insulin immunoreactivities by specific antibodies (Figure 2). This suggests that the insulin-related material of this protozoa may be composed of several peptides, one with common epitopes with mammalian type insulin and other(s) that are immunologically closer to hystricomorph-type insulins. Taken together, these data suggest that insulin antedates the pancreas phylogenetically.

3. Mollusca, Arthropoda, Echinodermata, and Tunicata

The first suggestion that insulin might be produced by invertebrates as well as vertebrates was as early as a brief report in 1923.[20] Methods developed for the isolation of insulin in semipure form from the pancreas of an ox were successfully applied to clam (*Mya arenaria*) tissue. This clam extract produced pronounced hypoglycemia and convulsions in a normal rabbit following injection. A subcutaneous injection of dextrose eliminated the convulsions. However, it was not until 1965 that the first invertebrate insulin-related material was isolated and characterized immunologically, histochemically, and biologically from echinoderm digestive tract tissue (the starfish *Pisaster ochraceous*).[21] Subsequent studies, although not all definitive, reported the presence of an insulin-related material in gastrointestinal tissues from mollusks, arthropods, and tunicates including the shore crab (*Carcinus maenas*), whelk (*Buccinum undatum*), octopus (*Eledone cirrosa*), scallop (*Pecten maximus*), and sea-squirt (*Ciona intestinalis*).[22] The finding of an insulin-like material in gastrointestinal tissues of several species is not surprising since embryological and comparative studies have shown that the pancreas originates from the endodermal lining of the gastrointestinal tract.

Chambers and de Pablo have performed indirect immunofluorescence studies using anti-pork insulin antiserum during development of two species of sea urchins, *Strongylocentrotus purpuratus* and *Lytechinus variegatus*. In both species, molecules that share epitopes with mammalian insulin were detected. Positive immunostain was diffuse in all cells during cleavage stages up to the blastula. In the pluteus larva, immunostain was localized to the gut epithelium (Figure 3).[155]

Recently Smit et al.[23] identified a molluscan insulin-related peptide precursor in neuroendocrine cells of *Lymnaea stagnalis*. By *in situ* hybridization, it was demonstrated that transcription of the gene for this molluscan insulin-related peptide (MIP) occurs in canopy cells and in cerebral light-green cells, neuroendocrine cells believed to regulate various growth processes in *Lymnaea*, such as soft body part and shell growth, glycogen metabolism, and ornithine decarboxylase activity. Although A and B chain amino acid sequence comparisons with vertebrate insulins did not show much similarity, elements important in the tertiary structure have been conserved, i.e., cysteine positions, interchain disulfide bridges, and hydrophobic core residues. The functions of MIP are believed to be different from an insulin-related peptide present in the digestive tract of *Lymnaea* involved in glucose metabolism. Thus, there are clear indications that insulin-related peptides are produced in invertebrates by both the brain and the digestive system; in vertebrates, it is still not known whether insulin is made by neurons in the brain. In addition, although MIP shares several characteristics in common with vertebrate insulins, IGF-I, IGF-II, and silkworm PTTH, it also possesses features not found in any of these peptides. This indicates that MIP is another branch of a relatively ancient superfamily of related hormones and growth factors regulating growth and metabolism.

C. VERTEBRATA
1. Agnatha

Insulin has been identified in all species of vertebrates studied, including the most primitive living vertebrate, the North Atlantic hagfish (*Myxine glutinosa*).[24] Hagfish and lampreys form the class of cyclostomes — jawless, cartilagenous vertebrates believed to be direct descendants of the Ostracoderms, the earliest vertebrates known from fossil records. This is also the first example in vertebrate evolution where a separate islet organ is formed before the development of an exocrine pancreas, thus forming a link between the gut-associated B-cells of invertebrates and the islets of Langerhans of higher vertebrates. From this phylogenetic point of view and from structural and functional points discussed below, hagfish insulin is of considerable interest.

Hagfish insulin has been isolated and purified from islet organs by acid-ethanol extraction, fractional precipitation, and gel filtration.[24] It contains 52 amino acids of which 18 differ from those found in mammalian insulin; however, the three-dimensional organization of the monomer is similar to porcine insulin.[1,2,24,25] Hagfish insulin has been reported to be 8% as potent as mammalian insulins in rat epididymal fat pad assays,[26] 5% as potent as mammalian insulins in

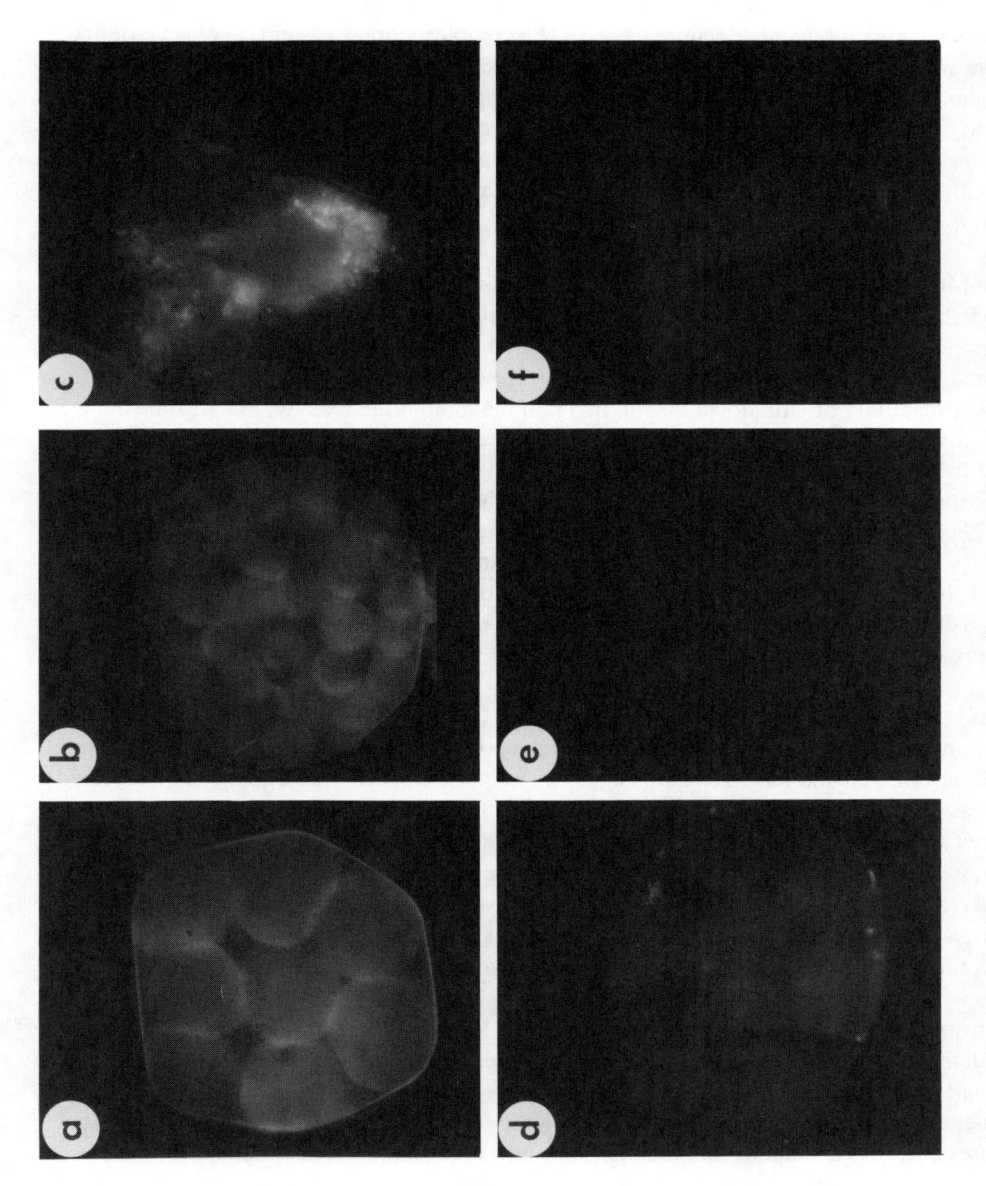

FIGURE 3. *L. variegatus* embryos were grown in artificial sea water and immunolocalization of insulin-related epitopes was carried out using the IgG fraction of a polyclonal anti-pork-insulin antiserum. Embryos at 8-cell (a, d), early blastula (b, e), and pluteus stage (c, f) were stained with either insulin antiserum (a, b, c) or control serum (d, e, f). The staining is diffuse in all cells up to the blastula, while it is preferentially localized in the gut region in the pluteus larva.

stimulating lipogenesis, 3-O-methyl glucose exchange, and glucose oxidation in isolated rat adipocytes,[27] and 5 to 10% as potent as porcine insulin in stimulating glucose oxidation and deoxyglucose transport in rat adipocytes.[28]

Among the many unique structural features of hagfish insulin is its ability to form dimers, but not hexamers, due to amino acid substitutions.[24,25] Preservation of the dimeric but not hexameric forms in hagfish insulin may be of fundamental evolutionary and/or functional significance to the endocrine system of cyclostomes. It is possible that hagfish have not evolved the granule storage mechanisms typical of mammals because a complex storage system is not yet required.

From amino acid analyses, Peterson et al.[24] concluded that hagfish insulin is as equally distant from mammalian insulins as it is from fish insulins, and that mammalian and fish insulins are closer to one another than is hagfish insulin to either one. This suggests that the cyclostomian evolutionary line probably diverged from that of more advanced vertebrates long before fish and mammalian evolution diverged. Nevertheless, despite its extraordinary evolutionary distance, hagfish insulin has retained intrinsic reactivity for mammalian receptors.

2. Amphibia

De Pablo et al. examined the insulin content of *Rana pipiens* oocytes and unfertilized eggs. Immunoreactive insulin recovered from acid-ethanol extraction and double gel-filtration was estimated to be 50 pg/g wet weight of whole egg. However, this number was not corrected for the lower reactivity of amphibian insulin in a porcine radioimmunoassay, nor was the oocyte insulin distinguished from any insulin that might be present in the jelly coating the eggs. (Unpublished observations, see Chapter 5.)

In *R. pipiens* embryos, insulin immunostain has been detected in the pancreatic anlage at Stage 22, which corresponds to $2\frac{1}{2}$ days of development (8 mm embryo).[29] Temporally, this is the first pancreatic hormone detected; glucagon cells are positive at Stage 24. In *Rana clamitans*, marked changes in the organization of the islet cells occur during metamorphosis.[30] The islets undergo differentiation from a larval to an adult type characterized by accumulation of insulin and reorganization of beta cells, the predominant cell type in amphibian islets.

3. Aves

Immunoreactive and bioactive insulin has been reported to be present in acid-ethanol extracts of both unfertilized and fertilized chick eggs.[31,32] In chick embryos, De Pablo et al.[31] detected insulin as early as Day 2 of embryogenesis, which is before the development of the endocrine pancreas and the beginning of hepatic metabolic functions. In whole embryos at Day 2, the immunoreactive insulin was estimated to be approximately 2 ng/g wet weight. A two- to three-fold increase in insulin content occurred between Days 4 and 6, coincident with differentiation of the pancreas. The physiologic function of chick embryonic insulin remains unknown; however, it is believed to be required for normal developmental pathways. (See Chapter 5.)

4. Mammalia

In all mammalian species studied, there is no evidence that free maternal insulin crosses the placenta.[33-40] Insulin-binding IgG antibodies have been detected, however, in newborn infants of diabetic mothers who had received insulin therapy.[41,42] Bauman and Yalow[43] have reported two cases of transplacental insulin transfer in insulin-requiring diabetic mothers. With the use of a human antiserum that permitted distinction between animal and human insulins, animal insulin was detected in the cord blood of the neonates. The higher the antibody titer of the mother, the greater was the total amount of insulin in the cord plasma and the greater was the fraction that was animal insulin. Thus, transplacental transfer occurred only when the insulin was complexed to antibody.

In fetal rats, pancreatic immunoreactive insulin has been detected on approximately the eleventh day of a 21-day gestation, coincident with the appearance of the pancreatic diverticu-

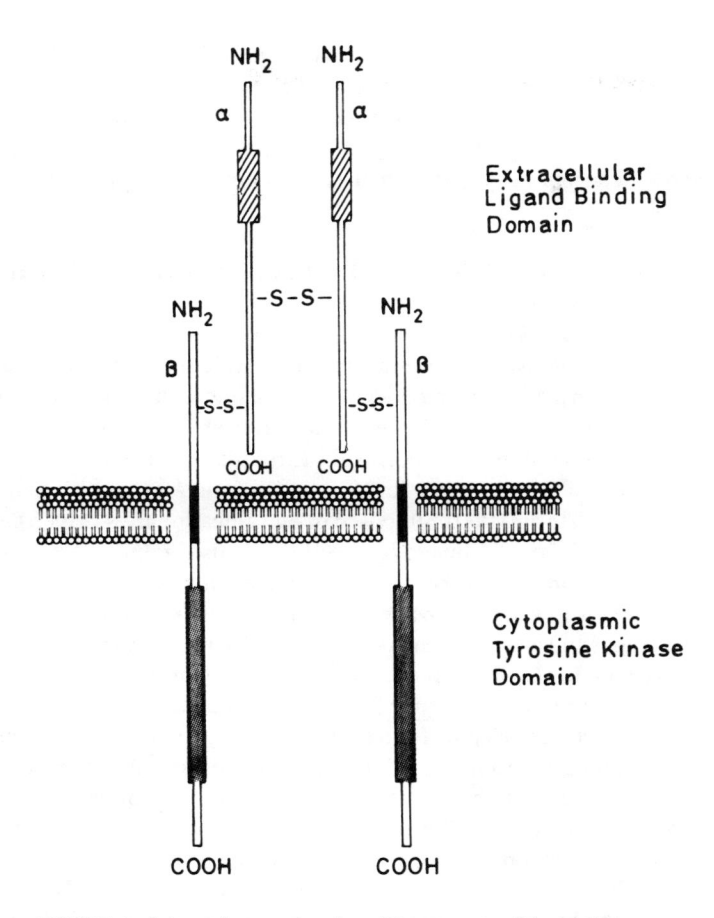

FIGURE 4. Schematic representation of the structure of the insulin receptor. The mature insulin receptor is an integral membrane heterotetrameric glycoprotein composed of two alpha and beta subunits linked by disulfide bonds. The alpha subunit binds insulin. The hatched boxes in the alpha subunit represent regions of high cysteine residue concentration. The transmembrane domains of the beta subunit are represented as solid boxes; the tyrosine kinase domains as stippled boxes.

lum.[44-46] Between Days 14 and 19, insulin-specific immunoreactivity increased more than 200-fold to a level characteristic of the fully differentiated pancreas. Kakita et al.[47] measured proinsulin mRNA in developing rat pancreata using Northern analysis. The pancreatic content of proinsulin mRNA increased 900-fold between Day 15 of gestation (undetectable levels) and 1 to 2 d after birth, coordinate with an increase in immunoreactive insulin. The concentrations of proinsulin mRNA and immunoreactive insulin in neonatal pancreata were more than eightfold higher than those found in the adult pancreas. In addition, these authors reported that glucose had a significant stimulatory effect on insulin biosynthesis as early as Day 15 of gestation, suggesting that the glucose regulatory control mechanisms for insulin biosynthesis are present and functional around the onset of endogenous insulin production in the fetus.

In contrast, in a study of bovine fetal pancreata, D'Agostino et al.[48] reported the synthesis of immunoreactive insulin during all three trimesters of gestation. Insulin was first detected during the first trimester with concentrations (units per g pancreas) increasing as development progressed. The profile of fetal pancreatic preproinsulin mRNA paralleled that of fetal insulin. Immunoreactive insulin was also detected in fetal sera from all three trimesters; however, serum concentrations remained constant throughout gestation.

Reports vary on the ontogeny of insulin production and secretion in the human.[49-52] The earliest detection of fetal pancreatic immunoreactive insulin was reported to be at 7 weeks of gestation. After its initial appearance, pancreatic insulin content was reported to increase in a linear fashion as a function of gestational age.[63] The appearance of immunoreactive insulin in the pancreas and sera during early gestation in a number of mammalian species suggests a role for insulin in fetal maturation.

III. INSULIN AND INSULIN-RELATED PEPTIDE BINDING

A. INSULIN RECEPTOR STRUCTURE

The biological effects of insulin are initiated by the interaction of insulin with its cell surface receptor. The mammalian insulin receptor derives from a single chain, glycosylated polypeptide precursor (M_r 180,000 to 210,000).[53,54] During transport from the endoplasmic reticulum to the Golgi and finally to the plasma membrane, disulfide bridges are formed and further glycosylation and cleavage occur. The mature insulin receptor is an integral membrane heterotetrameric glycoprotein (M_r 350,000 to 400,000) composed of two alpha subunits (M_r 120,000 to 135,000) and two beta subunits (M_r 90,000) linked by disulfide bonds (Figure 4).[55-59] The alpha subunit of the receptor binds insulin.[59-62] The beta subunit traverses the plasma membrane, contains the ATP-binding site, and is a protein tyrosine kinase that catalyzes phosphorylation of the beta subunit as well as exogenous peptides and proteins.[63-68] The beta subunit protein kinase domain exhibits homology with the EGF receptor and the *src* family of tyrosine kinases.[69] (Discussed in more detail in Chapters 3 and 7 through 9 of this volume.)

The insulin-like growth factors IGF-I and IGF-II are polypeptide hormones that show greater than 40% sequence similarity in amino acid structure with insulin, and are potent mitogens both *in vivo* and *in vitro*.[70-72] The receptors for insulin and IGF-I share a number of properties.[73-77] Like the insulin receptor, the mature IGF-I receptor is processed from a large precursor and exists as a membrane tetrameric glycoprotein consisting of two alpha and two beta subunits connected by disulfide bonds. The IGF-I receptor has more than 50% amino acid sequence similarity with the insulin receptor; the highest degree is in the tyrosine kinase domain. Upon binding to the alpha subunit, each ligand stimulates beta subunit tyrosine-specific protein kinase activity with subsequent beta subunit autophosphorylation. Both receptors can bind each other's ligand, although with an approximately 100-fold weaker affinity. In contrast to these two receptors, the putative IGF-II receptor consists of a single chain polypeptide (M_r 250,000) and has not been found to possess intrinsic tyrosine kinase activity or to undergo tyrosine autophosphorylation.[73,74,78] Neither the extracellular or intracellular domains of the IGF-II receptor share homology with the insulin or IGF-I receptors. The IGF-II receptor binds both IGF-I and IGF-II, although IGF-II is bound with a higher affinity.[79,80] It does not bind insulin. Although the *in vivo* and *in vitro* mitogenic functions of insulin and IGF-I are well documented, the physiological role of IGF-II and its receptor remains unknown. Recently, the IGF-II receptor has been found to be related, if not identical, to the cation independent mannose-6-phosphate receptor which has been proposed to participate in the lysosomal targeting of lysosomal proteins.[81-83]

B. INVERTEBRATA
1. Insecta

The structure of the insulin receptor appears to be highly conserved in vertebrate and in invertebrate cells. *Drosophila melanogaster* has proven to be a valuable organism with which to study insulin action and the insulin receptor due to its well-defined genetics and the isolation of an insulin-like peptide (see Section II.B.1). Moreover, mammalian insulin has been shown to be required for the survival and differentiation of cultured *Drosophila* cells from gastrula stage embryos.[84] To date, the putative *D. melanogaster* insulin receptor has been the only insulin receptor other than the human whose structure has been analyzed by use of site-specific antibodies and molecular cloning. These studies have shown that the structural features and intrinsic functions of the insulin receptor have been highly conserved despite the evolutionary divergence of vertebrates and invertebrates over 800 million years ago. The *D. melanogaster* insulin receptor homologue (dIRH) is similar to the human insulin receptor in deduced amino acid sequence, subunit structure, and ligand-stimulated protein kinase activity. Like the human insulin receptor, the mature dIRH is derived by processing from a proreceptor (M_r 200,000 to 210,000).[85] The dIRH is a cell-surface glycoprotein composed of insulin-binding alpha subunits (M_r 110,000 to 120,000) and protein tyrosine kinase transmembrane beta subunits (M_r 95,000).[85-87] However, unlike the mammalian receptor, the dIRH oligomer can also include a 170 kD receptor component in addition to the alpha and beta subunits.[85] This component, which appears to be largely incorporated into the receptor oligomer, is covalently linked to the alpha subunit by disulfide bonds, does not appear to bind insulin, and is phosphorylated on tyrosine residues in response to insulin. The significance of the subunit is not known; however, it may represent an incompletely processed proreceptor molecule.

Fernandez-Almonacid and Rosen[85] have demonstrated that the ligand-dependent protein tyrosine kinase activity of the dIRH beta subunit of embryos and cell lines is stimulated only by insulin and not by IGF-I, IGF-II, EGF, or silkworm 4K-PTTH-II. These features distinguish the dIRH from the *D. melanogaster* binding protein reported by Thompson et al. that has dual binding specificity for both insulin and EGF.[88] Of interest is the observation that IGF-I, even at very high concentrations (750 nM) which activate the human insulin receptor, did not activate the dIRH. This suggests that either there is no IGF-I receptor in *D. melanogaster* or that, if present, it is less similar to the insulin receptor than has been reported in humans.

The function of the *D. melanogaster* insulin receptor is not known. Of particular interest to the topic of this book is the suggestion that it may play a role during development. In 1985 Petruzzelli et al.[86] reported the first characterization of the membrane-associated, high affinity, insulin receptor in adult *D. melanogaster*. Glycoprotein-enriched fractions of adult *D. melanogaster* heads and bodies containing the insulin binding activity were shown to exhibit high levels of protein kinase activity that was not, however, significantly stimulated by insulin. In a subsequent study, insulin-stimulated tyrosine kinase activity was identified during embryogenesis.[87] Activity first appeared at 6 to 12 h of embryogenesis, peaked during 12 to 18 h, and fell to low levels in the adult. In addition, a *Drosophila* genomic sequence homologous to the kinase domain of the human insulin receptor was isolated which hybridized to an 11-kb mRNA. Although this mRNA was expressed throughout the *Drosophila* life cycle, it was predominant during 8 to 12 hr of embryogenesis, consistent with the stage-specific expression of insulin-dependent tyrosine kinase activity.[89]

In a recent study, insulin receptor transcripts were localized in *D. melanogaster* tissues during development.[90] Relative abundancies of two principal mRNAs of 11- and 8.6-kb were shown to be sex-specific and developmentally regulated. By Northern analysis, it was shown that, whereas both transcripts were present in similar amounts in unfertilized eggs, the 8.6-kb species was preferentially accumulated after 2 to 3 h and then became undetectable in later developmental stages. Levels of the 11-kb species began to rise after 4 h and reached a peak at 8 to 12 h, which corresponds temporally with the growth of the embryonic nervous system.[88,89] Adult female flies expressed both mRNAs, but adult males contained only the 11-kb species. *In situ* hybridization revealed that dIRH transcript expression was ubiquitous during early embryogenesis, consistent with a general metabolic role for insulin in cell growth. After mid-embryogenesis, higher levels were detected in the developing nervous system. In larvae, dIRH transcript expression was most prominent in the imaginal disks and the nervous system. The imaginal disks, which give rise to the adult epidermis, are composed of continuously dividing diploid cells, again consistent with a role for insulin in anabolic processes required for cell growth. The localization of transcripts in the nervous system suggests a role for insulin in neuronal growth. In adult flies, the highest levels of transcript expression were in the nervous system and in the ovaries. In the ovaries, hybridization was seen in the nurse cell cytoplasm of nurse cell-oocyte complexes as well as in mature oocytes. No hybridization was seen in follicle cells. Muscle, gut, and other abdominal tissues did not appear to hybridize significantly to the mRNA probe. Thus, the stage- and tissue-specific expression of dIRH mRNA is consistent with a role for the insulin receptor in development of *Drosophila* embryos.

2. Echinodermata

In 1976, using biochemical and immunocytochemical methods, Jeanmart et al.[91] reported insulin binding to sea urchin (*Paracentrotus lividus*) unfertilized and fertilized egg plasma membranes. Binding of ^{125}I-insulin to egg plasma membrane extracts was competitively inhibited by coincubation with an excess of unlabeled insulin, and was correlated with an increase in the total cAMP content of the eggs. The authors suggest that the presence of these receptors may represent specific markers of cell membrane differentiation during embryonic development. In addition, since insulin-related epitopes have been detected in sea urchin eggs and embryos (Section II.B.3), the insulin receptors may serve a developmental function.

C. VERTEBRATA

Muggeo et al.[92] systematically compared the characteristics of insulin-receptor interactions in a variety of vertebrate species from bony fish to humans. High affinity receptors capable of binding porcine insulin were detected in the erythrocytes or liver membranes in these species. Their data indicate that the binding site of the insulin receptor has been well conserved over a period of evolution of 400 million years. Properties of insulin binding, including pH and temperature dependence, specificity, affinity, and negative cooperativity, were virtually identical among all the species. Any receptor structural changes that have occurred, as suggested by differences in immunoreactivity to insulin receptor antibodies, appeared to be in regions not critical for insulin receptor binding.

Stuart[93] evaluated hepatic insulin receptors from a more diverse series of vertebrate species to determine their immunologic cross-reactivities with the human insulin receptor as well as their abilities to bind human insulin. Confirming previous work, he found that the binding site of the insulin receptor is highly conserved during evolution. Human insulin bound to hepatic membranes with remarkably consistent affinity. In an immunoassay specific for the human insulin receptor, he showed that the amount of shared antigenic determinants differed widely among the species tested and, in general, the farther the animal was from the human on the phylogenetic tree, the less was the recognition of the insulin receptor. Thus, despite considerable changes in the primary structure of the insulin receptor, the binding function has been highly conserved during evolution.

1. Agnatha

As discussed previously (Section II.C.1), insulin has been identified in the North Atlantic hagfish, the most primitive living vertebrate.[28] In the same study it was reported that the hagfish insulin receptor is even more highly conserved than insulin. The hagfish erythrocyte insulin receptor showed the negative cooperativity, time, temperature, and pH dependences of binding characteristic of all other insulin receptors. Moreover, the receptor recognized all of the bioactive insulins tested and retained the same affinities and rank order of preference for insulins and insulin analogues found with other vertebrate receptors. One unique binding feature observed was that hagfish insulin was more potent in binding to hagfish receptors than to other mammalian receptors. This is in contrast to the finding that the affinity of a given vertebrate receptor for a given insulin is independent of the nature of the endogenous insulin. Taken together, these data indicate that the insulin receptor is more highly conserved functionally than is insulin.

2. Chondrichthyes

A very recent report has characterized a novel dimeric insulin receptor from stingray (*Dasyatis americana*) liver.[94] Unlike the heterotetrameric structures of the insulin and IGF-I receptors containing two alpha and two beta subunits linked by disulfide bonds, the stingray insulin receptor consists of two noncovalently linked, identical subunits (apparent M_r 210,000) which contain both specific insulin binding and insulin-stimulated tyrosine kinase functions (Figure 4). The subunit structure of the stingray insulin receptor is more similar to that of the EGF receptor. The pH dependency and negative cooperativity characteristics of insulin binding in more advanced species were found to be similar in the stingray. Binding specificity studies showed that both IGF-I and IGF-II were four- to fivefold less potent than insulin. There was no evidence for IGF-I or IGF-II receptors. This report is the first description of an insulin receptor consisting of only one type of subunit which possesses both binding and tyrosine kinase functions. It is suggested that the stingray insulin receptor may represent a phylogenetic position prior to the evolutionary divergence of insulin and IGFs.

3. Osteichthyes

Transmembrane potentials in the presence of a range of insulin concentrations have been measured across loach (*Misgurnus fossilis*) embryos cultured during the first 6 h of development.[95] Physiologic concentrations of insulin (4 ng/ml) induced concentration-dependent hyperpolarization of the embryo membrane.

4. Amphibia

During oogenesis in the amphibian, oocytes are physiologically arrested in prophase of the first meiotic division (late G2 phase of the cell cycle). During the next several months or years, the oocytes grow without cell division to a diameter of 1.1 to 1.4 mm (Stage VI) as a result of deposition of vitellogenin. Upon exposure to progesterone, oocytes are released from this physiological arrest and resume meiosis. Maturation of the oocyte to a fertilizable egg involves breakdown of the germinal vesicle (GVBD), chromosome condensation, completion of meiosis I, and arrest at metaphase II. Although progesterone is probably the major physiological hormone involved in triggering resumption of meiosis, this arrest can also be released by a number of agents, including insulin, that work through several different signal transduction pathways.[96-104] These extracellular signals all elicit the activation of maturation promoting factor (MPF), a cytoplasmic factor responsible for the late meiotic events of the cell cycle. (For a review, see References 105 and 106.)

The *Xenopus* oocyte system has developed as a model for the study of not only the regulatory events of oocyte maturation and early development but also the mechanisms of the mitogenic signaling and protein phosphorylation actions of insulin. The advantages of this system include the physiological arrest of the oocyte in the late G2 phase of the cell cycle, the synchrony of the oocyte population after release from this block, and the extremely large cell size of the *Xenopus* oocyte (greater than 1 mm) which facilitates the use of semi-quantitative microinjection techniques and single cell analyses.

a. Amphibian Oocyte Maturation Studies

Addition of insulin to cultures of *Rana pipiens*[99-101] and *Xenopus laevis*[102-104] oocytes has been shown to induce meiotic maturation and cell division. Scatchard plot analysis of insulin binding to *Xenopus* oocytes revealed curvilinear kinetics as seen in other insulin binding systems, and indicated that the amphibian oocyte possesses a remarkably large number (4×10^7) of high affinity insulin receptors per cell. The observations that a high concentration of insulin was required to induce GVBD ($5 \times 10^{-8}M$), whereas purified IGF-I stimulated oocyte maturation at a low physiological concentration ($4 \times 10^{-10}M$), suggested that the effects of insulin on meiotic

TABLE 1
Comparison of the Regulation of *Xenopus* Oocyte Maturation by Insulin and Progesterone

Regulatory pathway	Insulin	Progesterone
Amphibian oocyte maturation cAMP system	+	+
Oocyte membrane adenylate cyclase	–	–
Hormone-induced GVBD	– by cholera toxin	– by cholera toxin
In vivo phosphodiesterase activity	+	n.e.
Phosphatidylinositol turnover		
Hormone-induced GVBD	+ by IP_3	+ by IP_3
Hormone-induced GVBD	+ by protein kinase C	n.e.
Hormone-induced GVBD	– by neomycin	n.e.
Protein phosphorylation		
Hormone-induced GVBD	– by protein tyrosine kinase Abs	n.e.
pHi		
Stage IV oocytes	+	n.e.
Stage VI oocytes	+	+
S6 phosphorylation		
Stage IV oocytes	+	n.e.
Stage VI oocytes	+	+
pHi (Stage VI)	n.e.	– by cholera toxin
S6 phosphorylation (Stage VI)	n.e.	– by cholera toxin
Hormone-induced GVBD	– by *ras* Ab	n.e.

Note: +, stimulates; -, inhibits; n.e., no significant effect GVBD, germinal vesicle breakdown; pHi, intracellular pH; IP_3, inositol-1,4,5-triphosphate; IP_3, protein kinase, and antibodies against protein tyrosine kinase domains and *ras* p 21 were microinjected into the oocytes.

maturation could be mediated via a low affinity interaction with IGF-I receptors, and not mediated by the high affinity insulin binding component. Moreover, anti-insulin receptor antibodies had no effect on the concentration-effect curve for induction of cell division, nor were anti-insulin receptor antibodies able to induce maturation, whereas in other systems treatment with anti-receptor antibodies mimics insulin action.[107]

Hormone-induced maturation of amphibian oocytes is a complex process that involves multiple cellular events and pathways. The studies described in the following paragraphs of this section are of particular interest to the theme of this book. They represent recent work that has characterized the differences in the regulation of these maturation events by insulin and progesterone, and they have contributed significantly to a better understanding of the mitogenic signalling pathways. These reports have focused on three signal transduction pathways, all of which involve the phosphorylation of proteins: (1) the cAMP system; (2) phosphatidylinositol turnover; and, (3) protein tyrosine kinase phosphorylation. The endpoint of these pathways is the activation of MPF. These studies are summarized in Table 1.

Insulin, IGF-I, and progesterone are capable of inducing GVBD in Stage VI *Xenopus* oocytes. A number of the hormone-induced biochemical events leading to GVBD have been identified, some of which are similar for these hormones. For example, one of the earliest events that is necessary to trigger the maturation response is a rapid decrease in basal cAMP levels, which has been demonstrated by intracellular microinjection of the regulatory and catalytic subunits of cAMP-dependent protein kinases.[108] This decrease in cAMP levels is due, in part, to an inhibition of oocyte membrane adenylate cyclase.[108,109] Insulin, IGF-I, and progesterone have been shown to inhibit oocyte adenylate cylase.[110,111] Further, cholera toxin, a potent irreversible activator of adenylate cyclase, inhibits GVBD induced by these hormones.[102,106,113,114] One of the late events in the maturation response is the appearance of MPF.[115] MPF has been detected in *Xenopus* oocytes after exposure to both insulin and progesterone.[102]

An example of the difference in the regulation of a key event in *Xenopus* oocyte maturation by insulin or progesterone is in the activation of phosphodiesterase. While previous studies have examined the role of adenylate cyclase inhibition in the decrease of cAMP levels, recent studies have focused on the possible involvement of phosphodiesterase activation. In contrast to the inhibition of adenylate cyclase activity induced by insulin, IGF-I, and progesterone in Stage VI oocytes, *in vivo* phosphodiesterase activity in intact oocytes was stimulated by insulin and IGF-I but was unaffected by progesterone.[112] The concentration of insulin and IGF-I required for phosphodiesterase activation was the same as the concentration required to inhibit adenylate cyclase *in vitro*, suggesting a concerted regulation of cAMP levels by insulin and IGF-I that affects both synthesis and degradation of cAMP. In this study, insulin was shown to be less potent than IGF-I in three activities: inhibition of oocyte membrane adenylate cyclase, stimulation of *in vivo* phosphodiesterase activity, and induction of oocyte maturation. These results support further the hypothesis discussed previously that insulin may stimulate oocyte meiotic maturation by interacting with IGF-I receptors. It is interesting to note that although progesterone alone had no effect on phosphodiesterase activity, low concentrations of progesterone combined with insulin greatly accelerated the time course of oocyte maturation which correlated with an increased stimulation of phosphodiesterase activity. This synergistic action of steroid plus insulin was specific for maturation-inducing steroids only.

An additional pathway by which maturation can be induced in *Xenopus* oocytes is via membrane phospholipid turnover and activation of protein kinase C. Studies of this pathway have provided further examples of differences between the mechanisms of action of insulin and progesterone.[116] Phosphatidylinositol breakdown produces two second messengers: diacylglycerol (DAG), which activates protein kinase C, and inositol-1,4,5-triphosphate (IP_3), which releases calcium from intracellular nonmitochondrial stores. Microinjection of IP_3 into oocytes did not induce maturation, but accelerated by approximately 25% the rate of GVBD induced by insulin or progesterone when injected just prior to hormone addition. Treatment of oocytes with 12-O-tetradecanoylphorbol 13-acetate (TPA), a DAG analogue and tumor promoter, induced GVBD in the absence of hormone treatment. This effect was inhibited by cholera toxin. Microinjection of highly purified protein kinase C into oocytes did not induce maturation but markedly accelerated the rate of insulin — but not progesterone — induced GVBD. Further, neomycin, a putative inhibitor of phosphoinositide turnover, reversibly inhibited insulin- but not progesterone-induced maturation. These results suggest that the products of phospholipid turnover are capable of exerting regulatory effects on oocyte maturation. These results further suggest that while insulin may, in part, act through activation of protein kinase C, progesterone uses another mechanism.

Amphibian oocytes have been shown to possess insulin receptors by binding studies[104] and by measurement of protein tyrosine kinase activity in response to insulin.[117] Microinjection of antibodies against the kinase domain into oocytes inhibited insulin- but not progesterone-induced maturation. A previous study has shown that microinjection of the *src* protein tyrosine kinase into oocytes accelerated the rate of progesterone-induced maturation.[118]

Two intermediate biochemical events, an increase in intracellular pH (pHi) and ribosomal protein S6 phosphorylation, have been induced in Stage VI oocytes by both insulin and progesterone to an equivalent extent and with the same time course.[119] However, further studies have indicated that the regulation of these events is different for the two hormones. For example, in the presence of cholera toxin, progesterone-induced changes in pHi and S6 phosphorylation were inhibited, whereas the action of insulin was virtually unaffected.[119] Since insulin-induced GVBD is inhibited by cholera toxin, these results suggest that neither increased pHi nor S6 phosphorylation is sufficient to induce GVBD. Another difference in the regulation of the events of meiotic maturation by these two hormones is their effect on small, Stage IV oocytes, which do not undergo GVBD in response to insulin or progesterone, but do undergo GVBD in response to microinjection of MPF. Both pHi and S6 phosphorylation increased significantly in Stage IV oocytes exposed to insulin; however, progesterone induced only a partial increase in pHi with no S6 phosphorylation.[119] The stimulation of S6 phosphorylation induced by insulin in Stage IV oocytes was less than the stimulation by insulin in Stage VI oocytes. Taken together, the results suggest that, although these insulin-induced responses are present in Stage IV oocytes, the magnitude may not be sufficient to induce GVBD, or that an event subsequent to S6 phosphorylation but prior to MPF appearance required for GVBD may not be present or functional in Stage IV oocytes. It has been shown that the insulin-dependent increase in S6 phosphorylation is due, in part, to activation of a ribosomal protein S6 kinase.[120,121]

Recently, Miller[122] has reported that microinjected insulin stimulated RNA and protein synthesis in Stage IV oocytes. RNA, protein, and glycogen synthesis were increased in Stage IV oocytes by external insulin. Interestingly, the effects of intracellular and external insulin were additive, and the author has suggested separate mechanisms of action. RNA synthesis was also stimulated by insulin in experiments with isolated nuclei.

As discussed previously, the rapid decrease in basal cAMP levels necessary to trigger the maturation response is due, in part, to an inhibition of membrane-associated adenylate cyclase activity. The mechanism by which progesterone inhibits oocyte membrane adenylate cyclase has been shown to involve a guanine nucleotide regulatory protein; however, studies suggest that this mechanism is novel and is not mediated via the inhibitory guanine nucleotide binding subunit, Gi, in the conventional manner.[123] Although oocytes possess functional Gs and Gi proteins, neither cholera nor pertussis toxin was able to block progesterone-induced cyclase inhibition. Interestingly, microinjection of the p21 *ras* protein has been shown to induce maturation in *Xenopus* oocytes[124,125] via production of MPF.[125] The *ras* family of proteins are membrane-bound, guanine nucleotide-binding proteins that have been implicated in mediating the effects of a variety of growth factors and hormones. Microinjection of H-*ras* p21 was not associated with any changes in intracellular cAMP concentrations, and the effects of the protein were only partially blocked by cholera toxin or a phosphodiesterase inhibitor.[124] Also, progesterone and p21 were found to act synergistically in the induction of maturation. These results suggest that not all, if any, of the effects of p21 in the *Xenopus* oocyte are mediated via adenylate cyclase. A subsequent study utilized microinjection of monoclonal antibody 6B7, an antibody directed against a synthetic peptide corresponding to a highly conserved region of p21 not found in other characterized GTP-binding proteins and required for p21 function.[126] Microinjection of this antibody inhibited p21-induced oocyte maturation, specifically inhibited insulin-induced GVBD in a concentration-dependent manner, and had no effect on progesterone-induced maturation. Recently, Lacal et al.[127] reported a fivefold, rapid increase in DAG levels, as well as increases in other products of phosphoinositide turnover, upon microinjection of H-*ras* p21 into *Xenopus* oocytes. Thus, p21 may be involved in mediating insulin-induced maturation in *Xenopus* oocytes.

Thus, it is clear that the regulation of amphibian oocyte maturation is a complex process under multifactorial control. These studies establish the *Xenopus* oocyte system as a useful model for investigating signal transduction pathways.

b. Amphibian Embryonic Studies

Investigations of later developmental stages of *Rana pipiens* have given ambiguous results. Subtotal pancreactectomy of *Rana* larvae did not inhibit growth nor did it cause a significant elevation of blood glucose.[128] Likewise, extirpation of the dorsal pancreatic rudiment of salamander (*Ambystoma punctatum*) larvae resulting in the total absence of pancreatic islets had no effect upon larval survival or the rate of growth and differentiation.[129] One interpretation of these results is that pancreatic insulin has no function during early larval development in these species. Alternatively, sources of insulin may exist in other regions of the larvae, and plasma glucose values may not be the main indicator of the function of embryonic insulin.

5. Aves

In addition to the presence of insulin immunoreactivity and bioactivity reported in Day 2 (20 to 30 somites) chick embryos (Section II.C.3), insulin receptors have also been detected at the same prepancreatic stage of development.[130-134] Receptors were reported to be widespread by Day 3, and binding to tissue preparations increased throughout embryogenesis following distinct and tissue-specific patterns. The detection limits of the binding assay do not exclude the possibility of an earlier appearance of the receptor. Functionally, insulin (ng/embryo doses) was found to have broad stimulatory effects on embryonic development by morphologic, biochemical, and metabolic parameters. In a previous report, high doses of insulin (μg/embryo) were found to have detrimental effects on chick embryo development.[135] Further, with the exception of muscle differentiation, embryonic growth and development has been shown to be retarded in the presence of anti-insulin antibodies.[136,137] (See Chapter 5 of this volume for a detailed review.)

As discussed previously, the general features of the insulin receptor have been well conserved. One tissue-related difference that has been reported, however, is in the electrophoretic mobility of the alpha subunit, which appears to be the consequence of variations in sialic acid content of the carbohydrate moieties of the receptor. Alpha subunit heterogeneity has been described among adult liver, brain, muscle, and erythrocyte insulin receptors of various species as well as between neonatal and late fetal rat brain and liver.[132,138-144] For example, the alpha subunit of the adult rat brain has been reported to have a lower apparent M_r compared to the liver.[136,137] In addition, the apparent alpha subunit M_r of Day 11.5 rat embryos is smaller than the yolk sac membrane subunit which is smaller than the adult liver subunit.[142] Bassas et al.[132] have shown that this tissue-specific heterogeneity is present early in chick organogenesis. Differences in apparent M_r ranging from 138 kDa to 129 kDa were detected in membranes of newly differentiated chick embryo tissues (Day 8 liver > Day 6 heart = Day 12 skeletal muscle > Day 6 brain). The alpha subunit M_r of the Day 2 whole chick embryo was identical to that of the Day 3 head and Day 6 brain. Desialylation of the insulin receptors by neuraminidase treatment produced significant changes in the alpha subunit M_r in liver and heart tissues but not in brain or whole Day 2 embryos. Previous to this experiment, the only cell type reported as being insensitive to neuraminidase treatment was the neuron. These results suggest that the insulin receptor predominant in the earliest stages of chick embryonic development is similar to the neuronal type. Moreover, receptor posttranslational modifications which could modulate the actions of insulin may occur during terminal differentiation of cardiac, skeletal, and hepatic tissues, but not in the receptors of neural cells or the earliest embryonic tissues. All of the tissue-specific features described for the insulin receptor were paralleled by the IGF-I receptor.

6. Mammalia

The insulin receptor is distributed ubiquitously in mammalian organisms. Insulin receptors have been characterized not only on the classic target cells, namely liver, muscle, and fat, but also on a variety of other tissues such as brain,[145] placenta,[146] lymphocytes,[147] and fibroblasts.[148] The physiological significance of the receptors in some of these nontraditional target cells is uncertain but will likely include a role in the control of cell proliferation and growth.

The nature and role of insulin and IGFs during early embryonic development represent areas of increasing interest to developmental biologists. Although considerable evidence suggests a role for insulin during fetal growth, few studies have examined the interactions of insulin with cells of early mammalian embryos, primarily due to the limited quantities of tissue available for study. Stage-specific insulin binding has been detected in the preimplantation mouse embryo by indirect immunofluorescence and light microscopic autoradiography.[149,150] Specific insulin binding was detected on morulae and blastocysts; no binding was demonstrated on oocytes or embryos throughout the eight-cell stage. In addition, insulin, IGF-I, and IGF-II binding was detected on cells of blastocyst outgrowths. The ontogeny of insulin receptors coincides with a switch in energy dependence of the embryo from pyruvate and lactate to glucose and an increase in protein synthetic activity.[151-154] (This topic is reviewed in detail in Chapter 6 of this volume.)

IV. CONCLUSIONS

Considerable evidence suggests that insulin is neither a unique product of pancreatic beta cells nor of vertebrate organisms. The structure of insulin is remarkably conserved throughout evolution. Natural selection has operated to preserve the tertiary and quaternary structure of the hormone. Phylogenetically, insulin antedates the pancreas.

The insulin receptor is more highly conserved functionally than insulin itself. The structural features and intrinsic functions are preserved throughout evolution. Any structural changes that have occurred appear to be in regions not critical for insulin binding.

Functional insulin receptors have been characterized on many nontraditional target cells. The available evidence suggests that the insulin receptor may function in the control of cell growth and proliferation. Recent studies have demonstrated a role for insulin in mediating oocyte maturation, embryonic development, and fetal growth.

Drosophila melanogaster and *Xenopus* oocyte systems have proven to be valuable models with which to study the insulin receptor and action and provide a unique opportunity to characterize signal transduction pathways.

ABBREVIATIONS

DAG	Diacylglycerol
dIRH	*Drosophila* insulin receptor homologue
EGF	Epidermal growth factor
GVBD	Germinal vesicle breakdown
IGF-I	Insulin-like growth factor I
IGF-II	Insulin-like growth factor II
IP_3	Inositol-1,4,5-triphosphate
MIP	Molluscan insulin-related peptide
MPF	Maturation promoting factor
pHi	Intracellular pH
PTTH	Prothoracicotropic hormone
TPA	12-O-Tetradecanoylphorbol 13-acetate

REFERENCES

1. **Blundell, T. L. and Wood, S. P.,** Is the evolution of insulin Darwinian or due to selectively neutral mutation?, *Nature (London),* 257, 198, 1975.

2. **Emdin, S. and Falkmer, S.,** Phylogeny of insulin, *Acta Paediatr. Scand.,* S270, 15, 1977.

3. **Steiner, D.,** Insulin today, *Diabetes,* 26, 322, 1977.

4. **Steiner, D. F., Peterson, J. D., Tager, H., Emdins, S., Ostberg, Y., and Falkmer, S.,** Comparative aspects of proinsulin and insulin structure and biosynthesis, *Am. Zool.,* 13, 591, 1973.

5. **Kimura, M. and Ohta, T.,** On some principles governing molecular evolution, *Proc. Natl. Acad. Sci., U.S.A.,* 71, 2848, 1974.

6. **Wyatt, G. R.,** The biochemistry of insect hemolymph, *Annu. Rev. Entomol.,* 6, 75, 1961.

7. **Chefurka, W.,** Some comparative aspects of metabolism of carbohydrates in insects, *Annu. Rev. Entomol.,* 10, 345, 1965.

8. **Bueding, E.,** Comparative aspects of carbohydrate metabolism, *Fed. Proc. Fed. Am. Soc. Exp. Biol.,* 21, 1039, 1962.

9. **Tager, H. S., Markese, J., Kramer, K. J., Speirs, R. D., and Childs, C. N.,** Glucagon-like and insulin-like hormones of the insect neurosecretory system, *Biochem. J.,* 156, 515, 1976.

10. **Kramer, K. J., Childes, C. N., Spiers, R. D., and Jacobs, R. M.,** Purification of insulin-like peptides from hemolymph and royal jelly, *Insect Biochem.,* 12, 91, 1982.

11. **El-Salhy, M., Falkmer, S., Kramer, K. J., and Spiers, R. D.,** Immunohistochemical investigations of neuropeptides in the brain, corpora cardiaca, and corpora allata of an adult lepidopteran insect, *Manduca sexta, Cell Tissue Res.,* 232, 295, 1983.

12. **Duve, H.,** The presence of a hypoglucemic and hypotrehalocemic hormone in the neurosecretory system of the blowfly, *Calliphora erythrocephala, Gen. Comp. Endocrinol.,* 36, 102, 1978.

13. **Duve, H., Thorpe, A., and Lazarus, N.,** The isolation of material displaying insulin-like immunological and biological activity from the brain of the blowfly, *Calliphora vomitoria, Biochem. J.,* 184, 221, 1979.

14. **Duve, H. and Thorpe, A.,** Immunofluorescent localization of insulin-like material in the median neurosecretory cells of the blowfly, *Calliphora vomitoria,* (Diptera), *Cell Tissue Res.,* 200, 187, 1979.

15. **Meneses, P. and De Los Angeles, M.,** A protein extract from *Drosophila melanogaster* with insulin-like activity, *Comp. Biochem. Physiol.,* 51A, 483, 1975.

16. **LeRoith, D., Lesniak, M. A., and Roth, J.,** Insulin in insects and annelids, *Diabetes,* 30, 70, 1981.

17. **Nagasawa, H., Kataoka, H., Isogai, A, Tamura, S., Suzuki, A, Mizoguchi, A., Fujiwara, Y., Suzuki, A., Takahashi, S. Y., and Isjizaki, H.,** Amino acid sequence of a prothoracicotropic hormone of the silkworm *Bombyx mori, Proc. Natl. Acad. Sci. U.S.A.,* 83, 5840, 1986.

18. **LeRoith, D., Shiloach, J., Roth, J., and Lesniak, M. A.,** Evolutionary origins of vertebrate hormones: substances similar to mammalian insulins are native to unicellular eukaryotes, *Proc. Natl. Acad. Sci. U.S.A.,* 77, 6184, 1980.

19. **De Pablo, F., Lesniak, M. A., Hernandez, E. R., LeRoith, D., Shiloach, J., and Roth, J.** Extracts of protozoa contain materials that react specifically in the immunoassay for guinea pig insulin, *Horm. Metab. Res.,* 18, 82, 1986.

20. **Collip, J. B.,** The demonstration of an insulin-like substance in the tissues of the clam (*Mya arenaria*), *J. Biol. Chem.,* 55, xxxix, 1923.

21. **Wilson, S. and Falkmer, S.,** Starfish insulin, *Can. J. Biochem.,* 43, 1615, 1965

22. **Davidson, J. K., Falkmer, S., Mehrotra, B. K., and Wilson, S.,** Insulin assays and light microscopical studies of digestive organs in Protostomian and Deuterostomian species and in Coelenterates, *Gen. Comp. Endocrinol.,* 17, 388, 1971.

23. **Smit, A. B., Vreugdenhil, E., Ebberink, R. H. M., Geraerts, W. P. M., Klootwijk, J., and Joosse, J.,** Growth-controlling molluscan neurons produce the precursor of an insulin-related peptide, *Nature (London),* 331, 535, 1988.

24. **Peterson, J. D., Steiner, D. F., Emdin, S. O., and Falkmer, S.,** The amino acid sequence of the insulin from a primitive vertebrate, the Atlantic hagfish (*Myxine glutinosa*), *J. Biol. Chem.,* 250, 5183, 1975.

25. **Peterson, J. D., Coulter, C. L., Steiner, D. F., Emdin, S. O., and Falkmer, S.,** Structural and crystallographic observations on hagfish insulin, *Nature (London),* 251, 239, 1974.

26. **Weitzel, G., Stratling, W. H., Hahn, J., and Martini, O.,** Insulin vom Schleimfisch (*Myxine glutinosa*; Cyclostomata), *Hoppe-Seyler's Z. Physiol. Chem.,* 348, 525, 1967.

27. **Emdin, S. O., Gammeltoft, S., and Gliemann, J.,** The degradation, binding affinity, and potency of insulin from Atlantic hagfish determined on isolated rat fat cells, *J. Biol. Chem.,* 252, 602, 1977.

28. **Muggeo, M., Van Obberghen, E., Kahn, C. R., Roth, J., Ginsberg, B. H., de Meyts, P., Emdin, S. O., and Falkmer, S.,** The insulin receptor and insulin of the Atlantic hagfish, *Diabetes,* 28, 175, 1979.

29. **Kaung, H.-L. C.,** Immunocytochemical localization of pancreatic endocrine cells in frog embryos and young larvae, *Gen. Comp. Endocrinol.,* 45, 204, 1981.

30. **Frye, B. E.,** Metamorphic changes in the blood sugar and the pancreatic islets of the frog *Rana clamitans, J. Exp. Zool.,* 155, 215, 1964.

31. **De Pablo, F., Roth, J., Hernandez, E., and Pruss, R. M.,** Insulin is present in chicken eggs and early chick embryos, *Endocrinology,* 111, 1909, 1982.

32. **Trenkle, A. and Hopkins, K.,** Immunological investigation of an insulin-like substance in the chicken egg, *Gen. Comp. Endocrinol.,* 16, 493, 1971.

33. **Davis, J. and Lacy, P. E.,** Observations on the failure of insulin to pass from the foetus to the mother in the rabbit, *Am. J. Obstet. Gynecol.,* 74, 514, 1957.

34. **Goodner, C. J. and Freinkel, N.,** Carbohydrate metabolism in pregnancy. IV. Studies on the permeability of the rat placenta to 131-I-insulin, *Diabetes,* 10, 383, 1961.

35. **Josimovich, J. B. and Knobil, E.,** Placental transfer of 131-I-insulin in the Rhesus monkey, *Am. J. Physiol.,* 200, 471, 1961.

36. **Buse, M. G., Roberts, W. J., and Buse, J.,** The role of the human placenta in the transfer and metabolism of insulin, *J. Clin. Invest.,* 41, 29, 1962.

37. **Seller, M. J.,** The effect of glucose and insulin on the pregnant rat and fetus, *J. Physiol.,* 172, 353, 1964.

38. **Adam, P. A. G., Teramo, K., Raiha, N., Gitlin, D., and Schwartz, R.,** Human fetal insulin metabolism early in gestation. Response to acute elevation of fetal blood glucose concentration and placental transfer of human 131-I-insulin, *Diabetes,* 18, 409, 1969.

39. **Wolf, H., Sabata, V., Frerichs, H., and Stubbe, P.,** Evidence for the impermeability of the human placenta for insulin, *Horm. Metab. Res.,* 1, 274, 1969.

40. **Posner, B. I.,** Insulin receptors in human and animal placental tissue, *Diabetes,* 23, 209, 1974.

41. **Spellacy, W. N. and Goetz, F. C.,** Insulin antibodies in pregnancy, *Lancet,* ii, 222, 1963.

42. **Isles, T. E. and Farquhar, J. W.,** The effect of endogenous antibody on insulin-assay in the newborn infants of diabetic mothers, *Pediatr. Res.,* 1, 110, 1967.

43. **Bauman, W. A. and Yalow, R. S.,** Transplacental passage of insulin complexed to antibody, *Proc. Natl. Acad. Sci. U.S.A.,* 78, 4588, 1981.

44. **Clark, W. R. and Rutter, W. J.,** Synthesis and accumulation of insulin in the fetal rat pancreas, *Dev. Biol.,* 29, 468, 1972.

45. **Sanders, T. G. and Rutter, W. J.,** The developmental regulation of amylolytic and proteolytic enzymes in the embryonic rat pancreas, *J. Biol. Chem.,* 249, 3500, 1974.

46. **Muglia, L. and Locker, J.,** Extrapancreatic insulin gene expression in the fetal rat, *Proc. Natl. Acad. Sci. U.S.A.,* 81, 3635, 1984.

47. **Kakita, K., Giddings, S. J., Rotwein, P.S., and Permutt, M. A.,** Insulin gene expression in the developing rat pancreas, *Diabetes,* 32, 691, 1983.

48. **D'Agostino, J., Field, J. B., and Frazier, M. L.,** Ontogeny of immunoreactive insulin in the fetal bovine pancreas, *Endocrinology,* 116, 1108, 1985.

49. **Steinke, J. and Driscoll, S. G.,** The extractable insulin content of pancreas from fetuses and infants of diabetic and control mothers, *Diabetes,* 14, 573, 1965.

50. **Rastogi, G. K., Letarte, J., and Fraser, T.R.,** Immunoreactive insulin content of 203 pancreases from fetuses of healthy mothers, *Diabetologia,* 6, 445, 1970.

51. **Kaplan, S. L., Grumbach, M. M. and Shephard, T. H.,** The ontogenesis of human fetal hormones, *J. Clin. Invest.,* 51, 3080, 1972.

52. **Schaeffer, L. D., Wilder, M. L., and Williams, R. H.,** Secretion and content of insulin and glucagon in human fetal pancreas slices *in vitro, Proc. Soc. Biol. Med.,* 143, 314, 1973.

53. **Deutsch, P. S., Wan, C. F., Rosen, O. M., and Rubin, C. S.,** Latent insulin receptor and possible receptor precursors in 3T3-L1 adipocytes, *Proc. Natl. Acad. Sci. U.S.A.,* 80, 133, 1983.

54. **Hedo, J. A., Kahn, C. R., Hayashi, M., Yamada, K. M., and Kasuga, M. J.,** Biosynthesis and glycosylation of the insulin receptor, *J. Biol. Chem.,* 258, 10029, 1983.

55. **Cuatrecasas, P. J.,** Interaction of Concanavalin A and wheat germ agglutinin with the insulin receptor of fat cells and liver, *J. Biol. Chem.,* 248, 3528, 1973.

56. **Massague, J., Pilch, P., and Czech, M. P.,** Electrophoretic resolution of three major insulin receptor structures with unique subunit stoichiometries, *Proc. Natl. Acad. Sci. U.S.A.,* 77, 7137, 1980.

57. **Hedo, J., Kasuga, M., Van Obberghen, E., Roth, J., and Kahn, C. R.,** Direct demonstration of glycosylation of insulin receptor subunits by biosynthetic and external labeling: evidence for heterogeneity, *Proc. Natl. Acad. Sci. U.S.A.,* 78, 4791, 1981.

58. **Massague, J., Pilch, P., and Czech, M. P.,** A unique proteolytic cleavage site on the beta subunit of the insulin receptor, *J. Biol. Chem.,* 256, 3182, 1981.

59. **Siegel, T., Ganguly, S., Jacobs, S., Rosen, O. M., and Rubin, C. S.,** Purification and properties of the human placental insulin receptor, *J. Biol. Chem.,* 256, 9266, 1981.

60. **Yip, C. C., Yeung, C. W. T., and Moule, M. C.,** Photoaffinity labeling of insulin receptors of rat adipocyte plasma membranes, *J. Biol. Chem.,* 253, 1743, 1978.

61. **Jacobs, S., Hazum, E., Schechter, Y., and Cuatrecasas, P.,** Insulin receptor: covalent labeling and identification of subunits, *Proc. Natl. Acad. Sci. U.S.A.,* 76, 4918, 1979.

62. **Pilch, P. F. and Czech, M.,** Interaction of cross-linking agents with insulin effector systems of isolated fat cells, *J. Biol. Chem.,* 254, 3375, 1979.

63. **Petruzzelli, L., Ganguly, S., Smith, C. J., Cobb, M. H., Rubin, C. S., and Rosen, O. M.,** Insulin activates a tyrosine-specific protein kinase in extracts of 3T3-L1 adipocytes and human placenta, *Proc. Natl. Acad. Sci. U.S.A.,* 79, 6792, 1982.

64. **Shia, M. A. and Pilch, P.,** The beta subunit of the insulin receptor is an insulin-activated protein kinase, *Biochemistry,* 22, 717, 1983.

65. **Stadtmauer, L. and Rosen, O. M.,** Phosphorylation of exogenous substrates by the insulin receptor-associated protein kinase, *J. Biol. Chem.,* 258, 6682, 1983.

66. **Zick, Y., Whittaker, J., and Roth, J.,** Insulin stimulated phosphorylation of its own receptor, *J. Biol. Chem.,* 258, 3431, 1983.

67. **Van Obberghen, E., Ross, B., Kowalski, A., Gazzano, H., and Ponzio, G.,** Receptor mediated phosphorylation of the hepatic insulin receptor: evidence that the M_r 95,000 receptor subunit is its own kinase, *Proc. Natl. Acad. Sci. U.S.A.,* 80, 945, 1983.

68. **Petruzzelli, L., Herrera, R., and Rosen, O. M.,** Insulin receptor is an insulin-dependent tyrosine protein kinase: copurification of insulin-binding activity and protein kinase activity to homogeneity from human placenta, *Proc. Natl. Acad. Sci. U.S.A.,* 81, 3327, 1984.

69. **Ullrich, A., Bell, J. R., Chen, E. Y., Herrera, R., Petruzzelli, L. M., Dull, T. J., Gray, A., Coussens, L., Liao, Y.-C., Tsubokawa, M., Mason, A., Seeburg, P. H., Grunfeld, C., Rosen, O. M., and Ramachandran, J.,** Human insulin receptor and its relationship to the tyrosine kinase family of oncogenes, *Nature (London),* 313, 756, 1985.

70. **Rinderknecht, E. and Humbel, R. E.,** The amino acid sequence of human insulin-like growth factor I and its structural homology with proinsulin, *J. Biol. Chem.,* 253, 2769, 1978.

71. **Rinderknecht, E. and Humbel, R. E.,** Primary structure of human insulin-like growth factor II, *FEBS Lett.,* 89, 283, 1978.

72. **Van Wyk, J. J.,** The somatomedins and their actions, in *Biochemical Actions of Hormones,* Vol. 4., Litwak, G., Ed., Academic Press, New York, 1978, 101.

73. **Kasuga, M., Van Obberghen, E., Nissley, S. P., and Rechler, M. M.,** Demonstration of two subtypes of insulin-like growth factor receptors by affinity cross-linking, *J. Biol. Chem.,* 256, 5305, 1981.

74. **Massague, J. and Czech, M. P.,** The subunit structures of two distinct receptors for insulin-like growth factors I and II and their relationship to the insulin receptor, *J. Biol. Chem.,* 257, 5038, 1982.

75. **Jacobs, S., Kull, F. C., Earp, H. S., Svoboda, M. E., Van Wyk, J. J., and Cuatrecasas, P.,** Somatomedin-C stimulates the phosphorylation of the beta-subunit of its own receptor, *J. Biol. Chem.,* 258, 9581, 1983.

76. **Sasaki, N., Rees-Jones, R. W., Zick, Y., Nissley, S. P., and Rechler, M. M.,** Characterization of insulin-like growth factor I stimulated tyrosine kinase activity associated with the beta subunit of Type I insulin-like growth factor receptors of rat liver cells, *J. Biol. Chem.,* 260, 9793, 1985.

77. **Ullrich, A., Gray, A., Tam, A. W., Yang-Feng, T., Tsubokawa, M., Collins, C., Henzel, W., Le Bon, T., Kathuria, S., Chen, E., Jacobs, S., Francke, U., Ramachandran, J., and Fujita-Yamaguchi, Y.,** Insulin-like growth factor I receptor primary structure: a comparison with insulin receptor suggests structural determinants that define specificity, *EMBO J.,* 5, 2503, 1986.

78. **Corvera, S., Whitehead, R. E., Mottola, C., and Czech, M. P.,** The insulin-like growth factor II receptor is phosphorylated by a tyrosine kinase in adipocyte plasma membranes, *J. Biol. Chem.,* 261, 7675, 1986.

79. **Ewton, D. Z., Falen, S. L., and Florini, J. R.,** The type II insulin-like growth factor (IGF) receptor has low affinity for IGF-I analogs: pleiotypic actions of IGFs on myoblasts are apparently mediated by type I receptor, *Endocrinology,* 120, 115, 1987.

80. **Rosenfeld, R. O., Conover, C. A., Hodges, D., Lee, P. D. K., Misra, P., Hintz, R. L., and Li, C. H.,** Heterogeneity of insulin-like growth factor-I affinity for the insulin-like growth factor-II receptor: comparison of natural, synthetic, and recombinant DNA- derived insulin-like growth factor I, *Biochem. Biophys. Res. Commun.,* 143, 199, 1987.

81. **Morgan, D. O., Edman, J. C., Standring, D. N., Fried, V.A., Smith, M. C., Roth, R. A., and Rutter, W. J.,** Insulin-like growth factor II receptor as a multifunctional binding protein, *Nature (London),* 329, 301, 1987.

82. **Roth, R. A.,** Structure of the receptor for insulin-like growth factor II: the puzzle amplified, *Science,* 235, 1269, 1988.

83. **Kiess, W., Blickenstaff, G. D., Sklar, M. M., Thomas, C. L., Nissley, S. P., and Sahagian, G. G.,** Biochemical evidence that the type II insulin-like growth factor receptor is identical to the cation-independent mannose 6-phosphate receptor, *J. Biol. Chem.,* 263, 9339, 1988.

84. **Seecof, R. L. and Dewhurst, S.,** Insulin is a *Drosophila* hormone and acts to enhance the differentiation of embryonic *Drosophila* cells, *Cell Differ.,* 3, 63, 1974.

85. **Fernandez-Almonacid, R. and Rosen, O. M.,** Structure and ligand specificity of the *Drosophila melanogaster* insulin receptor, *Mol. Cell. Biol.,* 7, 2718, 1987.

86. **Petruzzelli, L., Herrera, R., Garcia, R., and Rosen, O. M.,** The insulin receptor of *Drosophila melanogaster, Cancer Cells 3/Growth Factors and Transformation,* Cold Spring Harbor Laboratory, New York, 2718, 1985.

87. **Petruzzelli, L., Herrera, R., Garcia-Arenas, R., and Rosen, O. M.,** Acquisition of insulin-dependent protein tyrosine kinase activity during *Drosophila* embryogenesis, *J. Biol. Chem.,* 260, 16072, 1985.

88. **Thompson, K. L., Decker, S. J., and Rosner, M. R.,** Identification of a novel receptor in *Drosophila* for both epidermal growth factor and insulin, *Proc. Natl. Acad. Sci. U.S.A.,* 82, 8443, 1985.

89. **Petruzzelli, L., Herrera, R., Arenas-Garcia, R., Fernandez, R., Birnbaum, M. J., and Rosen, O. M.,** Isolation of a *Drosophila* genomic sequence homologous to the kinase domain of the human insulin receptor and detection of the phosphorylated *Drosophila* receptor with an anti-peptide antibody, *Proc. Natl. Acad. Sci. U.S.A.,* 83, 4710, 1986.

90. **Garofalo, R. S. and Rosen, O. M.,** Tissue localization of *Drosophila melanogaster* insulin receptor transcripts during development, *Mol. Cell. Biol.,* 8, 1638, 1988.

91. **Jeanmart, J., Uytdenhoef, P., DeSutter, G., and Legros, F.,** Insulin receptor sites and membrane markers during embryonic development, *Differentiation,* 7, 23, 1976.

92. **Muggeo, M., Ginsberg, B. H., Roth, J., Neville, D. M., De Meyts, P., and Kahn, C. R.,** The insulin receptor in vertebrates is functionally more conserved during evolution than insulin itself, *Endocrinology,* 104, 1393, 1979.

93. **Stuart, C. A.,** Phylogenetic distance from man correlates with immunologic cross-reactivity among liver insulin receptors, *Comp. Biochem. Physiol.,* 84B, 167, 1986.

94. **Stuart, C. A.,** Characterization of a novel insulin receptor from stingray liver, *J. Biol. Chem.,* 263, 7881, 1988.

95. **Kusen, S. I., Sanagursky, D. I., Murashchick, I. G., and Goyda, E. A.,** Transmembrane potential changes in the developing loach embryos under the influence of insulin and inhibition of transcription and translation, *Biofizika,* 25, 658, 1980.

96. **Brachet, J., Baltus, E., Pays-DeSchutter, A., Hanocq-Quertier, J., Hubert, E., and Steinert, G.,** Induction of maturation (meiosis) in *Xenopus laevis* oocytes by three organomercurials, *Proc. Natl. Acad. Sci. U.S.A.,* 72, 1574, 1975.

97. **Schorderet-Slatkine, S., Schorderet, M., and Baulieu E. E.,** Initiation of meiotic maturation in *Xenopus laevis* oocytes by lanthanum, *Nature (London),* 262, 289, 1976.

98. **Baulieu, E. E., Godeau, F., Schorderet, M., and Schorderet-Slatkine, E.,** Steroid-induced meiotic division in *Xenopus laevis* oocytes: surface and calcium, *Nature (London),* 275, 593, 1976.

99. **Lessman, C. A. and Schuetz, A. W.,** Role of follicle wall in meiosis reinitiation induced by insulin in *Rana pipiens* oocytes, *Am. J. Phys.,* 241, E51, 1981.

100. **Lessman, C. A. and Schuetz, A. W.,** Insulin induction of meiosis of *Rana pipiens* oocytes: relation to endogenous progesterone, *Gamete Res.,* 6, 95, 1982.

101. **Lessman, C. A. and Marshall, W. S.,** Electrophysiology of *in vitro* insulin- and progesterone-induced reinitiation of oocyte meiosis in *Rana pipiens, J. Exp. Zool.,* 231, 257, 1984.

102. **El-Etr, M., Schorderet-Slatikine, S., and Baulieu, E. E.,** Meiotic maturation in *Xenopus laevis* oocytes initiated by insulin, *Science,* 205, 1397, 1979.

103. **El-Etr, M., Schorderet-Slatkine, S., and Baulieu, E. E.,** The role of zinc and follicle cells in insulin-initiated meiotic maturation of *Xenopus laevis* oocytes, *Science,* 210, 928, 1980.

104. **Maller, J. L. and Koontz, J. W.,** A study of induction of cell division in amphibian oocytes by insulin, *Dev. Biol.,* 85, 309, 1981.

105. **Maller, J. L.,** Regulation of amphibian oocyte maturation, *Cell Differer.,* 16, 211, 1985.

106. **Maller, J. L.,** Mitogenic signalling and protein phosphorylation in *Xenopus* oocytes, *J. Cyclic Nucleotide and Protein Phosphorylation Res.,* 11, 543, 1987.

107. **Baldwin, D., Terris, S., and Steiner, G.,** Characterization of insulin-like actions of anti-insulin receptor antibodies, *J. Biol. Chem.,* 255, 4028, 1980.

108. **Maller, J. L. and Krebs, E. G.,** Progesterone-stimulated meiotic cell division in *Xenopus* oocytes: induction by regulatory subunit and inhibition by catalytic subunit of adenosine 3'5'-monophosphate-dependent protein kinase, *J. Biol. Chem.,* 252, 1712, 1977.

109. **Huchon, D., Ozon, R., Fischer, E. H., and DeMaille, J. G.,** The pure inhibitor of cAMP-dependent protein kinase initiates *Xenopus laevis* oocyte maturation: a 4-step scheme for meiotic maturation, *Mol. Cell. Endocrinol.,* 22, 211, 1981.

110. **Sadler, S. E. and Maller, J. L.,** Progesterone inhibits adenylate cyclase in Xenopus oocytes: action on the guanine nucleotide regulatory protein, *J. Biol. Chem,* 256, 6368, 1981.

111. **Sadler, S. E. and Maller, J. L.,** Inhibition of *Xenopus* oocyte adenylate cyclase by progesterone and 2'5'-dideoxyadenosine is associated with slowing of guanine nucleotide exchange, *J. Biol. Chem.,* 258, 7935, 1983.

112. **Sadler, S. E. and Maller, J. L.,** *In vivo* regulation of cyclic AMP phosphodiesterase in *Xenopus* oocytes: stimulation by insulin and insulin-like growth factor, *J. Biol. Chem.,* 262, 10644, 1987.

113. **Schorderet-Slatkine, S., Schorderet, M., Boquet, P., Godeau, F., and Baulieu, E. E.,** Progesterone-induced meiosis in *Xenopus laevis* oocytes: a role for cAMP at the "maturation-promoting factor" level, *Cell,* 15, 1269, 1978.

114. **Maller, J. L., Butcher, F. R., and Krebs, E. G.,** Early effects of progesterone on levels of cyclic adenosine 3'5'-monophosphate in *Xenopus* oocytes, *J. Biol. Chem.,* 254, 579, 1979,

115. **Masui, Y. and Markert, C. L.,** Cytoplasmic control of nuclear behavior during meiotic maturation of frog oocytes, *J. Exp. Zool.,* 177, 129, 1971.

116. **Stith, B. J. and Maller, J. L.,** Induction of meiotic maturation in *Xenopus* oocytes by 12-O-tetradecanoylphorbol 13-acetate, *Exp. Cell Res.,* 169, 514, 1987.

117. **Morgan, D. O., Ho, L., Korn, L. J., and Roth, R. A.,** Insulin action is blocked by a monoclonal antibody that inhibits the insulin receptor kinase, *Proc. Natl. Acad. Sci. U.S.A.,* 83, 328, 1986.

118. **Spivack, J. G., Erikson, R. L., and Maller, J. L.,** Microinjection pp60-v-*src* into *Xenopus* oocytes increases phosphorylation of ribosomal protein S6 and accelerates the rate of progesterone-induced meiotic maturation, *Mol. Cell. Biol.,* 4, 1631, 1984.

119. **Stith, B. J. and Maller, J. L.,** The effect of insulin on intracellular pH and ribosomal protein S6 phosphorylation in oocytes of *Xenopus laevis, Dev. Biol.,* 102, 79, 1984.

120. **Maller, J. L., Pike, L. J., Freidenberg, G. R., Cordera, R., Stith, B. J., Olefsky, J. M., and Krebs, E. G.,** Increased phosphorylation of ribosomal protein S6 following microinjection of insulin receptor-kinase into *Xenopus* oocytes, *Nature (London),* 320, 459, 1986.

121. **Stefanovic, D., Erikson, E., Pike, L. J., and Maller, J. L.,** Activation of a ribosomal protein S6 kinase in Xenopus oocytes by insulin and insulin-receptor kinase, *EMBO J.,* 5, 157, 1986.

122. **Miller, D. S.,** Stimulation of RNA and protein synthesis by intracellular insulin, *Science,* 240, 506, 1988.

123. **Sadler, S. E., Maller, J. L., and Cooper, D. M. F.,** Progesterone inhibition of *Xenopus* oocyte adenylate cyclase is not mediated via the *Bordetella pertussis* toxin substrate, *Mol. Pharmacol.,* 26, 526, 1984.

124. **Birchmeier, C., Brook, D., and Wigler, M.,** *Ras* proteins can induce meiosis in *Xenopus* oocytes, *Cell,* 43, 615, 1985.

125. **Deshpande, A. K. and Kung, H.-F.,** Insulin induction of *Xenopus laevis* oocyte maturation is inhibited by monoclonal antibody against p21 *ras* proteins, *Mol. Cell. Biol.,* 7, 1285, 1987.

126. **Korn, L. J., Siebel, C. W., McCormick, F., and Roth, R.,** *Ras* p21 as a potential mediator of insulin action in *Xenopus* oocytes, *Science,* 236, 840, 1987.

127. **Lacal, J. C., de la Pena, P., Moscat, J., Garcia- Barreno, P., Anderson, P. S., and Aaronson, S. A.,** Rapid stimulation of diacylglycerol production in *Xenopus* oocytes by microinjection of H-*ras* p21, *Science,* 238, 533, 1987.

128. **Frye, B. E.,** Metamorphic changes in the blood sugar and pancreatic islets of the frog *Rana clamitans, J. Exp. Zool.,* 155, 215, 1964.

129. **Frye, B. E.,** Extirpation and transplantation of the pancreatic rudiments of the salamanders *Ambystoma punctatum* and *Eurycea bislineata, Anat. Rec.,* 144, 97, 1962.

130. **Doetschman, T. C., Havaranis, A. S., and Herrmann, H.,** Insulin binding to cells of several tissues of the early chick embryo, *Dev. Biol.,* 47, 228, 1975.

131. **Hendricks, S. A., de Pablo, F., and Roth, J.,** Early development and tissue specific patterns of insulin binding in chick embryo, *Endocrinology,* 115, 1315, 1984.

132. **Bassas, L., de Pablo, F., Lesniak, M. A., and Roth, J.,** The insulin receptors of chick embryo show tissue-specific structural differences which parallel those of the insulin-like growth factor I, *Endocrinology,* 121, 1477, 1987.

133. **Bassas, L. L., Lesniak, M. A., Girbau, M., and De Pablo, F.,** Insulin-related receptors in the early chick embryo: from tissue patterns to possible function, *J. Exp. Zool.* (Suppl. 1), 299, 1987.

134. **Girbau, M., Gomez, J. A., Lesniak, M. A., and de Pablo, F.,** Insulin and insulin-like growth factor I both stimulate metabolism, growth, and diffeentiation in the postneurula chick embryo, *Endocrinology,* 121, 1477, 1987.

135. **De Pablo, F., Hernandez, E., Collia, F., and Gomez, J. A.,** Untoward effects of pharmacological doses of insulin in early chick embryos: through which receptors are they mediated?, *Diabetologia,* 28, 308, 1985.

136. **De Pablo, F., Girbau, M., Gomez, J. A., Hernandez, E., and Roth, J.,** Insulin antibodies retard and insulin accelerates growth and differentiation in early embryos, *Diabetes,* 34, 1063, 1985.

137. **Girbau, M., Lesniak, M. A., Gomez, J. A., and De Pablo, F.,** Insulin action in early embryonic life: anti-insulin receptor antibodies retard chicken embryo growth, but not muscle differentiation *in vivo, Biochem. Biophys. Res. Commun.,* 153, 142, 1988.

138. **Heidenreich, K. A., Zahniser, N. R., Berhanu, P., Brandenburg, D., and Olefsky, J.,** Structural differences between insulin receptors in the brain and peripheral target tissues, *J. Biol. Chem.,* 258, 8527, 1983.

139. **Hendricks, S. A., Agardh, C. D., Taylor, S. I., and Roth, J.,** Unique features of the insulin receptor in the rat brain, *J. Neurochem.,* 43, 1302, 1984.
140. **McElduff, A., Comi, R. J., and Grunberger, G.,** Structural difference of the insulin receptors from circulating monocytes and erythrocytes, *Biochem. Biophys. Res. Commun.,* 133, 1175, 1985.
141. **Simon, J. and LeRoith, D.,** Insulin receptor of chicken liver and brain. Characterization of alpha and beta subunit properties, *Eur. J. Biochem.,* 158, 125, 1986.
142. **Burant, C. F., Treutelaar, M. K., Block, N. E., and Buse, M.G.,** Structural differences between liver- and muscle-derived insulin receptors in rats, *J. Biol. Chem.,* 261, 14361, 1986.
143. **Lowe, W. L., Jr., Boyd, F. T., Clarke, D. W., Raizada, M. K., Hart, C., and LeRoith, D.,** Development of brain insulin receptors: structural and functional studies of insulin receptors from whole brain and primary cell cultures, *Endocrinology,* 119, 25, 1986.
144. **Unterman, T., Goewert, R. R., Baumann, G., and Freinkel, N.,** Insulin receptors in embryo and extraembryonic membranes of the early somite rat conceptus, *Diabetes,* 35, 1193, 1986.
145. **Havranakova, J., Roth, J., and Brownstein, M.,** Insulin receptors are widely distributed in the central nervous system, *Nature (London), 272,* 827, 1978.
146. **Goodner, C. J. and Freinkel, N. F.,** Studies on the permeability of rat placenta to 131-I-insulin, *Diabetes,* 10, 383, 1961.
147. **Galbraith, R. A., Buse, M. G., and Marchalonis, J. J.,** Insulin binding to cultured B- and T-lymphocytes, *Immunol. Lett.,* 4, 141, 1982.
148. **Leed, B. C. and Lane, D.,** Insulin receptor synthesis and turnover in differentiating 3T3-L1 preadipocytes, *Proc. Natl. Acad. Sci. U.S.A.,* 77, 285, 1980.
149. **Rosenblum, I. Y., Mattson, B. A., and Heyner, S.,** Stage-specific insulin binding in mouse preimplantation embryos, *Dev. Biol.,* 116, 261, 1986.
150. **Mattson, B. A., Rosenblum, I. Y., Smith, R. M., and Heyner, S.,** Autoradiographic evidence for insulin and insulin-like growth factor binding to early mouse embryos, *Diabetes,* 37, 585, 1988.
151. **Biggers, J. D. and Borland, R. M.,** Physiological aspects of growth and development of the preimplantation mammalian embryo, *Annu. Rev. Physiol.,* 38, 95, 1976.
152. **Wales, R. G., Khurana, N. K., Edirisinghe, W. R., and Pike, I. L.,** Metabolism of glucose by preimplantation mouse embryos in the presence of glucagon, insulin, epinephrine, cAMP, theophylline, and caffeine, *Aust. J. Biol. Sci.,* 38, 421, 1985.
153. **Kaye, P. L., Pemble, L. B., and Hobbs,** Protein metabolism in preimplantation mouse embryos, *Prog. Clin. Biol. Res.,* 217B, 103, 1986.
154. **Harvey, M. B. and Kaye, P. L.,** Insulin stimulates protein synthesis in compacted mouse embryos, *Endocrinology,* 122, 1182, 1988.
155. **De Pablo, F., Chambers, S. A., and Akira, O.,** Insulin-related molecules and insulin effects in the sea urchin embryo, *Dev. Biol.,* 130, 304, 1988.

Chapter 5

INSULIN AND INSULIN-LIKE GROWTH FACTORS (IGFs) IN AVIAN DEVELOPMENT

Flora de Pablo

TABLE OF CONTENTS

I. OVERVIEW

Fundamental questions remain regarding the signals that control growth and differentiation in early life, and they are as poorly answered in nonmammals as in mammals. However, the search for these factors in nonmammalian species has been facilitated by the advantage of dealing with developmental systems in which there is no placental barrier, and further, a system in which the maternal influences are relatively clearly separated from embryonic contributions.

Insulin can be considered a potentially important developmental factor because it stimulates not only metabolic and growth processes *in vivo* but, in addition, growth and differentiation in a number of embryonic systems *in vitro*. Insulin-like growth factors (IGFs), closely related to insulin in structure and function, have been considered even stronger candidates for a role in development because they have been shown to be more potent stimulators of growth and differentiation than insulin in many systems.

The demonstration of hormonal action *in vivo* requires the presence of the hormone, its receptor, and evidence of post-binding effects at physiological concentrations. Therefore, to address the role of insulin and IGFs in early embryogenesis, the simplest approach is to demonstrate the presence of the hormone(s) and functional receptors and to document the effects of insulin and IGFs in early development. In this brief review, I will summarize studies that have taken these approaches and show that the results indicate that insulin is a developmentally important hormone in a nonmammalian vertebrate embryo, the chick. In addition, I will describe a few observations in amphibians, suggesting possible functions of insulin in oocytes as well as early embryos.

I will review the scant information that is available on the effects of IGFs in nonmammalian developmental systems separately from those of insulin when possible. However, all the peptides in the insulin/IGF family will be discussed together when the data indicate blurred boundaries of specificity; for example, the effects of insulin and IGFs overlap to a significant extent in many systems, as illustrated in Figure 1.

One reason why the IGFs, rather than insulin, have been proposed more widely as important growth factors in development is because insulin was thought to be exclusively the product of the endocrine pancreas. Since endocrine glands are not well-defined anatomically until late in development, traditional hormones were precluded from a role in early developmental stages.[1] In contrast, the peptides IGF-I and IGF-II,[2,3] and more recently, the messenger RNAs for IGF-I and IGF-II, have been demonstrated in a wide range of tissues when studied at a variety of embryonic stages.[4-6] It should be emphasized that while insulin and the IGFs are related polypeptides, the exquisite specificity of the radioimmunoassays used in these studies allows discrimination of their separate identities.

The storage of insulin in the egg and the biosynthesis of insulin in embryonic tissues will be addressed first, followed by the evidence that supports the presence of IGFs in nonmammalian embryos. The effects of insulin and IGFs in oocytes and embryos will be discussed next, starting with a discussion of the experiments that have elucidated the effects of antibodies directed against insulin on development, and including experiments in which the target ligand was added to the developmental system. In addition to the effects of insulin, insulin analogues, and IGFs *in vivo*, I will comment on a few examples of effects *in vitro* using embryonic tissue and organ cultures. No attempt will be made, however, to mention all the reports that have described the effects of insulin and the IGFs in cell cultures.

Finally, I will focus on the ontogeny of insulin receptors and IGF receptors. The insulin receptor and the IGF-I receptor are very similar structurally[7] and are only discriminated biochemically by the relative binding affinities for their own ligands. In contrast, the IGF-II receptor has a quite different structure;[8] therefore, it is easier to determine its presence or absence.

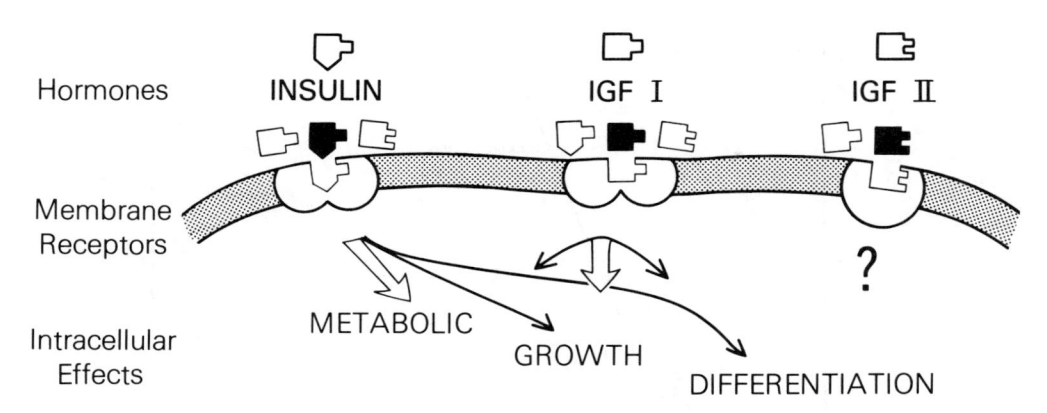

FIGURE 1. Scheme of the plasma membrane receptors for insulin, IGF-I, and IGF-II, and intracellular effects mediated by them. Insulin and IGF-I receptors are glycoproteins composed of alpha (binding) and beta (tyrosine kinase) subunits. Both types are found in nonmammalian vertebrates. The IGF-II is a monomeric transmembrane glycoprotein with no intrinsic tyrosine kinase activity. While IGF-II binds to IGF-I receptor and insulin receptors with lower affinity than their homologous ligands, typical IGF-II receptors have not been identified in nonmammalian vertebrates so far. Insulin and IGF-I receptors both mediate metabolic, growth, and differentiation effects with different potency rations depending upon each tissue. The function of the IGF-II receptor remains an unsolved puzzle. (From Bassas et al., *J. Exp. Zool.*, Suppl. 1, 299, 1987.)

II. INSULIN AND INSULIN-LIKE GROWTH FACTORS (IGFs)

Insulin, IGF-I, and IGF-II are structurally related polypeptides with overlapping biological effects and probably a common phylogenetic origin. The synthesis, processing and secretion of insulin in the pancreas has been well studied. IGF-I and IGF-II are much more widely distributed, being synthesized and released from a number of tissues; however, the mechanisms governing their synthesis and release are not yet understood.[2-6]

III. BIOSYNTHESIS OF INSULIN AND IGFs

A. INSULIN IN THE EGG

The finding of maternal insulin stored in the eggs of a few species suggests that this phenomenon may be relatively widespread and could be a mechanism for providing the early embryo with the hormone before it starts to synthesize its own. In chicken eggs, both the yolk and the white contain insulin-related immunoactivity as reported in an early study by Trenkle and Hopkins.[9] More recently, both insulin immunoreactivity and bioactivity have been found in partially purified extracts of unfertilized and fertilized eggs (Figure 2).[10] Whether this maternal insulin is synthesized in the oocyte or accumulated by the egg as it travels through the oviduct is not known. One possibility is that plasma insulin may be internalized by a receptor-mediated mechanism into the oocyte, or alternatively, into the cells of the oviduct. Another possibility is that the hormone is the product of local tissue synthesis in the oviduct.

The frog is a well-characterized developmental model, producing large numbers of eggs with medium amounts of yolk. Because of the large numbers of mesolecithal eggs, the frog is more appropriate than the chicken to address the question of insulin accumulation in oocytes. In a pilot experiment to test whether insulin immunoreactivity is detectable in frog eggs, a mixture of oocytes and eggs was obtained from *Rana pipiens* ovaries. After acid-ethanol extraction and double gel-filtration, a value of 50 pg/g wet weight of insulin immunoreactivity was obtained (Figure 3). Whether this "egg insulin" is identical in structure to the circulating plasma insulin or whether it is a somewhat modified form for storage purposes is not known. Nor is it known

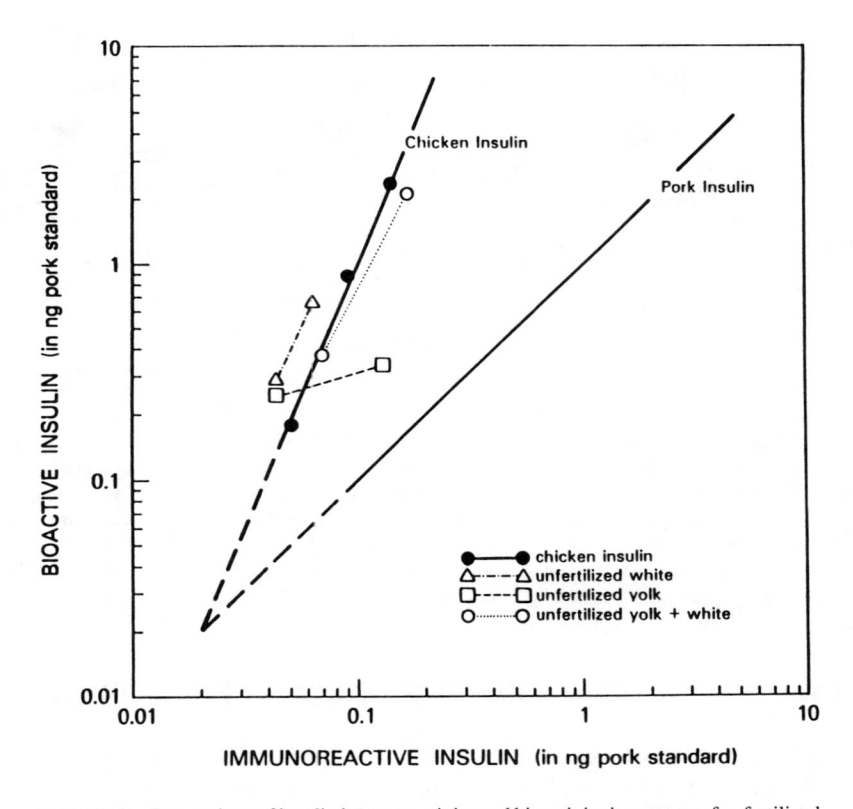

FIGURE 2.   Comparison of insulin immunoactivity and bioactivity in extracts of unfertilized chicken eggs. Using purified pork insulin as standard, pancreatic chicken insulin and extracts of egg yolk and white were tested in a radioimmunoassay and a lipogenesis bioassay. Two dilutions of each extract were analyzed and the ratio of biological to immunological activities found was typical of chicken insulin and quite distinct from pork insulin. (From De Pablo et al., *Endocrinology*, 111, 1909, 1982.)

whether insulin action is required for the maturation of the oocyte, or for post-fertilization events in this species (see Chapter 4 for further discussion).

B. INSULIN AND INSULIN MESSENGER RNA IN EMBRYONIC TISSUES

Initial investigators working with chicken embryos assumed that the pancreas is the sole source of insulin. Indeed, the hormone has been measured in chicken embryo plasma and pancreas extracts after Day 5 of embryogenesis.[11-13] Using immunocytochemical methods, pancreatic beta cells were first detected at 3 1/2 d, a few h later than the first glucagon cells.[12] However, when whole chicken embryo extracts were partially purified, low levels of immunoreactive and bioactive insulin were detected at Days 2 and 3,[10] as shown in Figure 4. This insulin-related material was demonstrated to have the immunoactivity to bioactivity ratio of authentic chicken insulin, indistinguishable from insulin extracted from chicken embryos at Day 8, which possess a fully functional pancreas. These data suggest that chicken embryos are capable of synthesizing insulin before the appearance of the pancreas. Although the argument can be made that the material represents a modified, insulin-like, growth factor, there are cogent reasons to believe that this is not the case. The main argument is that fully processed IGF-I and IGF-II from mammalian adult species are not recognized by anti-insulin antibodies; therefore, an IGF should not be detected by an insulin radioimmunoassay.[14] Nevertheless, we cannot exclude the possibility that an as yet unidentified, embryonic, avian IGF might be recognized by the anti-insulin antibodies. Further, it is of interest that the "head" portion of chicken embryos

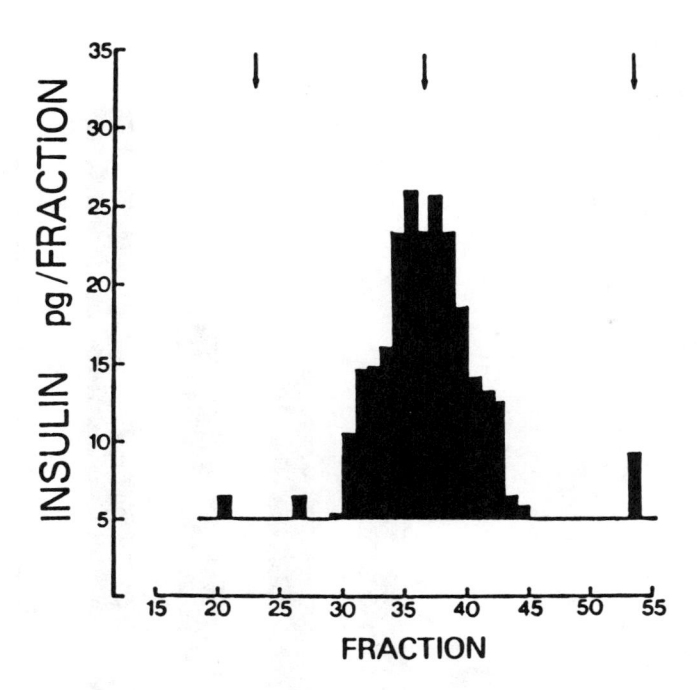

FIGURE 3. Chromatographic profile of insulin immunoactivity in amphibian oocytes and eggs. *Rana pipiens* oocytes and unfertilized eggs (100 gm obtained from 20 frogs by manual dissection from the ovaries) were extracted in acid-ethanol and the extract was filtered twice on Sephadex® G-50. The immunoactivity detected in a porcine radioimmunoassay coeluted with authentic insulin (middle arrow). The position of the void volume (left arrow) and salt (right arrow) are also indicated. Approximately 50 pg of insulin/g wet weight of eggs were recovered (probably an underestimation) based on the crossreactivity of amphibian insulin an a pork radioimmunoassay.

at Days 3 to 5 also contained this insulin-related material. Thus, the sites of insulin synthesis in the chicken embryo may include tissues other than the differentiated pancreatic beta cells.

To resolve these questions, of when and where the embryonic insulin gene is expressed, it was important to localize insulin messenger RNA. The insulin transcript (approximately 600 bases) detected by specific oligonucleotide probes (based on the chicken insulin genomic sequence) is not very abundant in the pancreas during mid-embryogenesis; there is at least 30-fold less in Day 10 embryos than in the post-hatched chick. A similarly low level of expression is detected in chick embryo liver; however, in contrast with the pancreas, the abundance does not increase between the second and third week of embryogenesis.[15] Therefore, it appears that there is differential gene expression which is developmentally regulated in tissues that express the insulin gene. Also, it appears that the regulation of insulin expression involves posttranscriptional control, since the changes in protein levels detected in pancreas, as well as in plasma, during the second half of embryogenesis[10-12] are much more modest than the corresponding changes in messenger RNA (mRNA). Using very large amounts of poly(A)+ RNA from the body region of Day 4 embryos and Day 2 whole embryos, a transcript that corresponds to the size of insulin mRNA can be detected in even lower abundance by Northern blot analysis.[64] It would be interesting to see if other, more sensitive techniques, e.g., *in situ* hybridization, confirm the presence of insulin mRNA prior to 3 1/2 d, before the differentiation of the embryonic pancreas.

C. IGF-I AND IGF-II

There are only fragmentary reports to indicate that molecules with immunological similarities to mammalian IGFs are present during nonmammalian early development. Liver cells from

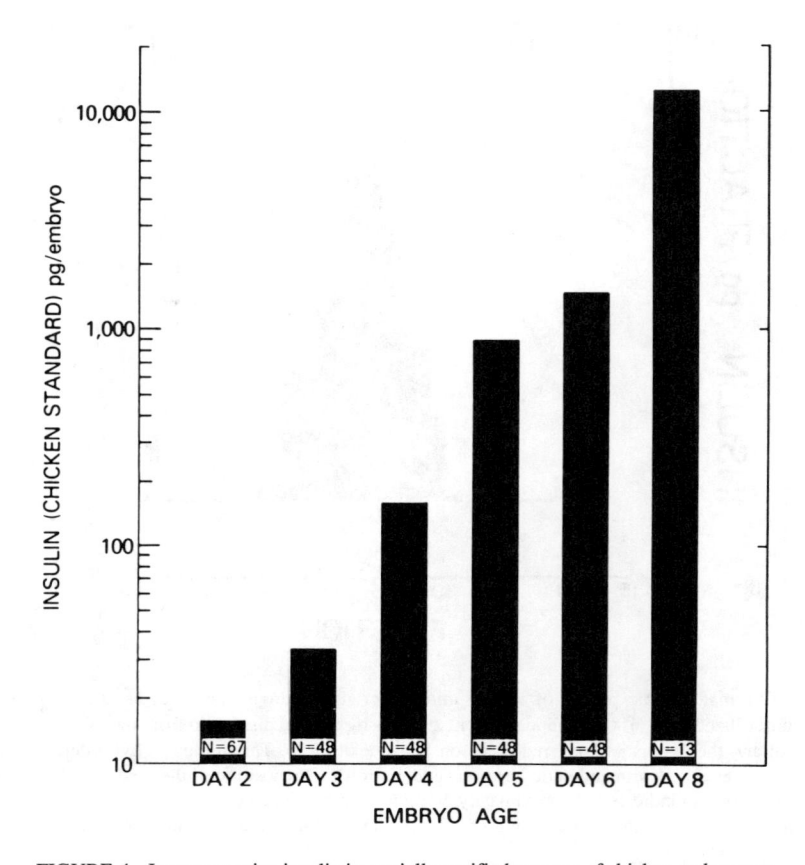

FIGURE 4.   Immunoractive insulin in partially purified extracts of chicken embryos at different ages. Groups of embryos (N = number) were pooled, homogenized and extracted with acid ethanol. The extracts were gel filtered on Sephadex® G-50. The material recovered in the elution region of insulin was concentrated and tested in an insulin radioimmunoassay using purified pancreatic chicken insulin as standard. (From De Pablo et al., *Endocrinology,* 111, 1909, 1982.)

Day 16 chick embryos in primary culture[16] and pelvic cartilage explants from Day 9 embryos[17] have been shown to synthesize a peptide which demonstrated cross-reactivity in a human IGF-I radioimmunoassay. It has been reported also that IGF-II messenger RNA is very abundant in Day 4 chick embryo limb buds, hearts, and eyes.[18] This report, however, is open to criticism; mRNA was detected in surprising abundance considering that heterologous, human, IGF-II probe was used. The transcript was a single mRNA of 1.9 kilobases, the same size as the ribosomal RNA, which differed significantly from that found in mammals. The abundance of the transcript in the three tissues was very similar. Preliminary findings suggest that IGF-I may also be stored in the chicken egg yolk, in even larger quantities than insulin itself.[65] However, until the genes of nonmammalian species have been cloned, it will be difficult to evaluate the developmental expression of IGF-I and IGF-II in other vertebrate species. Thus far, only mammalian IGF-I[19-21] and IGF-II[22-24] cDNAs have been described.

IV. ACTION OF INSULIN AND IGFs IN EMBRYOGENESIS

The scheme shown in Figure 5 summarizes the 21 d of chicken embryo development until hatching. Also shown in this scheme are the experiments designed to determine whether

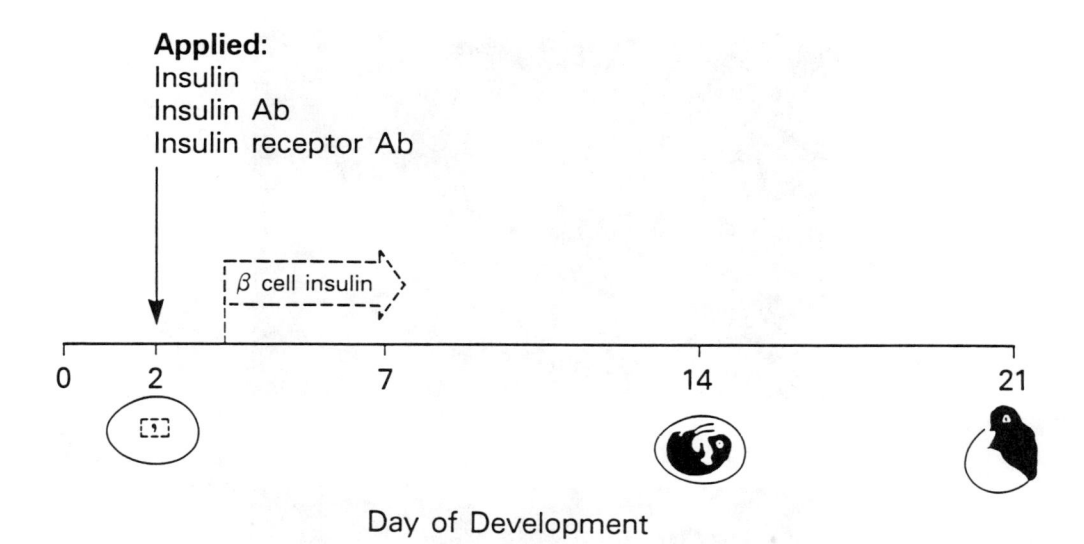

FIGURE 5. Experimental design of insulin and antibody treatments in chicken embryos during early embryogenesis. Chicken embryos, which hatch after 21 d, develop pancreatic beta cells in the dorsal pancreatic rudiment by 3.5 d. On Day 2, immediately after neurulation is completed, a window was created on the egg shell and materials, either insulin, insulin antibodies, insulin receptor antibodies, or others (see text) were applied over the blastoderm disc. Note that the embryos are drawn to scale compared to the egg.

"prepancreatic" insulin-related material has any developmental function. At Day 2, when neurulation is completed and organogenesis begins, (approximately 20 somite stage),[25] a "window" can be created in the eggshell over the embryo, and materials can be administered directly to the blastoderm. This technique results in very little disruption of development due to the procedure itself and allows subsequent monitoring of embryonic development for approximately 3 d, until Day 5, mid-organogenesis. Although this approach to studying the whole embryo *in vivo* has not provided details on insulin action in any particular cell type at any precise developmental stage, it does allow for testing of physiological effects induced by insulin action in chicken embryogenesis.

A. EFFECTS OF ANTIBODIES

The chicken embryo apparently needs insulin for normal growth during early embryogenesis. Treatment of the embryo with anti-insulin antibodies produces dramatic growth inhibition, presumably, due to neutralization of the embryo's endogenous insulin.[26] The administration of an affinity purified polyclonal, anti-porcine insulin antibody killed a significant percentage of embryos. Those embryos that survived were morphologically (Figure 6) and biochemically (Table 1) retarded. The antibody-treated embryos showed a decrease in weight, total protein, total DNA and RNA, triglycerides, and creatine kinase. Interestingly, the MB isozyme of creatine kinase, the marker of muscle differentiation, also was decreased. In contrast, all biochemical parameters that were decreased by the antibody treatment were increased, in an almost mirror image fashion, by exogenous application of moderate doses of insulin, as shown in Table 1. These results provide support for the specificity of action of anti-insulin antibodies. Since the results with anti-insulin antibodies were so dramatic, antibodies to other hormones, including antinerve growth factor and anti-epidermal factor, were also tested in a similar manner. Neither of these antibodies caused morphological alterations that were macroscopically detectable, nor any changes in weight, total protein, creatine kinase, or MB isozyme activities. Further histological evaluation of the embryos which might have revealed more subtle

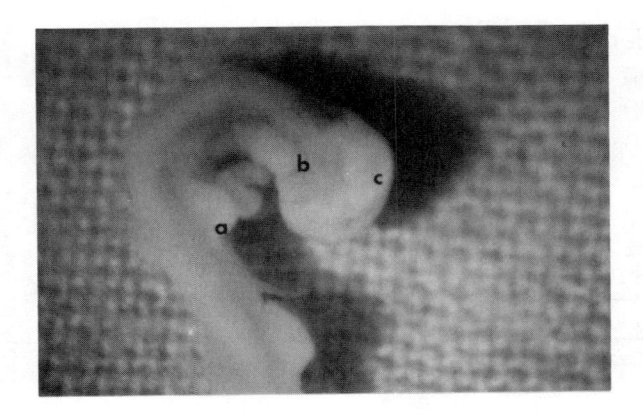

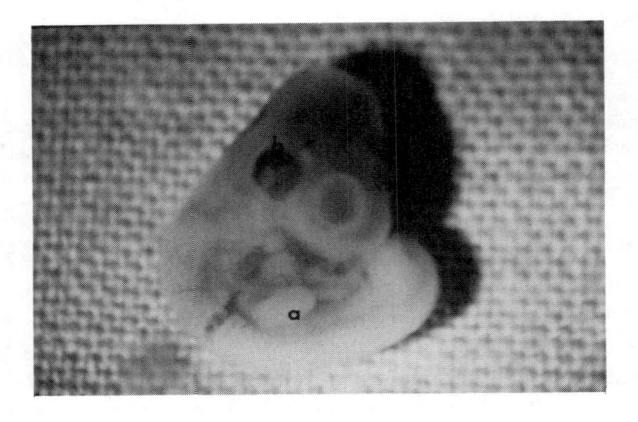

FIGURE 6. Effect of insulin antibodies on morphological development. Embryos were injected on Day 2 with control IgG (lower panel) or anti-insulin IgG (upper panel). On Day 4, these embryos were evaluated macroscopically according to the criteria of Hamburger and Hamilton[25]. The main features defining development at this stage include the limb buds, the eye, and the brain. The control embryo is at Stage 24. The anti-insulin treated embryo is at Stage 19, as indicated by much shorter limb buds, unpigmented, smaller eye, and less-defined brain vesicles. The magnification of the anti-insulin antibody-treated embryo is 1.5 × higher than the control.

TABLE 1
Effects of Adding Insulin Antibodies (Ab-I), Insulin Receptor Antibodies
(Ab-IR), and Insulin to Day 2 Chick Embryos

Parameters is day 4 embryos	Ab-I	Ab-IR	Insulin
Weight	Decreased	Decreased	Increased
Protein	Decreased	Decreased	Increased
DNA	Decreased	Decreased	Increased
RNA	Decreased	Decreased	Increased
Triglycerides	Decreased	Decreased	Increased
Total creatine kinase	Decreased	Decreased	Increased
Isozyme creatine kinase-MB	Decreased	Unchanged	Increased

abnormalities, however, was not carried out. Although antibodies to IGF-I would be very interesting to evaluate, the amounts required for *in vivo* experiments precluded their use.

Although it was clear that insulin antibodies affect growth and differentiation in the embryo, it is possible that metabolic pathways are simultaneously disturbed, contributing to the retarded growth of the embryo. The relationship between effects of insulin on metabolism, growth, and differentiation is not clear in this *in vivo* model; however, this relationship remains unclear even in much more simple *in vitro* systems.[27]

In contrast to the results from *in ovo* experiments, the role of insulin in the regulation of carbohydrate metabolism in chickens has been demonstrated clearly during the second and third weeks of embryogenesis. When anti-insulin antiserum was injected into chicken embryos between Days 9 and 16, an increase in blood glucose and a decrease in liver glycogen content occurred within 24 h. The resultant condition was described by the investigators as a "diabetic state".[13]

B. EFFECTS OF INSULIN, INSULIN ANALOGUES, AND IGFs

In addition to its classical role as a metabolic hormone, insulin has been shown in the past few years to exert growth regulatory activity and to influence differentiation, both actions reflecting rapid effects of insulin on gene expression. The growth promoting effects of insulin have been studied extensively in cultured cells and discussed in detail in several recent reviews.[28,29] The most thoroughly characterized effects of insulin on gene expression have been studied using mammalian cells[30] or explants.[31] Other studies have shown that morphological and biochemical differentiation, as well as growth, are also induced by insulin in nonmammalian systems and accompanied by concomitant changes in gene expression. The effects of the IGFs have been studied in parallel with those of insulin in some but not all cases.

Chicken embryonic muscle was the first tissue in which insulin-induced differentiation was described.[32] Recent studies have shown that insulin and IGF-I concentrations in the nanomolar range are capable of inducing myoblast differentiation.[33,34] In another study employing a different target organ, cultures of hepatocytes from Day 16 chicken embryos were shown to respond to 35 nM insulin by increasing albumin messenger RNA content.[35] In similar cultures, insulin was found to be equipotent with IGF-I in stimulating ^{3}H-uridine incorporation into RNA, and ^{3}H-valine incorporation into protein.[36] Ornithine decarboxylase activity increased in organ cultures of pelvic cartilage explants from Day 11 chicken embryos treated with 100 ng/ml insulin, which presumably exerted its effect at a posttranscriptional level.[37] Studies on the chicken embryo lens have shown that high concentrations of insulin (1 μg/ml) caused an increase in transcription and translation of delta crystallin mRNA and also stimulated epithelial cells to differentiate into fiber cells.[38] In a pure neuronal culture obtained from embryonic chick brain and grown in a defined, serum-free environment, insulin was the only hormone found to enhance the growth of neurons during the early proliferative phase.[39] In a recent elegant study, very low concentrations of insulin (1 ng/ml) have been shown to be sufficient to stimulate acetyltransferase activity in cultured retinal neurons obtained from Day 7 chicken embryos. This finding suggests that insulin may play a role in mediating cholinergic differentiation in the embryonic retina.[40]

Chicken embryos *in ovo* provide a convenient model to study hormonal influences in early organogenesis in vertebrates. The effects on early chicken development of insulin, proinsulin, desoctapeptide insulin, and IGF-I, after their addition to Day 2 chicken embryos *in ovo,* were evaluated at Day 4 of development.[41] Insulin, at concentrations ranging from 10 to 100 ng per embryo, increased the total protein content, creatine kinase activity, creatine kinase MB isozyme (Figure 7), triglycerides, cholesterol, phospholipids, DNA, and RNA in a dose-dependent manner. IGF-I was less potent than insulin in stimulating both metabolic and growth indices and

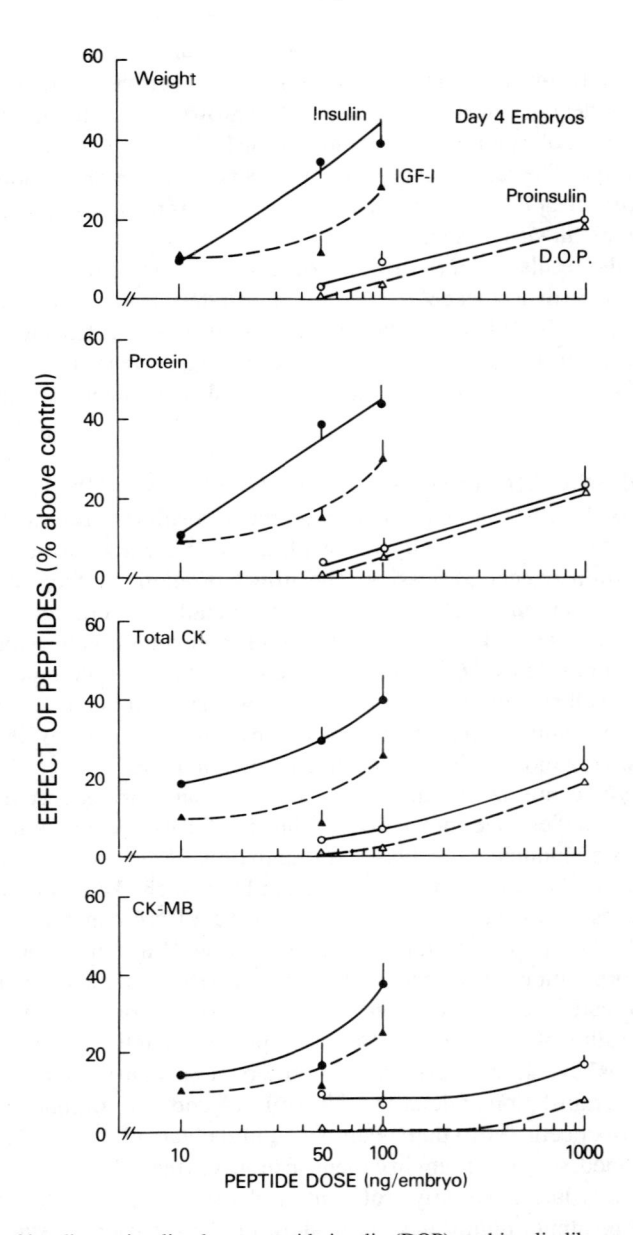

FIGURE 7. Effect of insulin, proinsulin, desoctapeptide insulin (DOP), and insulin-like growth factor I(IGF-I) on several parameters of general development (weight, total protein, total creatine kinase) and on creatine kinase MB (CK-MB), an index of differentiation. Day 2 chicken embryos were treated with various concentrations of peptides in 50 ml volume. At Day 4 of development, embryos were homogenized and the biochemical parameters measured. All values are expressed as the percentage above control obtained from embryos to which vehicle alone was applied. Vertical lines represent the standard errors, shown only for the two higher doses of insulin, IGF I, and proinsulin. (From Girbau et al., *Endocrinology,* 121, 1477, 1987.)

was nearly equipotent in stimulating the creatine kinase MB, a marker of muscle differentiation. By comparing the relative potencies of several insulin analogues, it was inferred that at low doses, insulin stimulates developmental processes mainly through the insulin receptor, with the possible exception of muscle differentiation.

The effects of exogenous IGF-I on embryo development were found to be similarly broad and appeared to overlap with the effects of insulin. These two peptides were not additive when

TABLE 2
Tissue-Specific Characteristics of Insulin and IGF-I Receptors
in Chick Embryos

Tissue	α Subunit M$_r$ (kDa)	Neuraminidase effect
Day 2 embryo	≈129	No
Day 3 head	≈129	No
Brain	≈129	No
Day 3 body	≈132	Yes
Muscle	≈131	Yes
Liver	≈138	Yes

injected together, and thus, it is possible that insulin and IGF-I receptors "cross talk" and stimulate common postbinding steps during embryogenesis. Pharmacological doses (μg per embryo) of insulin and proinsulin, however, have been shown to cause abnormal growth, accompanied by a decrease in protein, DNA, and RNA content of the embryos at Day 4,[42] resulting in teratogenic effects, and in some cases, mortality. These teratogenic effects occurred in spite of injection of glucose at the same time as the insulin. Since proinsulin in that study had a potency of approximately 60% that of insulin, the authors suggested that at high concentrations, insulin interferes with embryonic development, probably through a nonmetabolic pathway, possibly via an IGF receptor.

V. INSULIN AND IGF RECEPTORS

A. RECEPTOR CLASSES

In mammals there are, in addition to insulin receptors, two well-characterized IGF receptors; IGF-I, also referred to as the Type I receptor, and IGF-II, referred to as the Type II receptor. Insulin and IGF-I receptors are very similar in structure, and both belong to the family of tyrosine kinases. The typical insulin receptor has a high affinity for insulin and lower affinities for IGF-I and IGF-II, respectively. The typical IGF-I receptor has a high affinity for IGF-I and a low affinity for IGF-II and insulin, respectively.[7,43,44] The IGF-II receptor is quite different in structure from the other two and is very similar to the mannose-6-phosphate receptor; it has equal or lower affinity for IGF-I than for IGF-II and no affinity for insulin.[8,45,46] A simple scheme of the receptors in the plasma membrane is shown in Figure 1. In the case of insulin and IGF-I receptors, ligand binding results in receptor activation, involving autophosphorylation of the beta subunit, and this event leads to the stimulation of a wide range of intracellular events.[47] The events occurring after binding of IGF-II to its receptor remain unclear.

B. RECEPTOR STRUCTURE STUDIES

The classical methodological approach for examining the structural characteristics of insulin and IGF receptors include the technique of affinity labeling, followed by SDS-polyacrylamide gel electrophoresis. [^{125}I]-labeled ligands can be chemically cross-linked to their respective membrane receptors and the receptor binding subunit plus ligand subsequently detected by autoradiography. Two features emerged when different tissues were studied using these methods during early chicken development. First, the insulin receptors and IGF-I receptors had identical tissue-specific binding patterns, as far as heterogeneity in molecular weight of their binding subunits among different tissues (Figures 8 and 9). Second, the desialytion of the receptors by treatment with neuraminidase produced similar changes in the apparent molecular weight of the binding subunit of the insulin receptor and the IGF-I receptor.[48] These data are shown in Table 2.

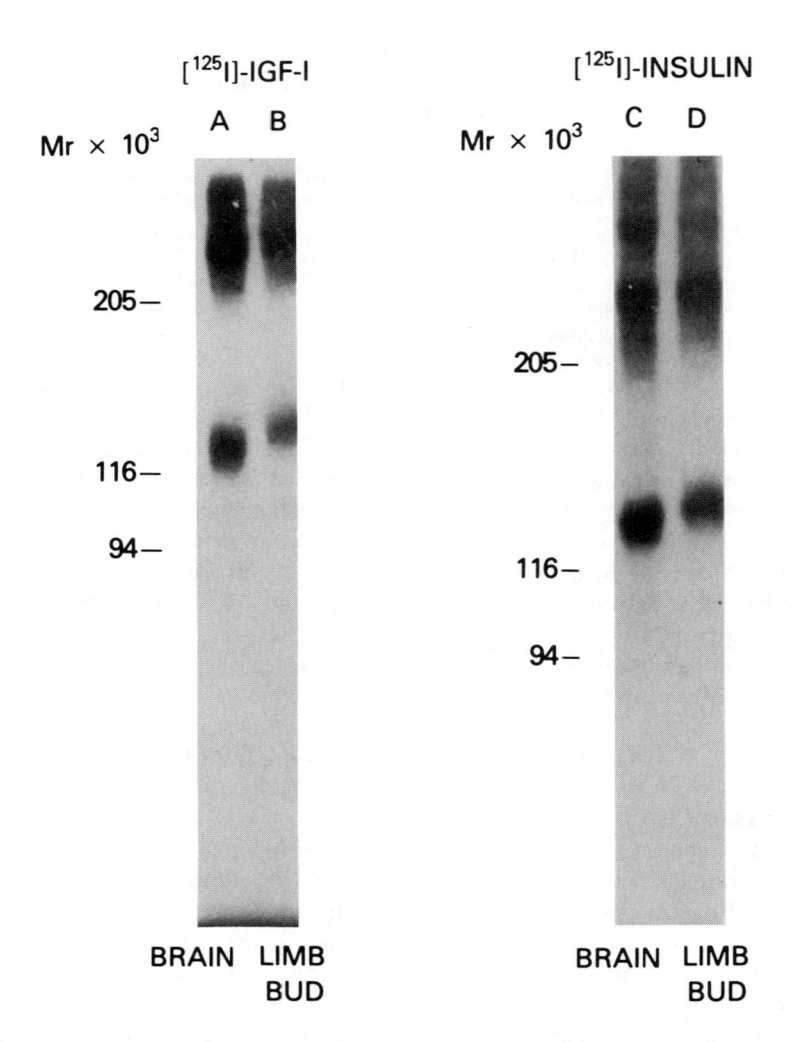

FIGURE 8.   Affinity labeling of IGF I receptors and insulin receptors in chicken embryo membranes from brain and limb buds. [^{125}I]IGF-I and [^{125}I]insulin were cross-linked to receptors on membranes from Day 6 embryo tissues. Shown is an autoradiogram of the polyacrylamide gel (6.5%) run under reducing conditions. The apparent molecular weight of the brain alpha subunit for both receptors was approximately 129 kDa, while the limb bud alpha subunit was approximately 133 kDa.

Tissue-dependent differences in the electrophoretic mobility of the alpha subunit of insulin receptors have been reported between adult liver, brain, and muscle of various vertebrate species.[49-52] This receptor heterogeneity, presumably due to differences in the carbohydrate moiety, could be demonstrated in the chicken embryo as soon as development had proceeded to the stage at which individual tissues could be dissected from the rest of the embryo.[48] Figure 8 shows that membranes from limb buds of chicken embryos have receptor subunits slightly larger than in brain, although they were smaller than in liver, as we have also observed in cardiac and skeletal muscle. Addition of sialic acid to the carbohydrate portion of the receptors during differentiation is one of the mechanisms by which this tissue heterogeneity is generated. Thus, chicken embryo liver and heart receptors have been shown to demonstrate increased electrophoretic mobility following neuraminidase digestion. In contrast, neither insulin nor IGF-I receptors in developing brain, nor Day 3 whole chicken embryo receptors have been demonstrated to be sensitive to neuraminidase action. Receptor structure appeared to be normal, since the alpha subunit of the receptors at this early embryonic stage had an apparent molecular weight

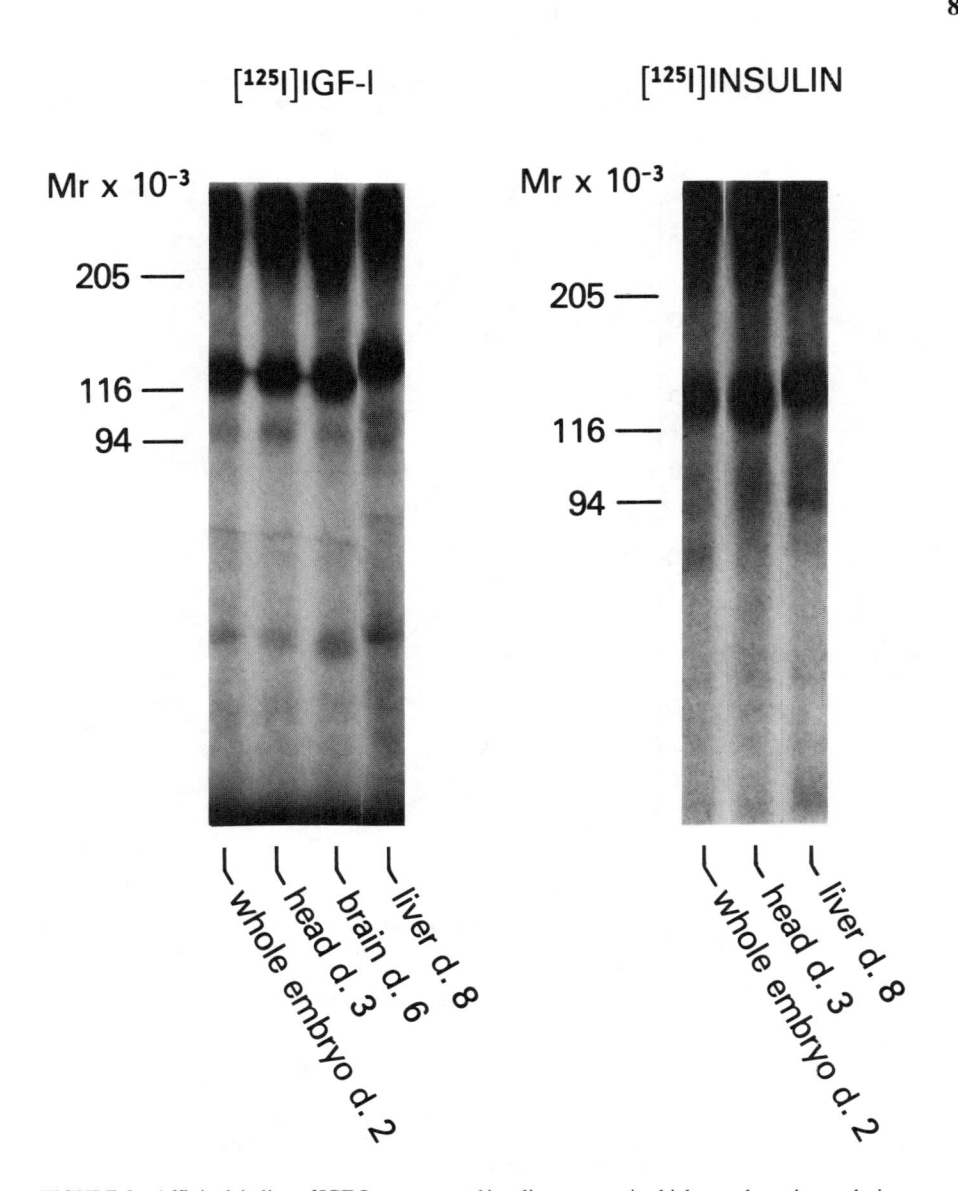

FIGURE 9. Affinity labeling of IGF-I receptors and insulin receptors in chicken embryo tissues during early organogenesis. [¹²⁵I]IGF-I was cross-linked to receptors on membranes from Day 2 whole embryos, Day 3 head portions, Day 6 brain, and Day 8 liver. Shown is an autoradiogram of the polyacrylamide gel run under denaturing conditions. A similar experiment was performed with [¹²⁵I]insulin cross-linked to aliquots of the same membrane preparations with the exception of brain. (From Bassas et al., *Endocrinology*, 121, 1468, 1987.)

similar to that of receptors isolated from the head at Day 3, and the brain at all stages after Day 6, as shown in Figure 9.

Membranes prepared from Day 2 chicken embryos, at approximately the 20 somite stage, have been shown to specifically bind [¹²⁵I]-IGF-I and [¹²⁵I]-insulin (Figure 10). The data from competitive binding experiments shown in Figure 10 confirmed the presence of distinct specific IGF-I and insulin receptors in these young embryos. Unlabeled IGF-I was 10-fold more potent than insulin in competing with [¹²⁵I]-IGF-I for membrane binding, providing evidence for the expression of IGF-I receptors, while insulin was almost a 100-fold more potent than IGF-I in competing with labeled insulin for binding to insulin receptors. Both IGF-I receptors and insulin

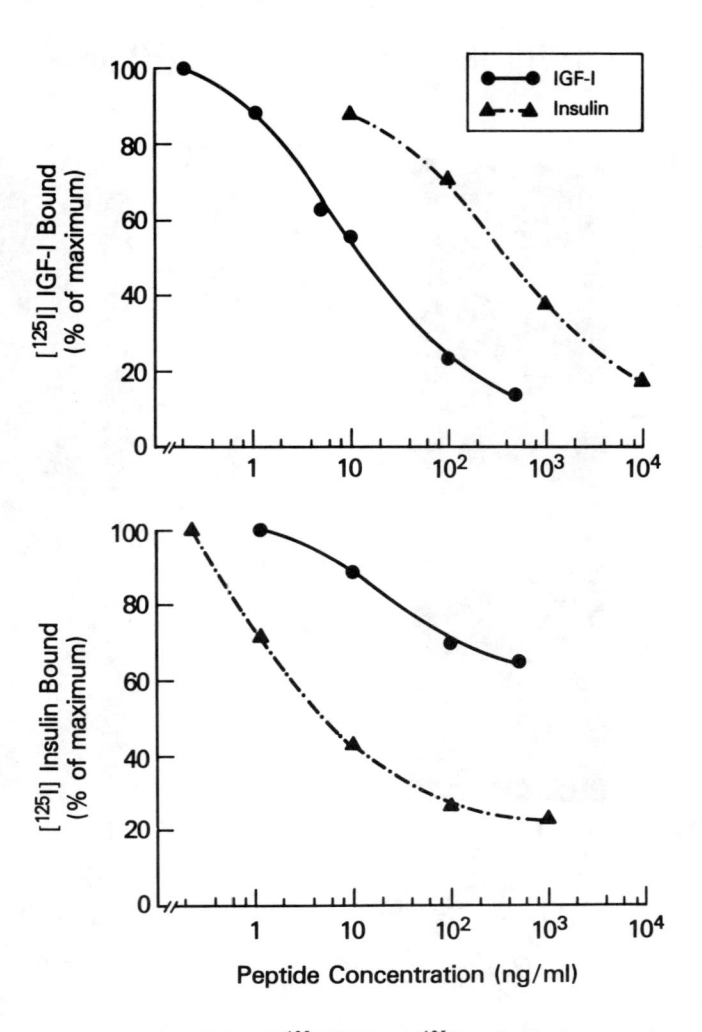

FIGURE 10.   Specificity of [^{125}I]IGF-I and [^{125}I]insulin binding to membrane receptors from whole Day 2 chicken embryos. Binding of [^{125}I]IGF-I (upper panel) and [^{125}I]insulin (lower panel) is expressed as a percentage of the maximum binding obtained in the absence of unlabeled peptides. Total specific binding of IGF-I was 4.2% and of insulin was 0.8%. (From Bassas et al., *Endocrinology,* 121, 1468, 1987.)

receptors from Day 2 whole embryos had the molecular weight characteristic of differentiated brain receptors, approximately 129 kDa, smaller than the receptors found in in liver, approximately 135 kDa. Thus, the predominant if not the only receptor detected in the youngest embryonic tissues is similar to the neuronal type. The functional implications of these tissue-specific differences in the alpha subunit of the insulin and IGF-I receptors are not known. The possible importance of this receptor feature is suggested by the fact that it is expressed very early in embryogenesis and is detected in both IGF-I and insulin receptors.

To study the ontogeny of receptors for insulin and IGFs, the chicken embryo is ideal, since each organ can be studied separately from the beginning of organogenesis, and whole embryos are accessible through gastrulation and neurulation.

Membrane-binding studies employing radio-labeled ligands or chemical cross-linking methods are not sufficiently sensitive to detect very low numbers of receptors and therefore have had limited application for studying receptor ontogeny in early development. However, by using

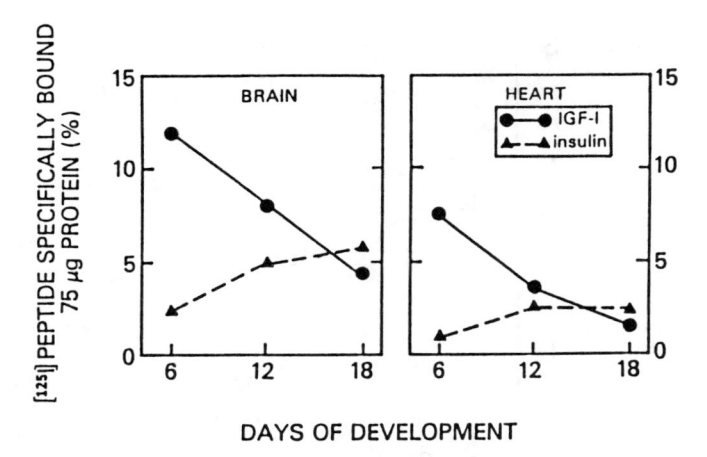

FIGURE 11. Specific binding of [^{125}I]insulin and [^{125}I]IGF-I to chicken embryo heart and brain membranes in early (Day 6), middle (Day 12), and late (Day 18) development. Specific binding was calculated by subtracting nonspecific binding from total binding.

a more sensitive autoradiographic technique applied to whole mounted chicken embryos, it is possible to detect ligand binding to specific insulin and also IGF-I receptors, which have been shown to be widespread throughout the blastoderm, before gastrulation is completed and the first somite formed.[53]

Tissue-specific binding patterns that are characteristic of specific developmental stages have been shown in brain,[54] eye lens,[55] heart, liver, and limb buds.[56] In every tissue studied, with the exception of liver, [^{125}I]-IGF-I binding was higher than [^{125}I]-insulin binding. In brain and heart, the [^{125}I]-IGF-I binding was highest during the first week of embryogenesis and decreased in the second and third weeks, as shown in Figure 11. Perhaps the abundance of IGF-I receptors is related to the growth rate of these organs, which is maximum between Days 3 and 6. Specific [^{125}I]-insulin binding in brain and heart, by contrast, was lower at early stages and increased in mid and late development, as shown in Figure 11. This pattern may be related to the stage of cellular differentiation.

There is evidence for developmental regulation of the expression of both IGF-I and insulin receptors in chicken embryo tissues. In the eye, the lens tissue showed a dramatic decrease in IGF-I receptors in the differentiated fiber cells between Day 6 and Day 19 of development, as shown in Figure 12. Scatchard analysis of radiolabeled insulin and IGF-I binding has demonstrated that the developmental changes in the binding of both ligands were due to changes in receptor concentration and not to modification in receptor affinity, supporting the notion of developmental regulation of expression.

[^{125}I]-IGF-II-specific binding has been evaluated in chicken embryo tissues, and the binding profile was shown to be remarkably parallel to [^{125}I]-IGF-I binding, although consistently lower. One interpretation of this result is that radiolabeled IGF-II is being bound to an IGF-I receptor; structural and competitive binding studies showed that this was the case.

To date, a receptor with the peptide binding and cross-linking characteristics of the mammalian IGF-II receptor has yet to be identified in any of the chicken embryo tissues studied, including limb buds, as well as in Day 1 post-hatched chicken liver and adult chicken adipose tissue.[56] In all studies reported so far, both IGF-I and IGF-II appeared to bind to an IGF-I receptor. This conclusion is based on several criteria. First, binding of [^{125}I]-IGF-II to brain, heart, and liver was always displaced to a greater extent by IGF-I than IGF-II, even when the relative impurity of the unlabeled IGF-II was taken into consideration (data shown in Figure 13). Second, insulin from several species, as well as proinsulin, competed for [^{125}I]-IGF-II binding,

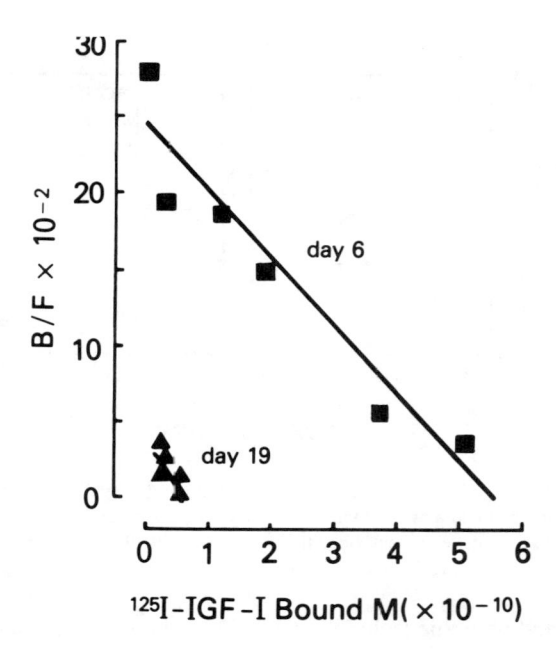

FIGURE 12.   Scatchard plot of [125I]IGF-I binding to Day 6 and Day 19 chicken embryo lens fiber cells. The specific binding expressed as the ratio of bound to free (B/F) of labeled IGF-I is plotted as a function of the total peptide bound to membrane receptors. Note the dramatic decrease in the number of receptors at Day 19 (as indicated by the imaginary intercept with the x-axis) with no significant change in receptor affinity (slope).

which is not expected for IGF-II receptors. Third, in every tissue examined, the specific binding profile of IGF-II parallels the binding of IGF-I, although it is 50% less than the IGF-I total binding per unit of protein. Fourth, the only receptor detected in cross-linking experiments using both radiolabeled IGF-I and IGF-II, had an $M_r \sim 130$ to 135 kDa, which corresponds in size to the characteristic alpha subunit of an IGF-I receptor. The cross-linked molecular species was fully inhibited by both IGFs and insulin. Thus, both specificity and size of this protein correlate to the IGF-I receptor. The absence of the IGF-II receptor however, does not necessarily imply that IGF-II is not present, since this ligand can bind to the IGF-I receptor with quite high affinity and mediate many of its biological effects.[57,58]

To clarify whether it is the insulin receptor that mediates the effects of insulin in embryonic life, chicken embryos at Day 2 of development have been exposed to insulin receptor antibodies *in ovo,* and the results evaluated at Day 4, as shown in Table 1. When compared to control IgG, the IgG fraction of a polyclonal human anti-receptor antibody (B10) induced perturbations of development. These included an increased death rate and production of retarded embryos, and in the survivors, decreased weight, total protein, total creatine kinase activity, DNA, and triglycerides. The effect on these parameters in embryos treated with antireceptor antibodies was even greater than in those embryos that had been treated with anti-insulin antibodies. Interestingly, despite retarded growth effects, creatine kinase MB, the marker of muscle differentiation, was not decreased in embryos treated with antireceptor antibodies.[59] In the chicken embryo, insulin appears to act preferentially through insulin receptors, influencing both growth and anabolic pathways. In specific tissues, such as muscle, it is probable that insulin binds in addition to IGF receptors to stimulate differentiation. Thus, even if insulin binding is lower than IGF-I binding in a number of chicken tissues, insulin receptors are functionally important for normal development.

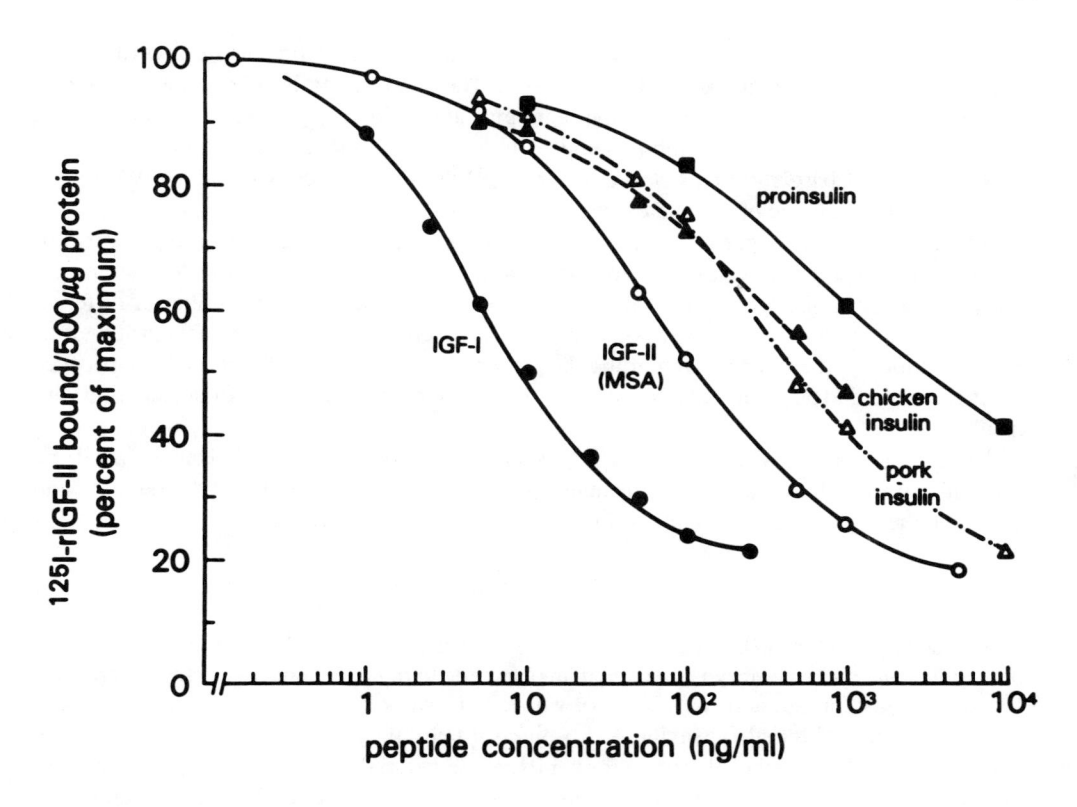

FIGURE 13. Specificity of ligand binding to IGF receptors in Day 8 chicken embryo brain membranes. [^{125}I]rIGF-II (MSA III-2, a gift of Dr. M.M. Rechler, NIDDK, NIH) binding is expressed as a percentage of the maximum binding obtained in the absence of unlabeled peptides. Membranes were incubated for 5 h at 15°C in the presence of various concentrations of peptide. Unlabeled rIGF-II.MSA-2 was a better competitive inhibitor of ^{125}I-rIGF-II binding compared to the less pure IGF-II.MSA purchased from Collaborative Research Inc. (data not shown). These data suggest that IGF-II binds to IGF-I receptors in membranes from Day 8 embryonic brain tissue. (Modified from Bassas et al., *Endocrinology*, 117, 2321, 1985.)

The study of receptors and insulin action in eggs and oocytes is difficult in the chicken, due to the extremely yolky egg. However, amphibian oocytes and developing eggs are readily accessible in large numbers. Insulin and possibly IGF-I receptors have been detected on full grown *Xenopus* oocytes[60,61] but have not been characterized extensively. Insulin at quite low concentrations reinitiates meiotic maturation in *Xenopus* oocytes. However, in the presence of anti-insulin receptor antibodies, there is no shift in the dose-response curve, despite partial inhibition of insulin binding. Thus, this particular effect of insulin is unlikely to be mediated by typical insulin receptors.[60-62] For a more detailed discussion, see Chapter 4.

VI. CONCLUSIONS

The majority of studies on insulin action in the past have focused on the effects of insulin on growth and metabolism; only a few have examined the effects of insulin on differentiation and gene expression. Although the results of *in vivo* studies suggest a significant role for insulin in chick embryogenesis, *in vitro* studies are needed to identify and characterize specific targets for insulin action during development. This chapter serves to review and discuss the significance of the recent intriguing evidence demonstrating the presence of specific receptors for insulin and IGFs in early stages (Day 2) of chicken embryogenesis.

Studies on the effects of insulin in embryogenesis do not have to be restricted to that period after the pancreas differentiates, since an insulin related material is stored in the eggs of several species, differing widely in size and yolk content. The chicken egg is approximately 6 cm in diameter, with a large yolk, while the amphibian egg is approximately 1,000 μm in diameter with little yolk; in contrast, the sea urchin is approximately 100 μm in diameter, with virtually no yolk[63] (see also Chapter 4). This stored maternal insulin hypothetically could be used by the embryo during cleavage or even later stages.

The confirmation of prepancreatic insulin gene expression in the embryo should also facilitate the acceptance of insulin's involvement in early embryogenesis. In addition, insulin receptors appear to be present in oocytes, eggs, and early embryos of a number of different species (see Chapter 4), indicating a potential for insulin action during development in numerous phyla. Although little is known regarding the presence of IGFs or their functional significance in other species and phyla, it is likely that these growth factors develop in close coordination with insulin.

Hopefully, this review will stimulate interest in the investigation of detailed biochemical and molecular events that underlie the mechanisms by which insulin and the IGFs regulate and stimulate the overall developmental program.

ACKNOWLEDGMENTS

I am deeply indebted to Dr. Jesse Roth for providing inspiration, guidance, and support during our studies on chicken embryos. Special thanks are due to my excellent co-workers, Drs. Lluis Bassas, Matias Girbau, and Jose Serrano, whose hard work has contributed greatly to these studies. I wish to thank also Maxine A. Lesniak for helpful discussions and criticisms of the manuscript and Mrs. Esther Bergman for her expert secretarial assistance. The partial support of our work by the U.S.-Spain Joint Committee for Scientific and Technical Cooperation is gratefully acknowledged.

REFERENCES

1. **Tata, J. R.,** The action of growth and developmental hormones, in *Biological Regulation and Development,* Vol. 3B, Goldberger, R.F. and Yamamoto K.R., Eds., Plenum Press, New York, 1984, 1.
2. **D'Ercole, A. J., Applewhite, G. T., and Underwood, L. E.,** Evidence that somatomedin is synthesized by multiple tissues in the fetus, *Dev. Biol,* 75, 315, 1980.
3. **Adams, S. O., Nissley, S. P., Handwerger, S., and Rechler, M. M.,** Developmental patterns of insulin-like growth factor-I and -II synthesis and regulation in rat fibroblasts, *Nature (London),* 302, 150, 1983.
4. **Brown, A. L., Graham, D. E., Nissley, S. P., Hill, D. J., Strain, A. J., and Rechler, M. M.,** Developmental regulation of insulin-like growth factor II mRNA in different rat tissues, *J. Biol. Chem,* 261, 13144, 1986.
5. **Han, K. M. V., D'Ercole, A. J., and Lund, P. K.,** Cellular localization of somatomedin (insulin-like growth factor) messenger RNA in the human fetus, *Science,* 236, 193, 1987.
6. **Beck, F., Samani, N. J., Penschow, J. D., Thorley, B., Tregean, G. W., and Goghlan, J. P.,** Histochemical localization of IGF-I and II mRNA in the developing rat embryo, *Development,* 101, 175, 1987.
7. **Ullrich, A., Gray, A., Yang-Feng, T., Tsubokawa, M., Collius, C., Henzel, W., LeBon, T., Kathuria, S., Chen, E., Jacobs, S., Francke, U., Ramachandran, J., and Fujita-Yamaguchi, Y.,** Insulin-like growth factor I receptor primary structure: comparison with insulin suggests structural determinants that define functional specificity, *EMBO J,* 5, 2503, 1986.
8. **Roth, R. A.,** Structure of the receptor for insulin-like growth factor II: the puzzle amplified, *Science,* 239, 1269, 1988.
9. **Trenkle, A. and Hopkins, K.,** Immunological investigation of an insulin-like substance in the chicken egg, *Gen. Comp. Endocrinol,* 16, 493, 1971.
10. **De Pablo, F., Roth, J., Hernandez, E., and Pruss, R. M.,** Insulin is present in chicken eggs and early chick embryos, *Endocrinology,* 111, 1909, 1982.

11. **Benzo, C. A. and Green, T. D.,** Functional differentiation of the chick endocrine pancreas: insulin storage and secretion, *Anat. Rec.,* 180, 491, 1974.

12. **Dieterlen-Lievre, F. and Beaupain, D.,** Immunocytochemical study of endocrine pancreas ontogeny in the chick embryo: normal development and pancreatic potentialities in the early splanchnopleure, in *The Evolution of Pancreatic Islets,* Adesanya, T., Grillo, I., Liebson, L. Epple, A., Eds. Pergamon Press, Oxford, 1976, 37.

13. **Liebson, L., Bondareva, V., and Soltitskaya, L.,** The secretion and the role of insulin in chick embryos and chickens, in *The Evolution of Pancreatic Islets,* Adesanya, T., Grillo, I., Liebson, L., and Epple, A., Eds., Pergamon Press, Oxford, 1976, 69.

14. **Rosenzweig, J. L., Lesniak, M. A., Samuels, B. E., Yip, C. C., Zimmerman, A. E., and Roth, J.,** Insulin in extrapancreatic tissues of guinea pig differs markedly from the insulin in their plasma and pancreas, *Trans. Assoc. Am. Physicians,* 93, 263, 1980.

15. **Serrano, J., Bevins, C. L., Young, W. S., Roth, J., and De Pablo, F.,** The insulin gene is differentially expressed in the pancreas and liver of developing chick embryos, *Endocrinology,* 120 (Suppl. 29), 1987.

16. **Haselbacher, G. K., Andres, R. Y., and Humbel, R. E.,** Evidence for the synthesis of a somatomedin similar to insulin-like growth factor I by chick embryo liver cells, *Eur. J. Biochem,* 111, 245, 1980.

17. **Burch, W. M., Weir, S., and Van Wyk, J. J.,** Embryonic chick cartilage produces its own somatomedin-like peptide to stimulate cartilage growth in vitro, *Endocrinology,* 119, 1370, 1986.

18. **Engstrom, W., Bell, K. M., and Schofield, P. N.,** Expression of the insulin-like growth factor II gene in the developing chick limb, *Cell Biol. Inst. Rep,* 11, 415, 1987.

19. **Jansen, M., Van Schaik, F. M. A., Ricker, A. T., Bullock, B., Wood, D. E., Gabbay, K. H., Nusbaum, A. L., Sussenbach, J. S., and Van Der Brande, J. L.,** Sequence of cDNA encoding human insulin-like growth factor I precursor, *Nature (London),* 306, 609, 1983.

20. **Bell, G. I., Stempien, M. M., Fong, N. M., and Rall, L. B.,** Sequences of liver cDNAs encoding two different mouse insulin-like growth factor I precursors, *Nucleic Acid Res.,* 14, 7873, 1986.

21. **Roberts, C. T., Lasky, S. R., Lowe, W. L., Seaman, W. T., and LeRoith, D.,** Molecular cloning of rat insulin-like growth factor I complementary deoxy-nucleic acids: differential messenger ribonucleic acid processing and regulation by growth hormone in extrahepatic tissues, *Mol. Endocrinol,* 1, 243, 1987.

22. **Bell, G. I., Merryweather, J. P., Sanchez-Pescador, R., Stempien, M. M., Priestly, L., Scott, J., and Rall, L. B.,** Sequence of a cDNA clone encoding human preproinsulin-like growth factor II, *Nature (London),* 310, 775, 1984.

23. **Dull, T. J., Grey, A., Hayflick, J. S., and Ullrich, A.,** Insulin-like growth factor II precursor gene organization in relation to insulin gene family, *Nature (London),* 310, 777, 1984.

24. **Stempien, M. M., Fong, N. M., Rall, L. B., and Bell, G. I.,** Sequence of a placental cDNA encoding the mouse insulin-like growth factor II precursor, *DNA,* 5, 357, 1986.

25. **Hamburger, V. and Hamilton, H. L.,** A series of normal stages in the development of the chick embryo, *J. Morphol.,* 88, 49, 1951.

26. **De Pablo, F., Girbau, M., Gomez, J. A., Hernandez, E., and Roth, J.,** Insulin antibodies retard and insulin accelerates growth and differentiation in early embryos, *Diabetes,* 34, 1063, 1985.

27. **Goldfine, I. D., Purrello, F., Vigneri, R., and Clawson, G. A.,** Direct regulation of nuclear functions by insulin: relationship to mRNA metabolism, in *Molecular Basis of Insulin Action,* Czech, M.P., Ed., Plenum Press, New York, 1985, 329.

28. **King, G. L. and Kahn, C. R.,** The growth-promoting effects of insulin, in *Growth and Maturation Factors,* Vol. 2, Guroff, G., Ed., John Wiley & Sons, New York, 1984, 224.

29. **Straus, D. S.,** Growth-stimulating actions of insulin in vitro and in vivo, *Endocr. Rev,* 15, 356, 1984.

30. **Granner, D., Andreone, T., Sasaki, K., and Beale, E.,** Inhibition of transcription of the phosphoenolpyruvate carboxykinase gene by insulin, *Nature (London),* 305, 549, 1983.

31. **Prosser, C. G., Sankaran, L., Hennighausen, L., and Topper, Y. J.,** Comparison of the roles of insulin and insulin-like growth factor I in casein gene expression and in the development of a-lactalbumin and glucose transport activities in the mouse mammary epithelial cells, *Endocrinology,* 120, 1411, 1987.

32. **De la Haba, G., Cooper, G. W., and Elting, V.,** Hormonal requirements for myogenesis in vitro: insulin and somatotropin, *Proc. Natl. Acad. Sci. U.S.A.,* 56, 1719, 1966.

33. **Ridpath, J. F., Huiatt, T. W., Trenkle, A. H., Robson, R. M., and Bechtel, P. J.,** Growth and differentiation of chicken embryo muscle cell cultures derived from fast- and slow-growing lines. Intrinsic differences in growth characteristics and insulin response, *Differentiation,* 26, 121, 1984.

34. **Schmid, C., Steiner, T., and Froesch, E. R.,** Preferential enhancement of myoblast differentiation by IGF-I and -II in primary cultures of chick embryonic cells, *FEBS Lett,* 116, 117, 1983.

35. **Plant, P. W., Deeley, R. G., and Grieninger, G.,** Selective block of albumin gene expression in chick embryo hepatocytes cultured without hormones and its partial reversal by insulin, *J. Biol. Chem.,* 258, 15355, 1983.

36. **Widmer, U., Schmid, C. H., Zapf, J., and Froesch E. R.,** Effects of insulin-like growth factors on chick embryo hepatocytes, *Acta Endocrinol. (Copenhagen),* 108, 237, 1985.

37. **Burch, W. M. and Lebovitz, H. E.,** Hormonal activation of ornithine decarboxylase in embryonic chick pelvic

cartilage, *Am. J. Physiol,* 241, E454, 1981.

38. **Milstone, L. M. and Piatigorsky, J.,** d-crystallin gene expression in embryonic chick lens epithelia cultured in the presence of insulin, *Exp. Cell Res.,* 105, 9, 1977.

39. **Aizenman, Y., Weichsel, M. E., and de Vellis, J.,** Changes in insulin and transferrin requirements of pure neuronal cultures during embryonic development, *Proc. Natl. Acad. Sci. U.S.A.,* 83, 2263, 1986.

40. **Kyriakis, J. M., Hausman, R. E., and Peterson, S. W.,** Insulin stimulates choline acetyltransferase activity in cultured embryonic chicken retinal neurons, *Proc. Natl. Acad. Sci. U.S.A.,* 84, 7463, 1987.

41. **Girbau, M., Gomez, J. A., Lesniak, M. A., and De Pablo, F.,** Insulin and insulin-like growth factor I both stimulate metabolism, growth and differentiation in the postneurula chick embryo, *Endocrinology,* 121, 1477, 1987.

42. **De Pablo, F., Hernandez, E., Collia, F., and Gomez, J. A.,** Untoward effects of pharmacological doses of insulin in early chick embryos: through which receptors are they mediated?, *Diabetologia,* 28, 308, 1985.

43. **Nissley, S. P. and Rechler, M. M.,** Somatomedin/insulin-like growth factor tissue receptors, *Clin. Endocrinol. Metab.,* 13, 43, 1984.

44. **Ullrich, A., Bell, J. R., Chen, E. Y., Herrera, R., Petruzzelli, L. M., Dull, A., Gray, A., Coussens, L., Liao, Y.-C., Tsubokawa, M., Mason, A., Seeburg, P. H., Grunfeld, C., Rosen, O., and Ramachandran, J.,** Human insulin receptor and its relationship to the tyrosine kinase family of oncogenes, *Nature (London),* 313, 756, 1985.

45. **Thibault, C. and Daughaday, W. H.,** Insulin-like growth factor II receptors, *J. Biol. Chem,* 259, 3361, 1985.

46. **Morgan, D. O., Edman, J. C., Standring, D. N., Fried, V. A., Smith, M. C., Roth, R. A., and Rutter, W. J.,** Insulin-like growth factor II receptor as a multifunctional binding protein, *Nature (London),* 329, 301, 1987.

47. **Rosen, O. M.,** After insulin binds, *Science,* 237, 1452, 1987.

48. **Bassas, L., De Pablo, F., Lesniak, M. A., and Roth, J.,** The insulin receptors of chick embryo show tissue-specific structural differences which parallel those of the insulin-like growth factor I receptors, *Endocrinology,* 121, 1468, 1987.

49. **Heidenreich, K. A., Zahniser, N. R., Berhanu, P., Brandenburg, D., and Olefsky, J. M.,** Structural difference between insulin receptors in the brain and peripheral target tissues, *J. Biol. Chem.,* 258, 8527, 1983.

50. **Shemer, J., Penhos, J. C., and LeRoith, D.,** Insulin receptors in lizard brain and liver: structural and functional studies of a- and b-subunits demonstrate evolutionary conservation, *Diabetologia,* 29, 321, 1986.

51. **Simon, J. and LeRoith, D.,** Insulin receptor of chicken liver and brain. Characterization of a and b subunit properties, *Eur. J. Biochem,* 158, 125, 1986.

52. **Lowe, W. and LeRoith, D.,** Insulin receptors from guinea pig liver and brain: structural and functional studies, *Endocrinology,* 118, 1669, 1986.

53. **Girbau, M., Bassas, L., and De Pablo, F.,** Autoradiographic detection of insulin and IGF I receptors in gastrula and neurula chick embryos, *Endocrinology,* 122 Suppl. A, 520, 1988.

54. **Bassas, L., De Pablo, F., Lesniak, M. A., and Roth, J.,** Ontogeny of receptors for insulin-related peptides in chick embryo tissues: early dominance of insulin-like growth factor over insulin receptors in the brain, *Endocrinology,* 117, 2321, 1985.

55. **Bassas, L., Zelenka, P. S., Serrano, J., and De Pablo, F.,** Insulin and IGF receptors are developmentally regulated in the chick embryo eye lens, *Exp. Cell Res,* 168, 561, 1987.

56. **Bassas, L., Lesniak, M. A., Serrano, J., Roth, J., and De Pablo, F.,** Developmental regulation of insulin and type I IGF receptors and absence of type II receptors in chick embryo tissues, *Diabetes,* 37, 637, 1988.

57. **Kin-Tak, Y. and Czech, M. P.,** The type I IGF receptor mediates rapid effects of MSA on membrane transport systems in rat soleus muscle, *J. Biol. Chem.,* 259, 3090, 1984.

58. **Mottola, C. and Czech, M. P.,** The type II insulin-like growth factor receptor does not mediate increased DNA synthesis in H-35 hepatoma cells, *J. Biol. Chem.,* 259, 12705, 1984.

59. **Girbau, M., Lesniak, M. A., Gomez, J. A., and De Pablo, F.,** Insulin action in early embryonic life: anti-insulin receptor antibodies retard chicken embryo growth, but not muscle differentiation *in vivo, Biochem. Biophys. Res. Commun.,* 153, 142, 1988.

60. **Maller, J. L. and Koontz, J. W.,** A study of the induction of cell division in amphibian oocytes by insulin, *Dev. Biol.,* 85, 309, 1981.

61. **El-Etr, M., Schorderet-Slatkine, S., and Baulieu, E. E.,** Meiotic maturation in *Xenopus laevis* oocytes initiated by insulin, *Science,* 205, 1397, 1979.

62. **El-Etr, M., Schorderet-Slatkine, S., and Baulieu, E. E.,** The role of zinc and follicle cells in insulin-initiated meiotic maturation of *Xenopus laevis* oocytes, *Science,* 210, 929, 1980.

63. **Alberts, B., Bray, D., Lewis, J., Raff, M., Roberts, K., and Watson, J. D.,** *Molecular Biology of the Cell,* Garland Publishing, New York, 1983.

64. **Serrano, J., Bevins, C. L., Young, S. W., and De Pablo, F.,** Insulin gene expression in chicken ontogeny: pancreatic, extrapancreatic, and prepancreatic, *Dev. Biol.,* 132, 410, 1989.

65. **Scavo, L. and De Pablo, F.,** unpublished.

66. **Bassas, L., Lesniak, M. A., Girbau, M., and De Pablo, F.,** Insulin-related receptors in the early chick embryo: from tissue patterns to possible function, *J. Exp. Zool.,* Suppl. 1, 1987.

Chapter 6

INSULIN AND INSULIN-LIKE GROWTH FACTORS IN MAMMALIAN DEVELOPMENT

S. Heyner, B. A. Mattson, R. M. Smith, and I. Y. Rosenblum

TABLE OF CONTENTS

I. INTRODUCTION

The earliest stages of mammalian development are characterized by rapid cell proliferation and the onset of differentiation. Developmental biologists have classically focused much of their attention on mechanisms that underly differentiation in an attempt to delineate the influences that guide totipotent cells into particular paths. However, the recent discoveries of the role of oncogenes in tumor biology have re-emphasized that the control of proliferation in development is also of crucial importance. Although development of the early embryo is controlled primarily by the embryonic genome, it is regulated also by various hormones and growth factors and by a continuing supply of energy. The involvement of specific growth-promoting substances during this period of development remains to be characterized. In this chapter, we will review studies showing that the early mammalian embryo expresses receptors that bind insulin and insulin-like growth factors (IGFs), discuss the evidence that there is autocrine production of these factors during early postimplantation developmental stages, and speculate on the role that these peptides may play in mammalian development.

II. INSULIN AND THE INSULIN-LIKE GROWTH FACTORS

It has been known for some time that a number of polypeptide hormones and growth factors found in blood and other tissues have the ability to stimulate the growth of mammalian cells *in vitro*.[1] Studies in a variety of systems have indicated that growth factors activate the production of proteins that are required for cell proliferation as well as for maintenance. While the use of cell lines maintained *in vitro* has led to an understanding of the interaction of endocrine and paracrine hormones and factors influencing growth, these mechanisms are poorly understood in the intact organism. This is perhaps best exemplified by insulin. Insulin is a 6 kilodalton (kDa) peptide hormone containing two polypeptide chains, A and B, linked by two invariant disulfide bonds. The two-chain hormone is derived biosynthetically from its immediate precursor, proinsulin, which consists of the A and B chains linked to a connecting, C-peptide. The initial translational product of insulin mRNA is preproinsulin, containing an N-terminal signal peptide of 24 amino acids linked to proinsulin. Preproinsulin is synthesized in the pancreatic beta cells, and in the pancreas, the fully processed insulin mRNA is 0.55 kilobase (kb) long.[2] Among the well-documented anabolic actions of insulin are the stimulation of amino acid and glucose transport[3] and DNA synthesis[4] of cells in culture.

The insulin-like growth factors (IGF-I and IGF-II) are single-chain peptide hormones that have a high degree of amino acid sequence homology with each other and with proinsulin; all three hormones appear to have a common ancestry.[5] Like insulin, the IGFs exert a number of biological effects on cells, including the induction of DNA synthesis and cell multiplication and the stimulation of several anabolic responses.

Insulin is one of the most important hormones influencing fetal growth and has classically been considered one of the most significant factors in the regulation of fetal metabolism. However, nearly all the data contributing to this view have been derived from pathological states in which there is either hypoinsulinemia or hyperinsulinemia. Thus, there are numerous reports on the effects of hypoinsulinemia or hyperinsulinemia on fertility and birth defects.[6] In addition, there is considerable information available concerning the distribution of insulin receptors in fetal tissues and the onset of fetal production of insulin.[7] However, there is very little known about the role of insulin in early mammalian embryogenesis, before the placenta is formed. This is of particular interest, since the placenta is believed to be impermeable to insulin and the conceptus does not produce pancreatic insulin until relatively late in gestation. Furthermore, glucose is a major metabolic requirement for mammalian embryos, and glucose imbalances are known to perturb normal developmental patterns.[7] Although the need for insulin or IGFs during mammalian prepancreatic development remains to be established, it is quite likely that such a requirement exists.

III. RECEPTORS FOR INSULIN AND INSULIN-LIKE GROWTH FACTORS

The biological effects of insulin and the IGFs are initiated by the interaction of the ligand with specific cell surface receptors. The mammalian insulin receptor is a complex integral membrane glycoprotein with intrinsic enzymatic activity. The basic structure appears to be the same in all vertebrates examined so far[8], consisting of a glycosylated tetramer made up of two alpha subunits that have the capacity to bind insulin and are linked by disulfide bonds to each other and to the two beta subunits. The beta subunits traverse the plasma membrane and possess an insulin-dependent protein tyrosine kinase. The IGF–I receptor is similar (but distinct), in structure and size of subunit components (see References 9 and 10 for reviews). From overlapping cDNA clones, the entire sequence of both the human insulin receptor mRNA[11,12] and the human IGF-I receptor mRNA[13] have been deduced. In both cases, the mRNA codes for a single polypeptide precursor that is about 1,370 amino acids long. The precursor includes a 27- to 30-residue signal peptide followed by the alpha receptor subunit of about 700 amino acids (M_r ~ 80,000) and the beta receptor subunit of about 625 amino acids (M_r ~ 70,000). The latter correspond to the mature 135,000 M_r (alpha) and 90,000 M_r (beta), fully glycosylated subunits. Despite functional and structural similarities, the receptors for insulin and IGF-I are likely to have different biological roles in development and in differentiated cells. Indeed, the two receptors are products of two distinct genes, located on separate chromosomes,[13] that may be controlled by different regulatory signals. Although the two receptors are structurally very similar, they may be discriminated by the affinity with which they bind ligands. Insulin binds with high affinity to the insulin receptor but also binds to the IGF-I receptor, with an approximately 100 times lower affinity. Similarly, the IGF-I receptor can bind both IGF-I and insulin, but the homologous ligand binds with approximately 100 times higher affinity.[14] Both receptors autophosphorylate in response to ligand binding, and the tyrosine kinase activity of the insulin receptor, and presumably the IGF-I receptor as well, appears to be essential for the ligand to stimulate its biological responses in cells.[15-17] The Type II receptor does not possess intrinsic kinase activity and is quite different structurally from the insulin and the Type I receptor. Although the Type II receptor binds IGF-II with high affinity, it has a lower affinity for IGF-I and does not bind insulin.[18,19] However, IGF-II itself can bind to both the IGF I and the insulin receptors, although the affinity with which it binds has been reported to vary considerably.[9] Recent isolation and sequencing of the IGF-II receptor cDNA predicts a structure with a single transmembrane region, an extracellular domain that comprises 93% of the receptor molecule and a relatively small cytoplasmic domain of M_r 18,000.[20] Neither domain of the Type II receptor shares homology with the insulin or the IGF-I receptor. On the other hand, the IGF-II receptor is now known to be identical to the mannose-6-phosphate (Man-6-P) receptor. This identity is based on several lines of evidence, including structural[21] (sequence homology of the purified human receptor), immunological cross-reactivity,[22,23] and ligand binding affinities and stoichiometry.[24] These intriguing data raise several important questions regarding the physiological role of the IGF-II receptor, which has never been proven to propagate a transmembrane signal following ligand binding. The question remains as to whether this membrane protein functions as a Man-6-P receptor to mediate the lysosomal targeting of proteins or as the IGF-II receptor which mediates genetic and/or metabolic responses involved in cell growth and differentiation. Can one receptor protein participate in, or mediate both, apparently unrelated processes? Data consistent with both roles have been reported in the literature.[25]

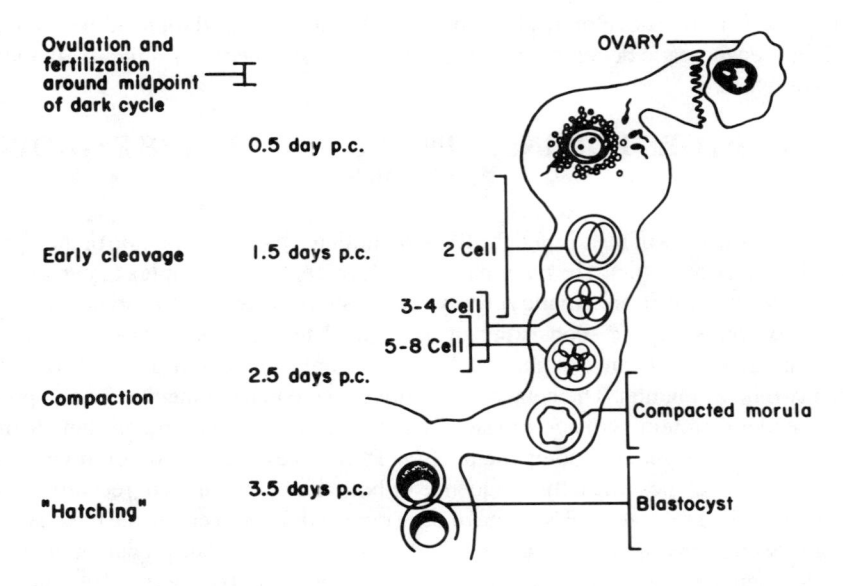

FIGURE 1. Schematic representation of preimplantation embryonic development in the mouse.

IV. THE MOUSE PREIMPLANTATION EMBRYO: A GENERALIZED PARADIGM

Recent discoveries in cellular and molecular biology, as well as immunology, have had profound effects on developmental biology. One of the major developments has been the application of molecular biological techniques, in an attempt to learn more about gene expression and control during embryogenesis. Although it is clear that there are differences in early developmental patterns between different mammalian species, comparative studies have shown that the mouse embryonic system provides a generalized paradigm for mammalian development.[26] The advantages offered by the early mouse embryo include a well-defined "window" in which the embryo can be recovered from the genital tract and cultured in a simple medium. The ability to superovulate the females allows the collection of large numbers of embryos at a defined stage of development. A schematic drawing of the preimplantation stages of mouse embryogenesis is shown in Figure 1.

The paucity of data concerning the role of insulin and IGFs in early developmental mammalian stages may be due in part to the difficulty of carrying out classical biochemical studies on the preimplantation mammalian embryo. The volume, dry weight, and total protein per embryo change very little between the fertilized ovum and the morula-early blastocyst. For example, the fertilized mouse ovum has a mean volume of 0.192 nl, while the morula-early blastocyst has a volume of 0.219 nl; similarly, the total protein content of the fertilized egg is 27.8 ng and that of the morula-early blastocyst is approximately 22 ng per embryo.[27] Inspection of these data will show immediately that the classic biochemical strategies of studying binding kinetics to assess receptor identity and function are simply not feasible in this system. One way around this problem is to utilize immunological probes. Antibodies provide high specificity and sensitivity and have been used to examine the expression of a wide range of surface molecules on early mammalian embryos.[28]

V. DETECTION OF STAGE-SPECIFIC INSULIN BINDING TO MOUSE EMBRYOS

Insulin binding to the cells of the preimplantation mouse embryo has been demonstrated by Rosenblum et al.[29] using a modification of the indirect immunofluorescent technique described by Majercik and Bourgignon.[30] Insulin binding to IM-9 lymphoblasts was incorporated as a positive control. Physiological levels of insulin were employed, and saturation appeared to occur over a concentration range between 4.0 ng/ml and 40 ng/ml. Using this experimental design, specific binding of insulin to the cells of the preimplantation embryo could be detected first at the morula stage, and the reactivity increased at the blastocyst stage as judged by increased immunofluorescent intensity, shown in Figure 2.

These experiments provided the first evidence that the cells of the preimplantation mouse embryo are capable of binding insulin. Because insulin binding occurred in a stage-specific manner, the results of these experiments suggested that there may be expression of the insulin receptor by cells of the morula-stage mouse embryo. However, it is known from the studies of Fleming and Pickering[31] that an extensive endocytotic system develops in the mouse embryo around the eight-cell stage. Therefore, these results could not be considered conclusive evidence that insulin binding was receptor-mediated until the possibility of insulin uptake by means of nonspecific endocytosis could be excluded.

VI. PREIMPLANTATION MOUSE EMBRYOS EXPRESS RECEPTORS THAT BIND INSULIN

Autoradiographic methods provide for specificity and also accommodate the use of very small tissue samples. In addition, they provide for an assessment of receptor-mediated uptake, by means of measuring the displacement of a labeled ligand by incubation in the presence of an excess of the unlabeled ligand. This approach was adopted by Mattson et al.[32] in order to confirm that insulin binding to the cells of the 16-cell mouse embryo was developmentally regulated and mediated via receptor(s).

Oocytes and preimplantation stages were incubated with physiological levels of ^{125}I-insulin, immobilized on coverslips using a modification of the method described by Wiley et al.,[33] sectioned and processed for autoradiography. Results confirmed that specific binding is detected first at the morula stage (Figure 3). Examination of autoradiographic grains on sections of blastocysts revealed grains overlying virtually all portions of the cellular structure, but not above the blastocoel cavity. This finding provided further support for the conclusion that insulin is bound specifically to insulin receptors in the membranes of the cells before being internalized by a receptor-mediated process. Uptake of ligand by the endocytotic system that develops around this time in the mouse embryo[31] is not a sufficient explanation of the phenomona described. First, the number of grains associated with morula and blastocyst stage embryos reached a plateau at near physiologic hormone levels and did not increase with further additions of insulin, as shown in Figure 4. If fluid-phase endocytosis were the operative mechanism, there would be a concentration-dependent increase in grains well beyond that which was observed. Second, co-incubation with an excess of unlabeled ligand significantly decreased the association of ^{125}I-insulin with morulae and blastocyst stages, again providing evidence for a receptor-mediated process.

As discussed above, insulin can be bound by the insulin receptor, but additionally, may be bound by the IGF-I receptor. In order to discriminate between these two possibilities, Mattson et al.[32] incubated oocytes and preimplantation embryos with oocytes and preimplantation

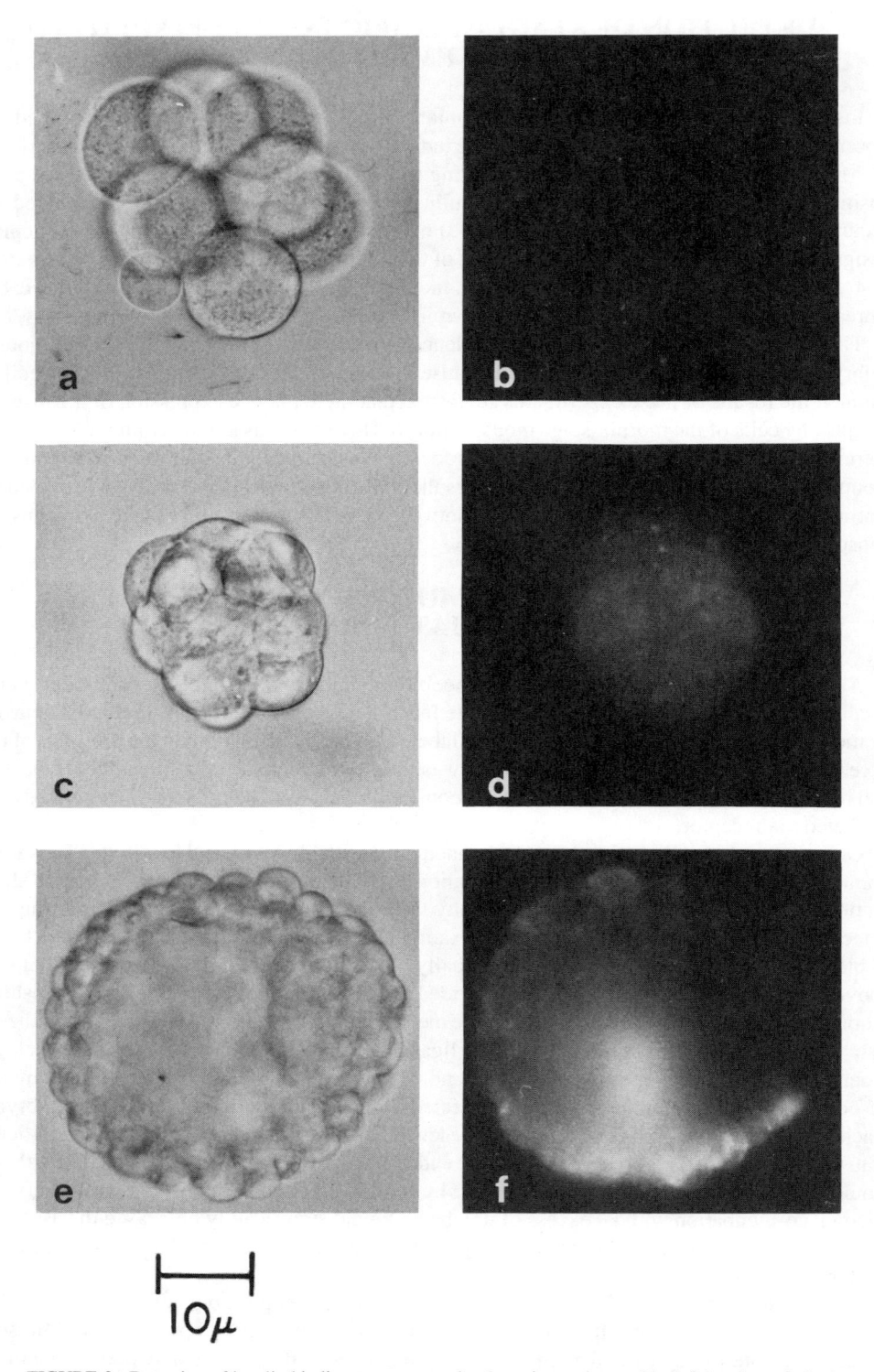

FIGURE 2. Detection of insulin binding to mouse preimplantation embryos. The left hand panel shows brightfield micrographs of (a) eight-cell, (c) morula, and (e) blastocyst stage embryos. Corresponding micrographs to demonstrate immunofluorescent localization of insulin are shown in the right panel, (b) eight-cell, (d) morula, and (f) blastocyst.

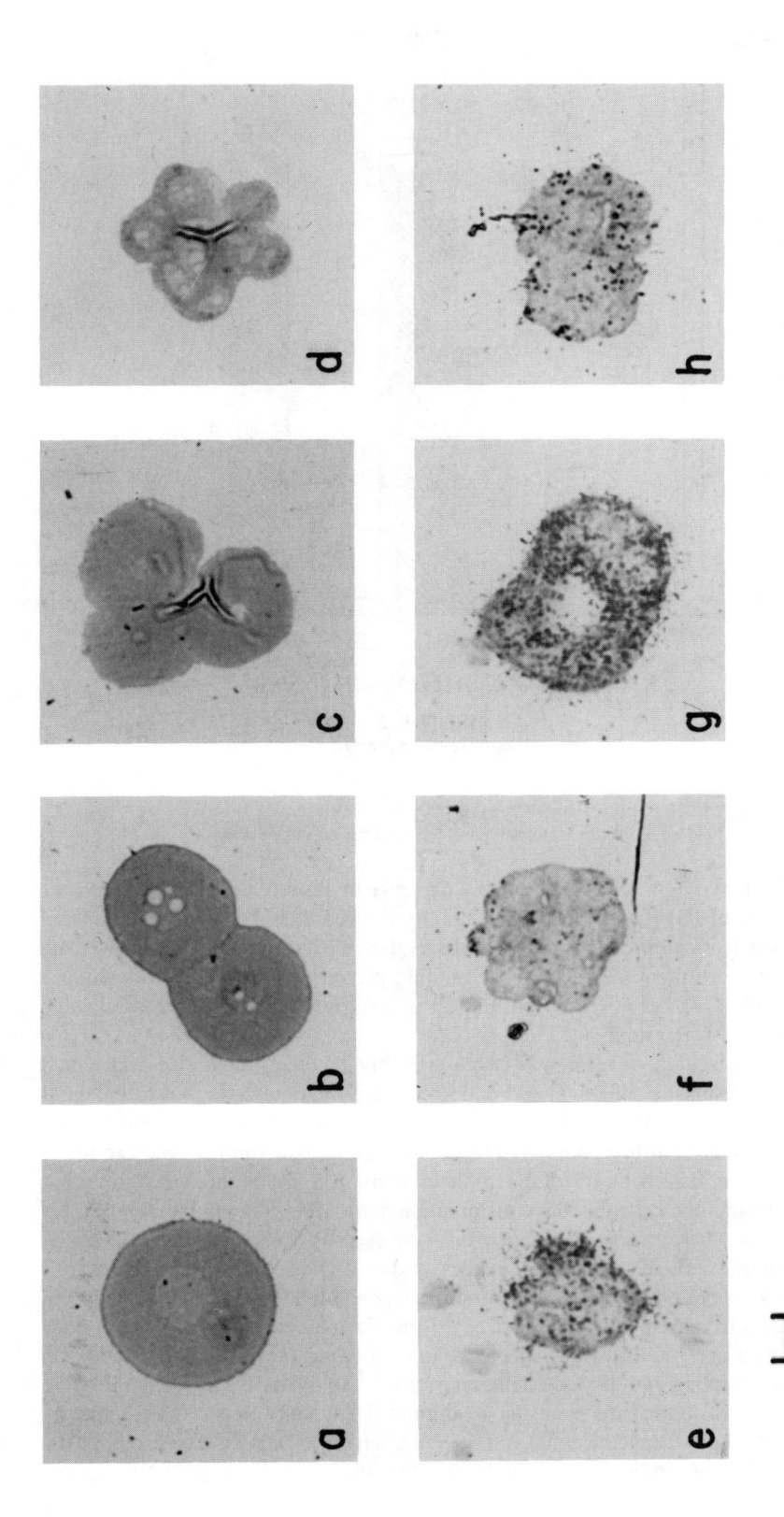

FIGURE 3. Light microscopic autoradiographs of 5-micron sections of (a) mouse oocyte, (b) two-cell, (c) four-cell, (d) eight-cell, (e and f) morula, and (g and h) blastocyst. All cells were incubated with 4 ng/ml ^{125}I-insulin. Controls consisted of incubation in the presence of 1000 × excess unlabeled ligand. When this was done for the (f) morula and (h) blastocyst, significant displacement of grains was noted.

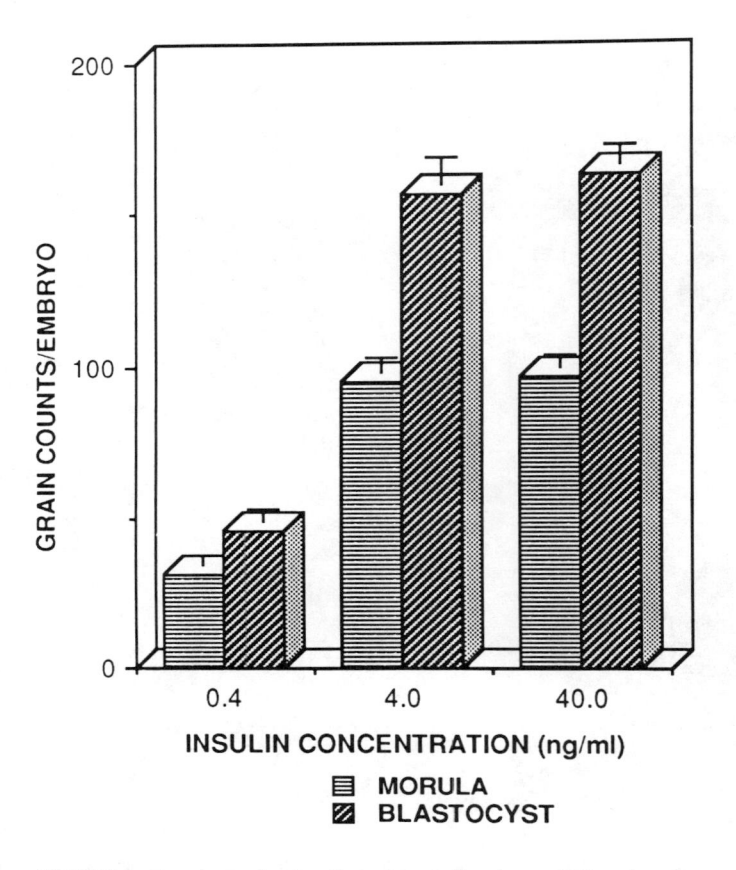

FIGURE 4. Summary of autoradiographic studies demonstrating dose-dependent binding of insulin to mouse morula and blastocyst stage embryos.

embryos incubated with 4.0 ng/ml ^{125}I-insulin in the presence or absence of a 1000-fold excess of insulin, or a mixture of IGF-I and IGF-II. The results showed that ^{125}I-insulin grains were displaced significantly from preimplantation embryos by native insulin or by IGFs. These data provide evidence for more than one receptor or binding protein expressed by early mouse embryos. This possibility was explored by the authors at a slightly later developmental stage using a blastocyst outgrowth system.

Murine blastocysts cultured *in vitro* will hatch from their zonae pellucidae and attach ("implant") on a glass or plastic substrate, and undergo differentiation analogous to the early stages of uterine implantation.[34] Cells of the inner cell mass may be distinguished readily from the trophectoderm cells by their morphology. They constitute a population of small rounded cells that overlie the larger flattened cells of the trophectoderm. It is important to recognize that blastocyst outgrowths include cells that have differentiated into more than two cell types, for example, ectoderm and endoderm. These outgrowths were used by Mattson et al.[32] to examine ^{125}I-insulin binding to the cells of the periimplantation embryo. The results, shown in Table 1, demonstrate that insulin binding is both concentration dependent and saturable. In addition, there was no apparent difference between binding of ^{125}I-insulin to the cells of the inner cell mass or the trophectoderm, as determined by through-focus light microscopic evaluation. In addition to insulin binding, the binding and displacement, respectively, of ^{125}I-IGF-I and ^{125}I-IGF-II, to the cells of the embryonic outgrowths were also examined. The results are shown graphically in Figure 5. These data provide evidence that at the periimplantation stage of development, the mouse embryo is capable of binding insulin. They also provide evidence not only that insulin

TABLE 1
Dose-Dependent Binding of ^{125}I-Insulin to Blastocyst Outgrowths

^{125}I-insulin (ng/ml)	Inner cell mass	Trophectoderm
	(Grains/μm²)	
0.04	19.2 ± 2.2	23.9 ± 1.5
4.0	60.0 ± 1.9	60.9 ± 1.2
40.0	62.3 ± 1.9	61.7 ± 1.7

Note: Blastocyst outgrowths were incubated with the concentrations of ^{125}I-insulin shown and prepared for autoradiography. Grain counts were estimated on three randon areas of outgrowth at each dose level.

and IGFs are bound in a receptor-mediated fashion to mouse early embryonic stages, but also that more than one type of receptor is expressed by this stage of development. This conclusion is based on the grain displacement counts which are shown in Figure 5. Since insulin also binds to the Type I IGF receptor, the binding of ^{125}I-IGF-I to its receptor will be displaced in the presence of either excess unlabeled IGF-I or insulin. However, greater displacement of the labeled ligand will result from co-incubation with excess unlabeled IGF-I. A similar argument applies to the displacement of insulin and IGF-II binding, respectively.

VII. PREIMPLANTATION MOUSE EMBRYOS EXHIBIT THE CLASSIC ENDOCYTOTIC PATHWAY FOR INSULIN UPTAKE AND INTERNALIZATION

Electron microscopy is a powerful tool for analyzing receptor-mediated binding. Furthermore, it allows elucidation of the intracellular pathway of a particular receptor-bound ligand. The "classic" receptor-mediated endocytotic pathway is shown in Figure 6. Numerous studies have shown that insulin may be visualized readily at the ultrastructural level by means of labeling with either ferritin or colloidal gold. In either case, the ligand was shown to retain biological activity identical to the native hormone.[35,36] Although ferritin is widely used, gold provides a more definitive image and has been used widely to visualize ligand binding and endocytosis.

Recently, we have used gold-labeled insulin in conjunction with high-resolution electron microscopy to visualize occupied insulin receptors in early mammalian embryos.[37,38] Data summarizing the location and the count of gold particles on preimplantation mouse embryos incubated in gold-labeled insulin and gold-labeled BSA, respectively, are presented in Table 2. These data show that the gold-labeled hormone is bound to the plasma membrane and microvilli, is concentrated in coated pits, and is internalized. In contrast, embryos incubated in gold-labeled BSA showed no evidence of gold uptake.

Electron micrographs (Figure 7) reveal details of the uptake and intracellular translocation of gold-insulin. First, aggregation of labeled ligand was visualized along a region of thickened plasma membrane, and this was followed by internalization of the labeled ligand via coated pits. Gold-labeled insulin was visualized in both coated and uncoated vesicles, while substantial amounts of the ligand accumulated in dense bodies and lysosomes. Controls consisting of gold-labeled BSA showed random binding patterns in the zona pellucida and glycocalyx; the label was not concentrated in coated pits or internalized. Since the gold-insulin complex is itself stabilized by BSA,[36] these results provide convincing evidence that the uptake of insulin was specific to the hormone itself.

Quantitative morphometry of the cell surface and gold-labeled insulin showed that approximately 11% of the cell-surface associated ligand was found in coated pits, which comprised only 3% of the cell surface area. These observations indicate a concentrative mode of uptake of insulin into the cells of the early mouse embryo. Similar quantitative morphometric studies using

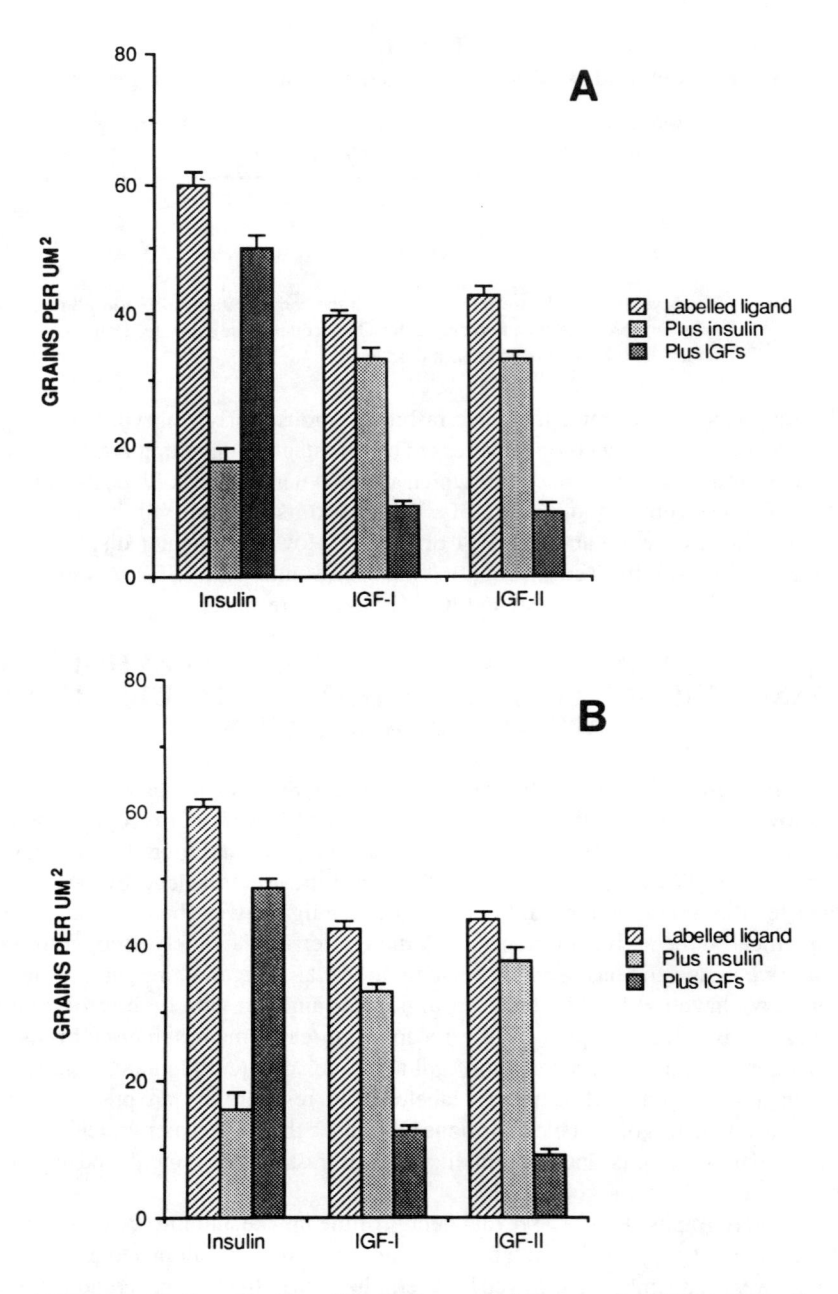

FIGURE 5. Binding of insulin and insulin-like growth factors to cells of (A) the inner cell mass and (B) trophectoderm of blastocyst outgrowths. Blastocyst outgrowths were incubated with labeled insulin or insulin-like growth factor I or II, in the presence or absence of excess unlabeled ligand. The blastocyst outgrowths were prepared for autoradiography, and the grain counts evaluated.

gold-labeled insulin and ferritin-labeled insulin have shown previously that endocytosis of insulin via coated pits does not occur on rat adipocytes, although this mode of uptake has been demonstrated on other cell types.[39] The results of these studies imply the importance of intracellular translocation of insulin in the rapidly proliferating early embryo.

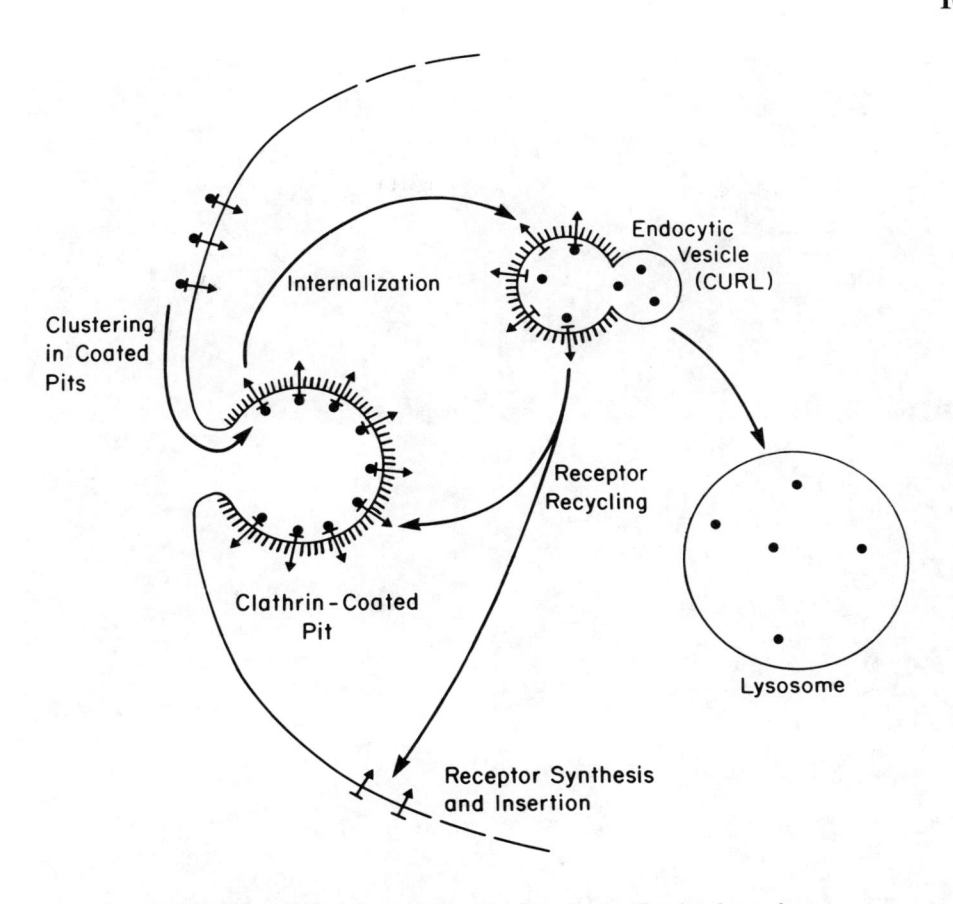

FIGURE 6. Schematic representation of receptor mediated endocytosis.

TABLE 2
Binding and Internalization of Gold-Labeled Insulin by Mouse Preimplantation Embryos

Stage (n)	Plasma membrane	Microvilli	Coated pit	Coated vesicle	Lysosome
			Gold-Insulin		
2—8 (10)	4.6 ± 1.2	2.0 ± 0.9	1.0 ± 1.0	0.6 ± 0.3	0.2 ± 0.1
Morula (8)	55.5 ± 5.8	16.1 ± 0.9	10.2 ± 2.3	9.2 ± 1.2	41.4 ± 5.5
Blatocyst (7)	155.4 ± 6.9	48.0 ± 4.3	24.4 ± 2.5	20.4 ± 1.5	51.4 ± 5.1
			Gold-BSA		
2—8 (4)	0.0 ± 0.0	0.1 ± 0.1	0.0 ± 0.0	0.2 ± 0.2	0.0 ± 0.0
Morula (6)	3.0 ± 0.9	1.7 ± 0.6	0.5 ± 0.2	1.7 ± 0.2	0.0 ± 0.0
Blastocyst (6)	1.8 ± 0.9	3.3 ± 1.0	0.5 ± 0.2	0.5 ± 0.2	0.0 ± 0.0

Note: Thin sections across whole embryos were evaluated for total gold particle counts. Controls consisted of incubating developmental stages with gold-labeled BSA under the same conditions. Particle counts represent mean ± SEM.

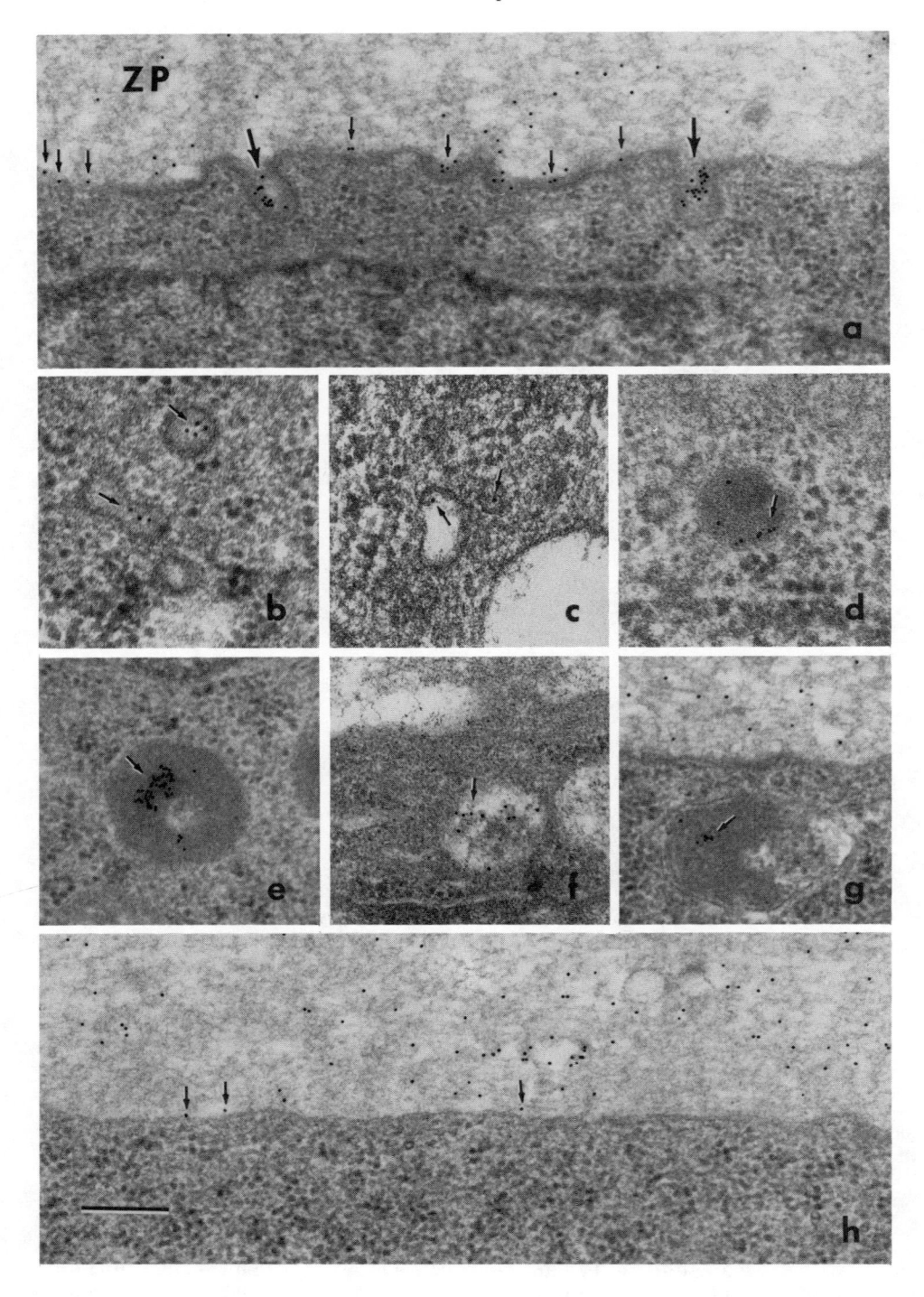

FIGURE 7. Electron microscopic visualization of gold-insulin uptake by cells of the blastocyst stage mouse embryo. (a) Gold-labeled insulin (small arrows) was found scattered throughout the zona pellucida (ZP) and bound in the glycocalyx of the plasma membrane in a widely distributed pattern. Receptor-mediated internalization of insulin was suggested by the large groups of particles observed in coated pits (large arrows). Internalized gold-insulin was found in (b) coated vesicles, (c) non-coated endosomes, (d and e) dense bodies of different sizes and granularity, (f) multivesicular bodies, and (g) lysosomes. When mouse embryos were incubated with gold-labeled BSA at the same concentration as gold-labeled insulin, gold particles were found scattered throughout the zona pellucida (h). A few particles were found in the glycocalyx, but no particles were seen in coated pits or intracellular structures (data not shown).

Autoradiographic studies at the light microscopic level did not reveal differential uptake of insulin between the cells of the trophectoderm and those of the inner cell mass. However, preliminary studies at the electron microscopic level have shown that there is a higher concentration of insulin-binding receptors on the cells of the trophectoderm vs. those of the inner cell mass. These observations suggest that insulin may be transported from the trophectoderm to the inner cell mass. Significantly fewer gold particles were visualized in cells of the inner cell mass, as compared to trophectoderm, and virtually no particles were seen binding to the cell membranes facing the blastocoel surface of either cell type. It is clear from autoradiographic studies on blastocyst sections and whole blastocyst outgrowths that the cells of the inner cell mass are capable of binding insulin. In order to confirm this at the ultrastructural level, expanded blastocysts were treated immunosurgically[40] and the isolated inner cell masses incubated with either gold-labeled insulin or gold-labeled BSA. The results of these experiments demonstrated that cells of the inner cell mass, isolated from expanded blastocysts, can bind and internalize gold-labeled insulin.[71]

VIII. INSULIN AND INSULIN-LIKE GROWTH FACTOR RECEPTORS AND LIGANDS IN POSTIMPLANTATION EMBRYOS

Recent studies have evaluated early mouse embryonic tissues for the presence of insulin, IGF-I, and IGF-II receptors.[41] Stages encompassing most of the period of organogenesis, from Day 9 (3 to 4 somites; neurulation) to vascularization of the yolk sac, (17 to 18 somites) through establishment of the chorioallantoic placenta (27 to 31 somites) and limb bud stages (45 somites, Day 12), were studied. The results of these studies indicate that Day-9 to -12 mouse embryonic tissues possess both Type I and II IGF receptors in greater abundance than the insulin receptor, but that the visceral yolk sac appears to have a greater abundance of insulin receptors. In addition, embryos contain extractable immunoreactive IGFs and IGF-binding proteins. These results are in general agreement with those obtained by Heath and Shi.[42] These investigators explanted extraembryonic tissues of the 9.5–d *post coitum* mouse embryo and metabolically radiolabeled the tissues with ^{35}S-methionine. Radiolabeled culture supernatants were immunoprecipitated with a rabbit anti-IGF-II antibody that recognized prohormone forms of IGF and the soluble IGF-binding protein. Results of these experiments showed that coexpression of four IGF-related proteins was detectable in amnion and extraembryonic yolk sac mesoderm, but not in either parietal endoderm or visceral endoderm. The authors concluded that IGF-like molecules and their cognate binding proteins are specifically expressed *in vitro* by tissues of extraembryonic mesodermal origin. Both IGF-I and IGF-II receptors have been found to be present in mouse limb buds cultured *in vitro* at all stages of differentiation from the blastema (Day 11) to well-differentiated cartilage (Day 19).[43] Further, the number of IGF-II receptors increased as differentiation occurred, although responsiveness to IGF II as measured by Alcian Blue uptake decreased.

Unterman and his colleagues have studied insulin receptor expression in rat embryos and extraembryonic membranes.[44] These investigators excised rat conceptuses on Day 9.5 of gestation and cultured them until either 10.4 or 11.6 d of development. At these two time points, individual components of the conceptus, including the embryo and the extraembryonic membranes, yolk sac, amnion, and allantoic placenta, were isolated and analyzed for insulin receptors by means of radiolabeled ligand binding analysis, photo-affinity labeling, and electrophoresis. The results of these studies demonstrated clearly that insulin-specific binding sites could be detected on all components of the rat conceptus by Day 10.4 of development. The inhibition of binding by anti-insulin receptor antiserum suggested that most, if not all, the insulin-specific binding was to the insulin receptor. Extraembryonic membranes also expressed insulin receptors, in fact, in greater concentration than the conceptus. When the yolk sac was separated from the other extraembryonic membranes, it was found to be rich in receptors that exhibited a slightly different molecular weight, and electrophoretic migration pattern, suggestive of a structural difference. The functional significance of this difference is not clear. However, the demonstration of high-affinity insulin receptors at this stage of development suggests that insulin plays an important role in early postimplantation mammalian development.

Insulin gene expression in the postimplantation rat conceptus has been investigated by Muglia and Locker.[45] Analysis of gene expression in the yolk sac at 14 d gestation revealed a 2.4 kb RNA species that hybridized to cloned insulin gene probes. This species increased throughout gestation. At 16 d, a second transcript of 0.72 kb was detectable, which became the predominant species by 18 d. Radioimmunoassay and gel filtration studies confirmed that there were approximately equal amounts of insulin and proinsulin in the 18-d yolk sac, suggesting that the transcripts detected were translated in this tissue.

Rat embryos at similar stages to those described above also have been examined histochemically for expression of IGF-I and -II mRNA.[46] IGF-I and -II mRNA were localized to tissues by means of *in situ* hybridization using oligodeoxynucleotide probes and the results confirmed by Northern blot analysis. These studies showed that IGF-II was predominant throughout development. It was strongly expressed in the liver and yolk sac and also expressed in a number of mesodermally derived structures in the process of differentiating; for example, proliferating chondrocytes were strongly labeled, while mature chondrocytes demonstrated only background levels. Probing for IGF-I failed to reveal evidence of its production at most of the sites that were positive for IGF-II. However, from Day 16.5 of gestation onward, faint positive hybridization signals were obtained in some mesodermally derived tissues. *In situ* hybridization with a probe for rat insulin also failed to reveal specific hybridization to cells of the perichondrium whereas 16.5–d fetal pancreatic cells gave a positive signal. In another study, Rotwein et al.[47] used a sensitive solution-hybridization assay to determine the steady state levels of IGF mRNAs during midgestation (Days 11 to 14) in the rat. IGF-I mRNA could be detected by Day 11, and rose approximately eightfold over the subsequent 48 h, whereas IGF-II mRNA, also detectable at Day 11, remained relatively constant over the ensuing 3 d.

Liu et al. have used an embryonic transplant system to examine the endocrine control of growth and differentiation in rat embryonic tissues.[48] Whole 10-d embryos were transplanted under the kidney capsule of intact hypophysectomized or diabetic juvenile syngeneic hosts and grown there for 3, 6, 9, or 12 d. Inhibition of growth in the experimental hosts occurred as early as Day 3 after transplantation but could be corrected by appropriate hormone replacement therapy of the hosts. The authors concluded that rat embryos/fetuses develop a dependence on insulin and a growth hormone-dependent factor, possibly IGF-I, before Day 13 of development.

In the rabbit, the occurrence and domain structure of the insulin receptor has been studied in liver and brown adipose tissue of developing fetuses.[49] Fetuses were examined on Days 20, 25, and 30 postcoitum. Photoaffinity labeling and immunoprecipitation were used to demonstrate that by Day 20, the insulin receptor was already present in liver and that addition of insulin stimulated phosphorylation of a 95 kDa protein. In brown adipose tissue, an insulin-stimulated phosphorylation of the beta subunit was also detected at all stages, starting from 20 d.

In the human, Scott et al. used dot blot and Northern blot analysis of RNA from human first trimester therapeutic abortions in order to examine the expression of IGF-II mRNA.[50] These investigators were in general agreement with the findings of Beck et al.[46] in the rat. A more recent investigation of IGF production during human development by Han et al. has utilized *in situ* hybridization with ^{32}P-labeled synthetic oligodeoxyribonucleotides complementary to portions of human IGF-I and IGF-II.[51] These authors found that IGF-I and IGF-II mRNAs were localized to connective tissues or cells of mesenchymal origin in 14 organs and tissues from human fetuses aged between 16 and 20 weeks gestation.

The placenta produces several peptides that have major influences on fetal growth and development. In addition to placental lactogen and chorionic gonadotrophin, insulin and the IGFs may be important growth-promoting peptides produced by the placenta before the

TABLE 3
Detection of Insulin and IGFs (Receptors and Ligands) in Postimplantation Mammalian Embryos

Target detected	Technique	Localization	Species/stage	Ref.
Insulin receptor	Radiolabeled ligand assay	Embryo and extraembryonic membranes	Mouse 9—12 d	41
Insulin receptor	Radiolabeled ligand assay	Embryo and extraembryonic membranes	Rat 10.4—11.6 d	44
Insulin receptor	Photoaffinity labeling	Liver, brown adipose tissue	Rabbit 20—30 d	49
Insulin	Northern blot RIA	Yolk sac	Rat 12—18 d	45
IGF-I receptor	Radiolabeled ligand assay	Embryo and yolk sac	Mouse 9—12 d	41
IGF-I receptor	Radiolabeled ligand assay	Limb bud	Mouse 9—19 d	43
IGF-I	RIA	Embryo	Mouse	41
IGF-I	*In situ* hybridization	Embryo and yolk sac	Rat 10.5—14.5 d	46
IGF-I	Solution hybridization	Embryo	Rat 11—14 d	47
IGF-I	Dot, Northern blots	Negative	Human 7—10 weeks	50
IGF-I	*In situ* hybridization	Connective tissues	Human 16—20 weeks	51
IGF-II receptor	Radiolabeled ligand assay	Limb bud	Mouse 9—12 d	41
IGF-II receptor	Radiolabeled assay	Limb bud	Mouse 9—19 d	43
IGF-II	Immuno-precipitation	Extraembryonic membranes	Mouse 9.5 d	41
IGF-II	Solution hybridization	Embryo	Rat 11—14 d	46
IGF-II	Dot, Northern blots	Connective tissues	Human 7—10 weeks	49
IGF-II	*In situ* hybridization	Connective tissues	Human 16—20 weeks	50

Note: Summary of studies reported in the literature on the detection of insulin and IGF receptors and their ligands on postimplantation mammalian embryos. References are numbered according to the textual citations and are listed in the reference section.

development of functional endocrine tissues in the fetus. Liu and colleagues[52] have shown that placental polyadenylated RNAs from the first and third trimester of normal pregnancies, as well as from term pregnancies of diabetic mothers, hybridize to a ^{32}P-labeled cDNA probe for an insulin-related sequence that is expressed in human fetal pancreas. Under the experimental conditions employed, only RNAs with substantial homology (> 80% sequence homology) would be expected to hybdridize. These results provide convincing evidence for expression of insulin-related genes in early pregnancy. In addition, the placental tissue from diabetic pregnancies contained quantitatively more of the sequences that hybridized strongly to the fetal insulin-related DNA probe.

Evidence supporting the expression of insulin and IGF receptors and the detection of their corresponding ligands during postimplantation development is summarized in Table 3.

IX. INSULIN AND INSULIN-LIKE GROWTH FACTORS AND THEIR RECEPTORS: EXPRESSION ON TERATOCARCINOMA CELLS

Teratocarcinomas in mice are malignant tumors that develop spontaneously in the gonads and can be produced experimentally from embryonic tissue and primordial germ cells. These tumors possess malignant stem cells (embryonal carcinoma cells) and the differentiated derivatives of these cells. Embryonal carcinoma (EC) cells can be maintained *in vitro* and induced to differentiate either by providing appropriate culture conditions, or by the addition of differentiation-inducing agents, for example, retinoic acid. These cells resemble cells of the Day 6 mouse embryo structurally[53] and have surface antigens in common with cleavage stage embryos.[54] Many investigators have considered that teratocarcinomas provide a paradigm for the early mammalian embryo, with the advantage of a vastly greater cell number. Although this premise has been questioned frequently, the amount of material available does allow for detailed biochemical analyses that are not possible with embryos. A number of investigators have examined different teratocarcinoma cell lines for the expression of insulin and other growth factor receptors. Detailed studies are described in Chapter 8. In the context of mammalian development, it is relevant to consider that high affinity receptors for insulin and multiplication-stimulating activity (MSA), the rat IGF-II have been identified on the mouse embryonal carcinoma cell line OTT-6050.[55] Similarly, F9 cells, originally derived from OTT-6050, also express receptors capable of binding MSA, and in these experiments, IGF-I and IGF-II both competed for MSA binding.[56] In a separate study, Nagarajan and Anderson have shown that insulin is capable of stimulating the growth of F9 EC cells under defined, serum-free conditions.[57] These investigators showed that antibodies to the insulin receptor did not alter the binding of radiolabeled MSA and that insulin stimulated the growth of F9 cells by acting directly through its own receptor and not through the MSA receptor, also present on these cells. In contrast to the studies described above, Heath et al.[58] were unable to demonstrate insulin-receptor binding activity or an insulin-stimulated growth response in PC13 teratocarcinoma cells. However, these cells expressed receptors for IGFs. When PC13 EC cells were induced to differentiate by administration of retinoic acid, the differentiated derivative cells expressed large numbers of insulin receptors, while the apparent number of unoccupied IGF receptors fell by about 60%. As a consequence of finding that IGF-II molecules were produced early in mouse development by extraembryonic membranes, the authors speculated that IGF-II may act in an autocrine manner.[58]

X. DISCUSSION

Although the roles of insulin and the IGFs await elucidation by means of functional experiments, it is clear that receptors for these ligands are expressed during early mammalian development. In the case of insulin, immunofluorescent, autoradiographic, and ultrastructural studies have provided confirmation that insulin is first bound at the morula stage of development. Further, autoradiographic studies have provided convincing evidence for the expression of receptors for not only insulin but also for IGFs at early stages.

High resolution electron microscopy has shown that insulin is concentrated and internalized by means of receptors that are concentrated in coated pits. Analysis of the cell surface and gold-labeled insulin distribution after 45 min incubation using quantitative morphometry reveals that there is an approximately fourfold concentration of the ligand in coated pits, compared to distribution on the plasma membrane. These data are evidence of a concentrative, receptor-mediated uptake process with subsequent intracellular translocation of the hormone by the rapidly dividing embryo. These observations imply a role for insulin at this stage of development. However, these findings should be interpreted with some caution, since the autoradiographic studies described earlier in this chapter provide evidence that receptors for IGF-I, capable of binding insulin, are expressed in a similar stage-specific manner.

At the morula stage, gold-labeled insulin was internalized by coated pits on the external surface of the cells and accumulated in coated and uncoated pits and in lysosomes. Gold particles were also visualized in intercellular spaces and in the cytoplasm of cells in the interior of the morula. In contrast, at the blastocyst stage, significantly fewer gold particles were seen in the cells of the inner cell mass as compared to trophectoderm, and virtually no particles were seen binding to the cell membranes facing the blastocoel surface of either cell type. These observations indicated that fewer receptors are expressed on the cells of the inner cell mass as compared to the trophectoderm. However, the presence of coated pits containing gold particles on the surface of cells of the inner cell mass adjacent to trophectoderm suggest that the trophectoderm at this stage may be acting as a transporting epithelium for the passage of insulin in a concentrative manner to the inner cell mass. This hypothesis is supported by the finding that relatively fewer gold particles accumulated in lysosomes of the trophectoderm, compared to lysosomes at the morula stage (Table 2), indicating passage of the ligand out of these cells.

The appearance of insulin receptors on cells of the trophectoderm of the mouse blastocyst may be highly significant in early development. Polar trophectoderm differentiates to groups of cells from which the placenta will develop.[60] The placenta is a vitally important endocrine tissue during pregnancy and is a rich source of insulin receptors. Studies that demonstrate the temporal appearance during fetal development of RNA sequences related to the insulin gene family suggest a possible extrapancreatic source of this hormone. Further, the occurrence of a higher number of these sequences in placentae from diabetic pregnancies correlates with the common clinical observation of diabetes-associated macrosomia of the newborn.[6] Macrosomia has been proposed to be a consequence of maternal hyperglycemia, resulting in fetal hyperglycemia and hyperinsulinemia. The experimental data suggest that fetal glucose metabolism may be influenced by extrapancreatic hormones as well as fetal pancreatic insulin.

Insulin receptors have been detected in the yolk sac of mouse postimplantation embryos,[42] whereas the rat yolk sac has been reported to contain both insulin and insulin receptors.[44,45] In addition to its function as a physical barrier between the maternal environment and the developing fetus, the yolk sac is responsible for nutritional requirements in early development. The observation of insulin receptor expression in the rat and the mouse suggest that the yolk sac may act as an extrapancreatic source of insulin during early development, before the fetal pancreas is functional. Clearly, more investigations are needed at different developmental periods and in a broader range of species to confirm this speculation.

Studies of energy metabolism in preimplantation mouse embryos have shown that pyruvate, but not glucose, meets the energy requirements of the embryo up to the eight-cell stage.[61] Glucose becomes the preferred energy substrate at the blastocyst stage and predominates during periimplantation development (see Chapter 2). Significantly, insulin binding to the preimplantation mouse embryo is apparent at the morula stage, just as glucose becomes the preferred substrate. Glucose transport across the plasma membrane is facilitated by insulin, and the hormone thus influences glycogen synthesis. The glycogen content of the freshly collected preimplantation mouse embryo decreases with development from the morula to the late blastocyst stage.[62,63] In contrast, the glycogen content of *in vitro* cultured embryos increases significantly between these two stages.[63] It is not known whether maternal hormones that govern glycogen metabolism are accessible to the embryo via the luminal secretions of the reproductive tract. On the other hand, it is known that several sources of bovine serum albumin, used in culture media, are contaminated with insulin, and it is possible that differences in the immediate environment may account for these experimental discrepancies. In an attempt to reconcile the environmental differences in embryonic glycogen synthesis, Wales et al. have investigated the effect of insulin on the uptake of glucose and the synthesis of glycogen during brief *in vitro* culture periods.[64] These authors found that insulin did not enhance the uptake of glucose or its utilization for the synthesis of glycogen. However, the hormone did stimulate the incorporation of labeled glucose into the nonglycogen, acid-insoluble fraction during both pulse and chase culture. Label in this fraction represented incorporation of substrate into macromolecules other than glycogen, and the investigators concluded that while glycogen metabolism appeared to be unresponsive to insulin, anabolic effects of the hormone were expressed in the embryo.

The role of IGFs in early development awaits clarification. Autoradiographic studies have provided evidence that receptors capable of binding IGFs are present by the periimplantation stage of mouse development. Studies in the postimplantation mouse embryo have shown a greater concentration of IGF-I receptors as compared with insulin receptors, although the significance of this observation is unclear. A number of studies in midgestational mammalian embryos have documented IGF-II receptor and ligand expression in tissues of mesodermal origin, notably in cartilage. However, there are a few discrepancies with respect to the ontogeny and level of expression of IGF-I vs. IGF-II, as outlined earlier. Some of the differences may be ascribed to differing techniques, and there also may be species differences.

In general, IGF I has been considered to act primarily in the regulation of stem-cell growth and proliferation. It is postulated that following cell differentiation, the rapidly proliferating tissues are assumed to become responsive to a variety of hormones, among them, insulin. Other investigators have proposed that IGFs may play a role in organogenesis but have not postulated a direct role for insulin in embryogenesis. Bhaumick and Bala have reported recently that IGF-II and insulin at a similar concentration (0.1 to 1 mg/ml) stimulated the uptake and incorporation of glucose into developing mouse limb buds *in vitro*, whereas IGF-I had little effect.[65] One of the difficulties involved in separating the roles of insulin and the IGFs is that many cells posses all three receptors. Indeed, it is not known which receptor is expressed first by the early mammalian embryo nor the receptor profile of cells in the pre- and periimplantation embryo.

An additional confounding factor has surfaced with the recent report that IGF-I binding proteins, but not the receptor, can be identified on membrane homogenates of mouse blastocysts.[41] IGF binding proteins have been detected on the surface of certain cells at times when those cells respond to IGF-I,[66,67] suggesting that the IGF-I binding protein may function as a serum carrier protein. Two classes of IGF-binding proteins have been identified in fetal blood,[9] supporting the suggestion that these proteins may serve a serum transport function. The production and secretion of IGF-binding proteins by cells in culture confuses the interpretation of binding data obtained in experiments that measured the uptake of radiolabeled IGF.[68] Biochemical analysis is required to discriminate between the IGF receptors and their binding proteins. Such experiments were not feasible with the small numbers of embryos used in our studies. Clearly, more studies are needed in order to establish the ontogeny of IGF vs. insulin receptor expression in mammalian embryos.

Although the precise roles of insulin and the IGFs during embryogenesis in the mouse and other mammals await elucidation by means of functional experiments, there are several lines of evidence that suggest a developmental role for insulin. First, the ultrastructural studies indicate clearly that there is a receptor-mediated mode of uptake for insulin. This suggests a requirement for the hormone in early development. Second, preliminary experiments have shown that physiological levels of insulin are capable of stimulating the incorporation of ^{3}H-uridine into total acid-insoluble RNA in the morula and blastocyst stage mouse embryo,[69] while others have reported that insulin stimulates protein synthesis in compacted mouse preimplantation embryos.[70] The experiments of Wales and his colleagues[64] provide further support for an anabolic effect of insulin in the preimplantation mouse embryo. On the other hand, a number of studies have shown that development of the mouse preimplantation embryo up to the blastocyst stage can take place in simple culture media and does not depend upon the presence of macromolecules or known exogenous growth factors. It is quite conceivable that the preimplantation embryo becomes capable of binding insulin and IGFs in readiness for implantation, when the requirements for embryo growth become more complex and when rapid proliferative growth is essential for successful establishment of pregnancy. Clearly, many more studies are needed in order to clarify the functional importance of these hormones during specific temporal phases of development.

This brief review has presented evidence for the expression of insulin and IGF receptors during initial stages of mammalian embryogenesis. In addition, scattered studies suggest that during early postimplantation developmental stages, insulin and IGF synthesis may occur in tissues other than those that are the major sites of hormone production in the adult. Taken together, these investigations provide persuasive evidence that the receptors for insulin and IGFs are developmentally regulated in mammals. The early expression of these receptors reflects significant roles for peptides of the insulin family in embryogenesis.

ACKNOWLEDGMENTS

We wish to acknowledge the skilled assistance of Neelima Shah, Susan Haydock, and Maria Wikarczuk during the course of these studies. This work was supported in part by grants from the National Institutes of Health, DK 19525 and HD 23511.

REFERENCES

1. **Gospodarowicz, D. and Moran, J. S.,** Growth factors in mammalian cell culture, *Annu. Rev. Biochem.*, 45, 531, 1976.
2. **Steiner, D. F., Chan, S. J., Welsh, J. M., and Kwok, S. C. M.,** Structure and evolution of the insulin gene, *Annu. Rev. Genet.,* 19, 463, 1985.
3. **Van Obberghen, E. and Gammeltoft, S.,** Insulin receptors: structure and function, *Experientia,* 42, 727, 1986.
4. **Richman, R. A., Claus, T. H., Pilkis, S. J., and Friedman, D. L.,** Hormonal stimulation of DNA synthesis in primary cultures of adult rat hepatocytes, *Proc. Natl. Acad. Sci. U.S.A.,* 73, 3589, 1976.
5. **Rinderknecht, E. and Humbel, R. E.,** Primary structure of human insulin-like growth factor II, *FEBS Lett.,* 89, 282, 1978.
6. **Hill, D. E.,** Effect of insulin on fetal growth, in *The Diabetic Pregnancy: A Perinatal Perspective,* Merkatz, R. and Adam, P. A. J., Eds., Grune and Stratton, New York, 1979.
7. **Gluckman, P. D.,** The role of pituitary hormones, growth factors and insulin in the regulation of fetal growth, *Oxford Rev. Reprod. Biol.,* 8, 1, 1986.
8. **Petruzzelli, L. M., Herrera, R., Garcia, R., and Rosen, O. M.,** The insulin receptor of *Drosophila melanogaster,* in *Growth Factors and Transformation; Cancer Cells,* Feramisco, J., Ozanne, B., and Stiles, C., Eds., Vol. 3, Cold Spring Harbor Laboratory, Cold Spring Harbor, NY, 1985.
9. **Froesch, E. R., Schmid, C., Schwander, J., and Zapf, J.,** Actions of insulin-like growth factors, *Annu. Rev. Physiol.,* 47, 443, 1985.
10. **Rechler, M. M. and Nissley, S. P.,** The nature and regulation of the receptors for insulin-like growth factors, *Annu. Rev. Physiol.,* 47, 425, 1985.
11. **Ullrich, A., Bell, J., Chen, E., Herrera, R., Petruzzelli, L. M., Dull, T. J., Gray, A., Coussens, L., Liao, Y., Tsubokawa, M., Mason, A., Seeburg, P., Grunfeld, C., Rosen, O., and Ramachandran, J.,** Human insulin receptor and its relationship to the tyrosine kinase family of oncogenes, *Nature (London),* 313, 756, 1985.
12. **Ebina, Y., Ellis, L., Jarnagin, K., Edery, M., Graf, L., Clauser, E., Ou, J., Masiarz, F., Kan, Y. W., Goldfine, I., Rother, R. A., and Rutter, W. J.,** The human insulin receptor cDNA: the structural basis for hormone-activated transmembrane signalling, *Cell,* 40, 747, 1985.
13. **Ullrich, A., Gray, A., Tam, A., Yang-Feng, T., Tsubokawa, M., Collins, C., Henzel, W., LeBon, T., Kathuria, S., Chen, E., Jacobs, S., Francke, U., Ramachandran, J., and Fujita-Yamaguchi, Y.,** Insulin-like growth factor I primary structure: comparison with insulin receptor suggests structural determinants that define functional specificity, *EMBO J.,* 5, 2503, 1986.
14. **Rechler, M. M., Zapf, J., Nissley, S. P., Froesch, E. R., Moses, A. C., Podskalny, J. M., Schilling, E. E., and Humbel, R. E.,** Interactions of insulin-like growth factors I and II and multiplication-stimulating activity with receptors and serum carrier proteins, *Endocrinology,* 107, 1451, 1980.

15. **Kasuga, M., Fujita-Yamaguchi, Y., Blith, D. L., White, M. F., and Khan, C. R.,** Characterization of the insulin receptor kinase purified from human placental membranes, *J. Biol. Chem.,* 558, 10973, 1983.

16. **Petruzzelli, L. M., Herrera, R., Rosen, O. M.,** Insulin receptor is an insulin-dependent tyrosine protein kinase: co-purification of insulin binding activity and protein kinase activity to homogeneity from human placenta, *Proc. Natl. Acad. Sci. U.S.A.,* 81, 3327, 1984.

17. **Rosen, O. M.,** After insulin binds, *Science,* 237, 1452, 1987.

18. **Ewton, D. Z., Falen, S. L., and Florini, J. R.,** The type II insulin-like growth factor (IGF) receptor has low affinity for IGF-I analogs: pleiotropic actions of IGF on myoblasts are apparently mediated by the type I receptor, *Endocrinology,* 120, 115, 1987.

19. **Rosenfeld, R. G., Conover, C. A., Hodges, D., Lee, T. D., Misra, P., Hintz, R. L., and Li, C. H.,** Heterogeneity of insulin-like growth factor I affinity for the insulin-like growth factor II receptor: comparison of natural, synthetic and recombinant DNA-derived insulin like growth factor I, *Biochem. Biophys. Res. Commun.,* 143, 199, 1987.

20. **Morgan, D. O., Edman, J. C., Standring, D. N., Fried, V. A., Smith, M. C., Roth, R. A., and Rutter, W. J.,** Insulin-like growth factor II receptor as a multifunctional binding protein, *Nature (London),* 329, 301, 1987.

21. **Roth, R. A., Stover, C., Hari, J., Morgan, D. O., Smith, M. C., Sara, V., and Fried, V. A.,** Interactions of the receptor for insulin-like growth factor II with mannose 6-phosphate and antibodies to the mannose 6-phosphate receptor, *Biochem. Biophys. Res. Commun.,* 149, 600, 1987.

22. **MacDonald, R. G., Pfeffer, S. R., Coussens, L., Tepper, M. A., Brocklebank, C. M., Mole, J. E., Anderson, J. K., Chen, E., Czech, M. P., and Ullrich, A.,** A single receptor binds both insulin-like growth factor II and mannose 6-phosphate, *Science,* 239, 1134, 1988.

23. **Tong, P. Y., Tollefsen, S. E., and Kornfeld, S.,** Cation-independent mannose 6-phosphate receptor binds insulin-like growth factor II, *J. Biol. Chem.,* 263, 2585, 1988.

24. **von Figuera, K. and Hasilik, A. A.,** Lysosomal enzymes and their receptors, *Annu. Rev. Biochem.,* 55, 167, 1986.

25. **Roth, R. A.,** Structure of the receptor for insulin-like growth factor II: the puzzle amplified, *Science,* 239, 1269, 1988.

26. **Johnson, M. H.,** The molecular and cellular basis of preimplantation mouse development, *Biol. Rev.,* 56, 463, 1981.

27. **Biggers, J. D. and Borland, R. M.,** Physiological aspects of growth and development of the preimplantation mammalian embryo, *Annu. Rev. Physiol.,* 38, 95, 1976.

28. **Wiley, L. M.,** The cell surface of the mammalian embryo during development, in *Ultrastructure of Reproduction,* Van Blerkom, J., and Motta, P. M., Eds., Martinus Nijhoff, Dordrecht, 1984.

29. **Rosenblum, I. Y., Mattson, B. M., and Heyner, S.,** Stage-specific insulin binding in mouse preimplantation embryos, *Dev. Biol.,* 116, 261, 1986.

30. **Majercik, M. H. and Bourgignon, L. W.,** Insulin receptor capping and its correlation with calmodulin-dependent myosin light chain kinase, *J. Cell. Physiol.,* 124, 403, 1985.

31. **Fleming, T. P. and Pickering, S. J.,** Maturation and polarization of the endocytotic system in outside blastomeres during mouse preimplantation development, *J. Embryol. Exp. Morphol.,* 89, 175, 1985.

32. **Mattson, B. A., Rosenblum, I. Y., Smith, R. M., and Heyner, S.,** Autoradiographic evidence for insulin and IGF binding to early mouse embryos, *Diabetes,* 37, 585, 1988.

33. **Wiley, L. M., Takaki, K. K., and Yamagata, M.,** Electron microscopy of mouse preimplantation embryos and isolated blastomeres immobilized on coverslips, *Gamete Res.,* 11, 51, 1985.

34. **Gwatkin, R. B. L.,** Amino acid requirements for attachment and outgrowth of the mouse blastocyst *in vitro, J. Cell. Physiol.,* 68, 335, 1966.

35. **Smith, R. M. and Jarett, L.,** Quantitative ultrastructural analysis of receptor-mediated insulin uptake into adipocytes, *J. Cell. Physiol.,* 115, 199, 1983.

36. **Smith, R. M., Goldberg, R. I., and Jarett, L.,** Preparation and characterization of a colloidal gold-insulin complex with binding and biological activities identical to native insulin, *J. Histochem. Cytochem.,* 36, 359, 1988.

37. **Heyner, S., Rosenblum, I. Y., Wikarczuk, M. L., and Smith, R. M.,** Receptor-mediated uptake of gold-labelled insulin by early mouse embryos, *J. Cell Biol.,* 105, 256a, 1987.

38. **Heyner, S., Rosenblum, I. Y., Mattson, B. A., Farber, M., and Smith, R. M.,** Ontogeny of insulin binding to the preimplantation mouse embryo, in *Growth Factors in Reproduction and Early Development,* Sadler, W.A., Ed., Plenum Press, New York, 1988.

39. **Goldberg, R. I., Smith, R. M., and Jarett, L.,** Insulin and alpha 2-macroglobulin-methylamine undergo endocytosis by different mechanisms in rat adipocytes. I. Comparison of cell surface events, *J. Cell. Physiol.,* 133, 203, 1987.

40. **Solter, D. and Knowles, B. B.,** Immunosurgery of mouse blastocysts, *Proc. Natl. Acad. Sci. U.S.A.,* 72, 5099, 1975.

41. **Smith, E. P., Sadler, T. W., and D'Ercole, D. J.,** Somatomedins/insulin-like growth factors, their receptors and binding proteins are present during mouse embryogenesis, *Development,* 101, 72, 1987.

42. **Heath, J. and Shi, W.-K.,** Developmentally regulated expression of insulin-like growth factors by differentiated murine teratocarcinomas and extraembryonic mesoderm, *J. Embryol. Exp. Morphol.,* 95, 193, 1986.

43. **Bhaumick, B. and Bala, R. M.,** Receptors for insulin-like growth factors I and II in developing embryonic mouse limb bud, *Biochim. Biophys. Acta,* 927, 117, 1987.

44. **Unterman, T., Goewert, T. T., Baumann, G., and Freinkel, N.,** Insulin receptors in embryo and extraembryonic membranes of early somite rat conceptus, *Diabetes,* 35, 1193, 1986.

45. **Muglia, L. and Locker, J.,** Extrapancreatic insulin gene expression in the fetal rat, *Proc. Natl. Acad. Sci. U.S.A.,* 81, 3655, 1984.

46. **Beck, F., Samani, N. J., Penschow, J. D., Thorley, B., Tregear, G. W., and Coghlan, J. P.,** Histochemical localization of IGF-I and -II mRNA in the developing rat embryo, *Development,* 101, 175, 1987.

47. **Rotwein, P., Pollock, K. M., Watson, M., and Milbrandt, J. D.,** Insulin-like growth factor gene expression during rat embryonic development, *Endocrinology,* 121, 2141, 1987.

48. **Liu, L., Russell, S. M., and Nicoll, C. S.,** Growth and differentiation of transplanted rat embryos in intact, diabetic, and hypophysectomized hosts: comparison with their growth *in situ, Biol. Neonat.,* 52, 307, 1987.

49. **Peyron, J. F., Samson, M., Van Obberghen, E., Brandenberg, D., and Fehlmann, M.,** Appearance of a functional insulin receptor during rabbit embryogenesis, *Diabetologia,* 28, 369, 1985.

50. **Scott, J., Cowell, J., Robertson, M. E., Priestley, L. M., Wadey, R., Hopkins, B., Pritchard, J., Bell, G. I., Rall, L. B., Graham, C. F., and Knott, T. J.,** Insulin-like growth factor-II gene expression in Wilms' tumour and embryonic tissues, *Nature (London),* 317, 260, 1985.

51. **Han, V. K. M., D'Ercole, A. J., and Lund, P. K.,** Cellular localization of somatomedin (insulin-like growth factor) messenger RNA in the human fetus, *Science,* 236, 193, 1987.

52. **Liu, K.-S., Wang, C.-Y., Mills, N., Gyves, M., and Ilan, J.,** Insulin-related genes expressed in human placenta from normal and diabetic pregnancies, *Proc. Natl. Acad. Sci. U.S.A.,* 82, 3868, 1985.

53. **Graham, C. F.,** Teratocarcinoma cells and normal mouse embryogenesis, in *Concepts in Mammalian Embryogenesis,* Sherman, M. I., Ed., MIT Press, Cambridge, MA, 1977.

54. **Gachelin, G.,** Le teratocarcinome experimental de la souris: une systeme modele pour l'etude des relations entre antigenes des surfaces cellulaires et differenciation embryonnaire, *Bull. Cancer,* 63, 95, 1976.

55. **Salomon, D.,** Correlation of receptors for growth factors on mouse embryonal carcinoma cells with growth in serum-free, hormone-supplemented medium, *Exp. Cell Res.,* 128, 311, 1980.

56. **Nagarajan, L., Nissley, S. P., Rechler, M. M., and Anderson, W. B.,** Multiplication-stimulating activity stimulates the multiplication of F9 embryonal carcinoma cells, *Endocrinology,* 110, 1231, 1982.

57. **Nagarajan, L. and Anderson, W. B.,** Insulin promotes the growth of F9 embryonal carcinoma cells apparently by acting through its own receptor, *Biochem. Biophys. Res. Commun.,* 106, 974, 1982.

58. **Heath, J., Bell, S., and Rees, A. R.,** Appearance of functional insulin receptors during differentiation of embryonal carcinoma cells, *J. Cell Biol.,* 91, 293, 1981.

59. **Smith, R. M., Cobb, M. H., Rosen, O. M., and Jarett, L.,** Ultrastructural analysis of the organization and distribution of insulin receptors on the surface of 3T3-L1 adipocytes: rapid microaggregation and migration of occupied receptors, *J. Cell Physiol.,* 123, 167, 1987.

60. **Rossant, J. and Papaioannou, V. E.,** The biology of embryogenesis, in *Concepts in Mammalian Embryogenesis,* Sherman, M. I., Ed., MIT Press, Cambridge, MA, 1977.

61. **Wordinger, R. J. and Brinster, R. L.,** Influence of reduced glucose levels on the *in vitro* hatching, attachment, and trophoblast outgrowths of the mouse blastocyst, *Dev. Biol.,* 53, 294, 1976.

62. **Stern, S. and Biggers, J. D.,** Enzymatic estimation of glycogen in the cleaving mouse embryo, *J. Exp. Zool.,* 168, 61, 1968.

63. **Ozias, C. B. and Stern, S.,** Glycogen levels of preimplantation mouse embryos developing *in vitro, Biol. Reprod.,* 8, 467, 1973.

64. **Wales, R. G., Khurana, N. K., Edirisnghe, W. R., and I. L. Pike,** Metabolism of glucose by preimplantation embryos in the presence of glucagon, insulin, epinephrine, cAMP, theophylline and caffeine, *Aust. J. Biol. Sci.,* 38, 421, 1985.

65. **Bhaumick, B. and Bala, R. M.,** Parallel effects of insulin-like growth factor-II and insulin on glucase metabolism of developing mouse embryonic limb buds in culture, *Biochem. Biophys. Res. Commun.,* 152, 359, 1988.

66. **Drop, S. L. S., Valiquette, G., Guyda, H. J., Corvol, M. T., and Posner, B. I.,** Partial purification and characterization of a binding protein for insulin-like activity in human amniotic fluid: a possible inhibitor of insulin-like activity, *Acta Endocrinol.,* 90, 505, 1979.

67. **Knauer, D. J. and Smith, G. L.,** Inhibition of biological activity of multiplication-stimulating activity by binding to its carrier protein, *Proc. Natl. Acad. Sci. U.S.A.,* 77, 7252, 1980.

68. **Nissley, S. P., Haskell, J. F., Sasaki, N., de Vroede, M. A., and Rechler, M. M.,** Insulin-like growth factor receptors, *J. Cell Sci.,* Suppl. 3, 39, 1985.
69. **Pritchard, M. L., Haydock, S. W., Wikarczuk, M. L., Farber, M., and Heyner, S.,** Effect of insulin on RNA synthesis on the preimplantation mouse embryo, *Biol. Reprod.,* 36 (Suppl. 1) 77, 1987.
70. **Harvey, M. B. and Kaye, P. L.,** Insulin stimulates protein synthesis in compacted mouse embryos, *Endocrinology,* 122, 1182, 1988.
71. **Heyner, S., Rao, L. V., Jarett, L., and Smith, R. M.,** Preimplantation mouse embryos internalize maternal insulin via receptor-mediated endocytosis: pattern of uptake and functional correlations, *Dev. Biol.,* in press.

Chapter 7

USE OF EMBRYONAL CARCINOMA CELLS TO STUDY GROWTH FACTORS DURING EARLY MAMMALIAN DEVELOPMENT

Angie Rizzino

TABLE OF CONTENTS

I. INTRODUCTION

Work with a wide variety of cells has established that growth factors regulate both growth and differentiation *in vitro*, yet relatively little is known about the roles of growth factors *in vivo*, especially during mammalian embryogenesis. Although many different growth factors have been detected in embryos from midstage and late stages of gestation, very little is known about the growth factors present during early development and even less is known about their functions. Nonetheless, there is good reason to believe that growth factors begin to exert effects on both growth and differentiation prior to the middle of gestation.

Efforts to understand the roles of growth factors during early development have been hampered by the time and expense of working with embryos and by the cellular complexities of embryos beyond the blastocyst stage. Consequently, a simpler model system was sought, one that closely parallels the cellular and biochemical events of mammalian embryogenesis. Since the mid-1970s, it has become clear that mouse embryonal carcinoma (EC) cell lines mimic, both morphologically and biochemically, important stages of early mammalian development.[1-4] Numerous mouse and human EC cell lines have been established and conditions for their long-term cultivation have been developed. Many EC cell lines reproducibly differentiate into specific cell types in culture, and several different EC cell lines can be directed, by manipulation of the culture medium, to differentiate into cells that exhibit properties of parietal or visceral extraembryonic endoderm. The latter EC cell lines are of particular interest because parietal and visceral extraembryonic endoderm are two of the first cell types to form during early mammalian development. Consequently, EC cell lines appear to provide a useful model system for investigating the effects of growth factors during early development.[5]

This chapter reviews our current understanding of growth factors and EC cells and attempts to relate this information to our current understanding of growth factors in the early embryo. For readers unfamiliar with specific EC cell lines, the first section briefly describes the EC cell lines most frequently used to study growth factors. Also covered are some of the issues to consider when interpreting results obtained with EC cells. This is followed by a review of studies in five areas: (1) growth factors produced by EC cells, (2) growth factors produced by differentiated cells derived from EC cells, (3) growth factor receptors expressed by EC cells and their differentiated cells, (4) possible roles of growth factors produced by EC cells and their differentiated cells, and (5) growth factors produced by early mouse embryos. The last section provides only a brief review of pertinent literature, since a detailed review of growth factors and early embryos was published recently.[6] Lastly, the discussion focuses on the significance of these studies and attempts to define some of the more important questions that need to be addressed in the near future. For readers unfamiliar with the specific growth factors mentioned, a brief description of three growth factor families is provided in the appropriate sections, as well as references to detailed reviews.

II. EC CELL LINES USED TO STUDY GROWTH FACTORS

A. PROPERTIES OF SPECIFIC EC CELL LINES

Although many EC cell lines have been used to study growth factors and EC cells, four different EC cell lines are used most frequently: F9, PC-13, OC-15, and PSA-1. These cell lines differ from one another in several respects, the most important is their capacity to differentiate. F9 and PC-13 EC cells appear to differentiate into relatively few cell types, and for this reason it has been suggested that these EC cell lines be referred to as restricted EC cell lines.[5] OC-15 EC cells can differentiate into more cell types than F9 and PC-13 EC cells, and PSA-1 EC cells appear to differentiate into still more cell types. Consequently, OC-15 and PSA-1 are often referred to as multipotent EC cell lines.

F9 and PC-13 are widely used because of the relative ease with which they can be maintained as undifferentiated populations *in vitro*.[7] Both EC cell lines form undifferentiated embryonal carcinomas when injected into suitable hosts, and they undergo very little spontaneous differentiation in culture. These EC cell lines are heavily used for a second reason. They can be readily induced to differentiate in culture. In 1978, Strickland and Mahdavi demonstrated that retinoic acid (RA) can induce monolayer cultures of F9 cells to irreversibly differentiate into cells that exhibit the properties of parietal extraembryonic endoderm, and further maturation of the differentiated properties occurs in the presence of dibutyryl cyclic AMP.[8,9] It is generally accepted that the majority of the differentiated cells exhibit the properties of parietal extraembryonic endoderm. However, more than one cell type may be present, since the differentiated cells exhibit more than one morphology.[10-12] RA can also induce PC-13 EC cells to differentiate, but the cells formed, which are sometimes referred to as END cells, appear to differ from those formed by F9.[13] Although there is some disagreement, the PC-13-differentiated cells may be related to visceral extraembryonic endoderm. Overall, both EC cell lines are relatively simple to use, since RA can induce virtually all cells in each population to differentiate. However, few cell types are formed and, in each case, they appear to be limited to cells related to extraembryonic endoderm. Consequently, it has been suggested that F9 and PC-13 EC cells may be closely related to primitive endoderm[14] which gives rise to parietal and visceral extraembryonic endoderm.

Although more studies have been conducted with F9 and PC-13 EC cells, several important studies have been conducted with multipotent EC cells. OC-15 EC cells form highly differentiated teratocarcinomas when injected into suitable hosts. In culture, OC-15 can be induced to differentiate into many different cell types and plating these cells at low cell density induces them to differentiate into cells that exhibit several properties of extraembryonic endoderm.[13] The differentiated cells formed under the latter conditions have not been fully characterized, but they appear to be similar to the cells formed by RA-treated PC-13 EC cells and, thus, are often referred to as END cells. OC-15 EC cells can also be induced to differentiate by plating them in serum-free medium.[15] The cells formed under these conditions are morphologically homogeneous and virtually all the cells produce tissue type plasminogen activator, which is a marker for parietal extraembryonic endoderm.

PSA-1 EC cells and other members of the PSA series of EC cells form highly differentiated teratocarcinomas when injected into suitable hosts. *In vitro,* PSA-1 EC cells are normally maintained as an undifferentiated population on a fibroblast feeder layer.[16] Removal of PSA-1 EC cells from its feeder layer and cultivation as cellular aggregates permits them to differentiate and mimic the development of periimplantation mouse embryos. Under these conditions, the cellular aggregates initially form structures referred to as embryoid bodies, which consist of an undifferentiated core of EC cells and an outer layer of endoderm-like cells that exhibit the properties of parietal or visceral extraembryonic endoderm.[16] Embryoid bodies formed under these conditions are strikingly similar to the embryoid body formed by the inner cell mass when it is devoid of trophectoderm. Subsequent development of the PSA-1-derived embryoid bodies gives rise to highly differentiated structures that exhibit important properties of mouse embryos at 6.5 to 8.5 d of gestation.[17]

Each of the four EC cell lines described in this section can be directed to differentiate into cells that exhibit properties of early embryonic cells. Although PSA-1 EC cells most closely mimic early development, these cells are more difficult to use because they require a feeder layer. In addition, PSA-1 EC cells grown on feeder layers and embryoid bodies formed by PSA-1 EC cells pose technical problems when studying growth factors and growth factor receptors. For readers interested in additional information on EC cells, several reviews are available.[1-4]

B. INTERPRETING OBSERVATIONS MADE WITH EC CELLS

Despite the apparent capacity of EC cells to mimic the early stages of mammalian development, observations made with EC cells must be interpreted with care and all conclusions drawn should be directly verified with embryos. Although it is suggested (Section VII) that EC cells and their nontransformed counterparts are likely to utilize growth factors in a very similar way, this is far from proven. Even if this is true, it may still be difficult to correctly infer the roles of growth factors during early development from studies with EC cells. To draw appropriate conclusions, one needs to know which embryonic cell is the closest relative of the EC cells employed, since different embryonic cells are likely to utilize growth factors differently. For instance, PSA-1 EC cells may be related to the cells of the inner cell mass just prior to formation of primitive extraembryonic endoderm, whereas F9 EC cells may be more closely related to a stem cell for extraembryonic endoderm.

There are other issues to be considered when EC cells are used to predict the roles of growth factors during early development and, ultimately, our efforts to define the roles of growth factors during early development must account for two important findings. First, mouse embryos can develop *in vitro* to the blastocyst stage in a very simple medium lacking growth factors, such as those supplied by serum. This suggests that either the early embryo does not require growth factors prior to this stage or the early embryo produces all of its required growth factors. Second, *in vitro* development of blastocysts can proceed to the early somite stage provided appropriate sera are added to the culture medium. This finding indicates that development to the early somite stage does not require unique maternal factors, but some of the embryonic requirements cannot be fulfilled by the embryo.

III. GROWTH FACTORS PRODUCED BY EC CELLS

A. EARLY INDICATIONS THAT EC CELLS PRODUCE GROWTH FACTORS

The first serum-free and defined media for EC cells were developed by Rizzino and co-workers.[18,19] This work demonstrated that: (1) F9 EC cells do not respond to growth factors added to their culture medium, and (2) growth of these EC cells in the absence of serum is density dependent. The latter finding suggested that EC cells condition their culture medium and also suggested that EC cells produce growth factors. The latter possibility was supported by the finding that F9 EC cells grown at high cell densities in defined medium produce a cell-associated factor that promotes growth of their differentiated cells cultured in defined medium.[20] The factor produced by the EC cells remained associated with the tissue culture substratum after lysis of the cells with a hypotonic solution. Although the factor was not characterized, growth promotion was observed when F9-derived differentiated cells were plated onto the conditioned tissue culture dishes in defined medium. In addition to confirming that EC cells produce growth-promoting factors, these findings suggested that paracrine growth control might operate during the early stages of mammalian development.

Soon after these findings were reported, Rizzino and co-workers determined that several different EC cell lines produce one or more growth factors that exhibit transforming growth factor (TGF) activity.[21,22] Initially, it was determined that EC cells release factors that are able to induce the soft agar growth of NRK-49F cells (a nontransformed rat kidney fibroblast cell line). The majority of this activity is heat- and acid-stable, but it is inactivated by trypsin.[23] Partial purification of the factor from EC cells demonstrated that it is unrelated to TGF-α or TGF-β, since it does not copurify with either growth factor. As described below (Section III.D.2), it now seems that the majority of the TGF activity released by EC cells is due to factors related to platelet-derived growth factor (PDGF) and fibroblast growth factor (FGF).

The finding that EC cells can condition their culture medium with factors able to influence growth of their differentiated cells was confirmed and extended by other investigators. Isacke and Deller observed that PC-13 EC cells and their differentiated cells plated at low serum concentrations influence the growth of one another.[24] Medium conditioned by PC-13 EC cells stimulated the labeling index of PC-13-derived differentiated cells (END cells). Similarly, growth of PC-13 EC cells at low cell densities in medium supplemented with low concentrations of fetal bovine serum was promoted by the presence of END cells. This work suggested the existence of reciprocal interactions between EC cells and their differentiated cells and, by analogy, the existence of similar interactions during the early stages of mammalian development.

Efforts to identify the types of factors produced by EC cells led to the identification of growth factors belonging to at least three different growth factor families. In the next three sections, a description of each growth factor family precedes the discussion of each type of growth factor produced by EC cells.

B. PRODUCTION OF GROWTH FACTORS BELONGING TO THE PDGF FAMILY

1. The PDGF Family of Growth Factors

Biologically active platelet-derived growth factor (PDGF) has been shown to exist as a dimer.[25] PDGF isolated from platelets is composed of two chains (A and B). PDGF exhibits a molecular weight of 27 to 31 kDa and the gene sequences that code for the two chains have been determined.[26-28] Interest in PDGF increased dramatically with the finding that simian sarcoma virus carries an oncogene, v-sis, that exhibits a very high degree of sequence identity with the gene (c-sis) for the PDGF B chain.[29,30] Cells infected with simian sarcoma virus are readily transformed and produce a biologically active growth factor that binds to PDGF receptors and is required for transformation.[31,32] Characterization of this factor demonstrated that it exists as a homodimer of the B chain and exhibits a molecular weight of 20 to 28 kDa.[33] More recently, it has been determined that some tumor cells produce a PDGF-like growth factor composed of a homodimer of the A chain.[34]

PDGF has been shown to affect the monolayer growth of many different cells.[35] Although it is widely assumed that PDGF only stimulates the growth of cells derived from the mesodermal germ layer, PDGF also stimulates the growth of some cells derived from other germ layers.[36] Besides its effects on monolayer growth, recent studies have shown that PDGF can induce the soft agar growth of several different nontransformed cell lines.[37] Thus, PDGF behaves as a TGF. Additional information relating to the PDGF family of growth factors is available in recent reviews.[35,38]

2. PDGF-Related Growth Factors Produced by EC Cells

Several years ago, a number of different EC cell lines were found to produce a growth factor related to PDGF. This factor was observed initially by Gudas et al. in conditioned medium prepared from PSA-1-G EC cells.[39] These cells had been adapted to grow without the feeder layer required by the parental PSA-1 EC cell line. Although this factor was not purified, it competed with radiolabeled PDGF for binding to 3T3 cells. Subsequent studies by Rizzino and Bowen-Pope demonstrated that the same factor, or a closely related one, is produced by F9 and

PC-13 EC cells.[40] The factor produced by F9 and PC-13 EC cells competed with radiolabeled PDGF for binding to human diploid fibroblasts. Moreover, the factor prepared from F9 cells was found to be immunologically related to, though not identical with, PDGF in whole mouse blood. Although these studies established that a PDGF-related growth factor is produced by several different EC cell lines, they did not establish whether the factor is derived from the B-chain gene (*c*-sis), the A chain gene or the gene for another, as yet undiscovered, member of the PDGF growth factor family. Recent studies by Tiesman and Rizzino indicate that mRNA isolated from F9 EC cells and mRNA isolated from PC-13 EC cells do not hybridize with a v-sis DNA probe.[41] This is consistent with an earlier claim that *c*-sis is not expressed by PC-13 EC cells.[42] However, EC cells do produce mRNA for the PDGF A-chain.[41] Thus, it appears that EC cells produce homodimers of the PGDF A-chain.

C. PRODUCTION OF GROWTH FACTORS BELONGING TO THE FGF FAMILY

1. The FGF Family of Growth Factors

The fibroblast growth factor (FGF) family comprises at least four different members and each exists as a single polypeptide chain. Two forms of FGF have been purified to homogeneity: a cationic form (FGFb), which exhibits a pI between 9 and 10, and an acidic form (FGFa), which exhibits a pI between 4 and 5. cDNA gene probes for both forms have been cloned and sequenced.[43,44]

FGFb is known by several other names, including heparin-binding growth factor β. The exact size of FGFb *in vivo* remains to be determined. FGFb has been isolated as a protein consisting of 146 amino acids.[45] However, smaller forms[46] and larger forms have been isolated, including one of 157 amino acids.[47] The 157 amino acid form was isolated from human term placentas. Based on the published cDNA sequence for human FGFb, one would predict that the full-length unprocessed protein should consist of 155 amino acids (17.2 kDa). The reason for the existence of an extended form of FGFb is currently unknown, but the shorter forms are likely to be due to proteolytic cleavage.

FGFa is also known by several names, including: endothelial cell growth factor, prostatropin, and heparin-binding growth factor α. Like FGFb, FGFa has been isolated in several sizes. Based on the published cDNA sequence for FGFa, the full-length unprocessed form of FGFa is also predicted to be 155 amino acids.[44] FGFa and FGFb exhibit approximately 55% amino acid sequence identity,[43,44] and have several properties in common. They bind to the same membrane receptors,[48] they bind to heparin,[49] and they are heat-labile (10 min incubation at 70°C destroys biological activity). In addition, both induce angiogenesis,[50-52] and both induce the soft agar growth of several different nontransformed cell lines.[53] The latter indicates that FGF behaves as a TGF.

Besides FGFa and FGFb, there appear to be several other members of the FGF family. Two putative oncogenes, hst and int-2, were recently cloned and sequenced.[54,55] The protein products of these oncogenes have not been isolated, but the amino acid sequences predicted from the cloned genes show significant amino acid sequence identity with FGFb and both should exhibit pI values equal to or greater than the pI of FGFb. int-2 and hst also exhibit significant amino acid sequence identity with FGFa. In regard to biological activity, recent studies indicate that cells transformed with hst release a growth factor that stimulates the soft agar growth of BHK-21 cells.[56] Thus, the hst gene product may exhibit FGF activity. However, at the time of this writing, it is unclear whether int-2 codes for a growth factor that can bind to the receptor for FGFa and FGFb. In this regard, interleukin 1β exhibits approximately 30% amino acid sequence identity with FGFa but does not exhibit biological activity in a bioassay for FGF.[53] Additional information relating to the FGF family of growth factors is available in recent reviews.[57,58]

2. FGF-Related Factors Produced by EC Cells

In 1984, Heath and Isacke reported the isolation and purification of a cationic protein from medium conditioned by PC-13 EC cells.[42] This protein exhibited a molecular weight of approximately 17.5 kDa. Although these properties are consistent with this factor being related to FGFb, insufficient information was reported to unambiguously assign this factor to the FGF growth factor family. In particular, it is unclear whether this factor binds to heparin, is heat-labile, and binds to FGF receptors. However, this factor stimulates the growth of END cells derived from PC-13 EC cells and stimulates the growth of other cells that are known to respond to FGF. More recently, van Veggel et al. isolated a heparin-binding protein from medium conditioned by PC-13 EC cells. This protein elutes from a heparin-Sepharose® column at a slightly lower salt concentration than required to elute FGFb.[59] Still more recently, Rizzino et al., using extracts of F9 EC cells, purified a heat-labile cationic heparin-binding protein that competes with radiolabeled FGFb for binding to cell surface receptors and is immunologically related to FGFb.[60] The heat lability of this factor was nearly identical to that of FGFb. It was also determined that this FGFb-like factor is produced by the human multipotent EC cell line NT2/D1.[60]

These results indicate that EC cells produce a growth factor closely related to FGFb. Recent efforts to identify the gene responsible for the FGFb-related growth factor strongly suggest that it is coded for by the *hst* proto-oncogene.[124] RNA isolated from F9 and PC-13 EC cells hybridizes with a probe for *hst*, but not with probes for FGFb or FGFa. In addition, int-2 does not appear to be expressed at significant levels in EC cells.[61]

D. PRODUCTION OF GROWTH FACTORS BELONGING TO THE TGF-β FAMILY

1. TGF-β Family of Growth Factors

There are at least five members of the type β transforming growth factor (TGF-β) family and two have been shown to be biologically active as disulfide-linked dimers that exhibit a molecular weight of approximately 25 kDa. Holley et al. in 1978 were the first to identify a member of the TGF-β family.[62] Several years later, TGF-β was isolated from several different sources[63,64] and was given the name TGF-β due to its ability to induce the soft agar growth of nontransformed cells. Interestingly, TGF-β has been found in virtually all normal cells,[63] and it is currently isolated from platelets.[65] Efforts to purify TGF-β from porcine platelets led to the discovery that two different, but closely related, forms of TGF-β exist.[66] Porcine TGF-β1 is a homodimer with an amino acid sequence identical to that of TGF-β isolated from human platelets. Porcine TGF-β2 also exists as a homodimer and it exhibits a high degree of sequence identity with TGF-β1 (approximately 70% in their amino-terminal halves). TGF-β1 and TGF-β2 were also isolated from demineralized bone by Seyedin et al. and were named cartilage-inducing factors (CIF) A and B.[67] CIF-A and CIF-B are identical to TGF-β1 and TGF-β2, respectively. Thus far, TGF-β1 appears to be the form present in most tumor cells. However, TGF-β2 has been isolated from a human prostatic adenocarcinoma cell line and a human glioblastoma cell line.[68,69] In most systems, TGF-β1 and TGF-β2 are functionally equivalent, but in at least two systems TGF-β1 and TGF-β2 are not equivalent.[70,71]

In 1984 it was determined that TGF-β and the growth inhibitor isolated by Holley et al. are closely related.[72] Until then, reports concerning TGF-β described its stimulatory effects. During the past 3 years, TGF-β has been shown to inhibit the growth of numerous cell lines, especially those of epithelial origin. Of even greater interest is the finding that TGF-β can inhibit or stimulate the differentiation of many cell types, including adipocytes,[73] chondrocytes,[74] bronchial epithelial cells,[75] myoblasts,[76] and mesenchymal stem cells.[77] In addition, TGF-β has been shown to stimulate production of extracellular matrices in a number of different systems.[78-81]

TGF-β also exhibits significant amino acid similarity with Mullerian-inhibiting substance (MIS) and the inhibins and activins.[82] MIS has only been found in significant amounts in the testes where it induces regression of Mullerian ducts during development of the male reproductive tract. MIS (approximately 140 kDa) is a larger homodimer than TGF-β. Interestingly, MIS has been found to inhibit the growth of tumors derived from tissues related to the Mullerian duct. However, TGF-β does not appear to mimic the effects of MIS. Additional information relating to the TGF-β family of growth factors is available in recent reviews.[83,84]

2. TGF-β Related Factors Produced by EC Cells

Recently, Rizzino and co-workers have determined that EC cells produce low levels of TGF-β. Initially, a radioreceptor assay was used to detect TGF-β in serum-free conditioned medium prepared from F9 and PC-13 EC cells.[85] Later, it was determined that RNA prepared from each of these EC cell lines hybridizes to a TGF-β1 cDNA probe.[124] RNA isolated from the multipotent human EC cell line, NT2/D1, also hybridizes to this probe. However, these results must be interpreted with caution. They could be due to hybridization of the probe with TGF-β1 mRNA or TGF-β2 mRNA, since it is not known whether the TGF-β1 DNA probe can hybridize with TGF-β2 mRNA under the conditions employed.

Given our current understanding of growth factors produced by EC cells, it is unlikely that the TGF activity originally detected in EC cells was due to TGF-β, even though EC cells produce TGF-β. This is suggested by four observations. First, NRK-49F cells were the indicator cells employed in the original study.[21,22] TGF-β does not induce the soft agar growth of NRK-49F cells in the absence of EGF,[37] and two different groups could not detect EGF or other members of the EGF growth factor family in conditioned medium prepared from EC cells.[22,86] Second, the TGF activity produced by EC cells does not copurify with TGF-β.[87] Third, the TGF activity produced by EC cells is likely to be due to the PDGF- and FGF-related growth factors produced by EC cells, since PDGF can induce the soft agar growth of NRK-49F cells in the absence of EGF and TGF-β, and FGF potentiates the soft agar response of these cells to PDGF.[37] Fourth, the soft agar response of NRK-49F cells to PDGF and FGF is reduced by TGF-β in the absence of EGF,[37] and TGF-β suppresses the soft agar growth response of NRK-49F cells to conditioned medium prepared from EC cells.[87] Together, these findings suggest that the TGF activity released by EC cells is unlikely to be due to TGF-β. In addition, the finding that TGF-β suppresses the soft agar growth response of NRK-49F cells to conditioned medium prepared from EC cells suggests that the conditioned medium contains very little, if any, EGF-related growth factors.

E. UNCHARACTERIZED GROWTH FACTORS PRODUCED BY EC CELLS

EC cells also release a number of different factors that have not been fully characterized. Jakobovits et al. have described three different activities in conditioned medium prepared from PSA-1 EC cells.[88] These factors are of particular interest because they are produced by multipotent EC cells. Although these factors were fractionated on the basis of size and classified by function, they were not purified. One factor behaves as an autocrine factor and stimulates thymidine incorporation of PSA-1 cells. The activity of this factor is matched by 10% fetal bovine serum but is not mimicked by EGF or PDGF. The second factor stimulates the proliferation of NIH/3T3 cells in both monolayer and soft agar, and the third factor stimulates thymidine incorporation by Friend erythroleukemia cells and inhibits DMSO-induced differentiation of these cells.

An important question raised by these findings is whether multipotent EC cells and restricted EC cells produce the same factors. More specifically, are the factors described by Jakobovits et al.[88] related to the growth factors produced by F9 and PC-13 EC cells? One can only speculate at this point, but it is possible that the three activities described above correspond to the factors identified from F9 and PC-13 EC cells. The autocrine factor may be related to the FGFb-related growth factor, since EC cells exhibit high affinity receptors for FGF (see Section V.B). Based on recent findings that PDGF and FGF can induce the soft agar growth of other 3T3 cell lines, the second factor described by Jakobovits et al. may also be related to the FGFb-related growth factor and/or the PDGF-related growth factors produced by F9 and PC-13 EC cells. The third factor described by these investigators could be related to TGF-β. Despite the current ambiguity, it is clear that both multipotent EC cells and restricted EC cells produce a number of different growth factors. This increases the likelihood that the cells of the early embryo, and cells derived from the inner cell mass in particular, produce many or all of the same growth factors. As discussed below (Section VII), this possibility should be considered seriously.

IV. GROWTH FACTORS PRODUCED BY DIFFERENTIATED CELLS DERIVED FROM EC CELLS

A. DIFFERENTIATION SUPPRESSES THE PRODUCTION OF SOME GROWTH FACTORS

The discovery that EC cells produce growth factors prompted an investigation of whether the differentiated cells derived from EC cells produce the same growth factors. This question was first addressed for the TGF activity produced by EC cells. Rizzino and co-workers demonstrated that differentiation of F9 and PC-13 EC cells reduces production of the TGF activity more than 90%.[22,23] The same result was observed later for the PDGF-related growth factor produced by EC cells. Unlike the conditioned media prepared from F9 and PC-13 EC cells, the conditioned media prepared from the differentiated cells derived from these two EC cell lines did not compete significantly with radiolabeled PDGF for binding to membrane receptors.[40] Precise estimates are difficult to make, but production of this factor is reduced at least 10-fold when F9 and PC-13 EC cells differentiate. The same picture emerges for the production of the FGFb-related growth factor. Both conditioned media and cell extracts prepared from F9- and PC-13-differentiated cells contain far less FGF activity than the conditioned media and cell extracts prepared from the parental EC cells.[60]

B. DIFFERENTIATION DOES NOT AFFECT THE PRODUCTION OF ALL GROWTH FACTORS EQUALLY

The findings described above indicate that production of several growth factors is dramatically reduced when EC cells differentiate. However, it appears that the production of at least one growth factor is not significantly suppressed when EC cells differentiate. Recent studies have determined that conditioned media prepared from F9- and PC-13-differentiated cells exhibit approximately the same capacity to compete with radiolabeled TGF-β for binding to membrane receptors as the conditioned media prepared from the parental EC cell lines.[85] Furthermore, Northern blot analysis indicates that mRNA from F9 EC cells and mRNA from F9-differentiated cells hybridize with a TGF-β1 cDNA probe.[41] Thus, the production of TGF-β does not appear to be reduced and, in fact, its production may be increased when EC cells differentiate. Recent studies by Heath and Shi indicate that EC-derived differentiated cells also produce growth factors that are not produced by the parental EC cells.[89] Conditioned medium prepared from PC-13-differentiated cells contained a factor that is immunologically related to IGF-II. In addition, this conditioned medium contained an IGF binding protein that exhibits a molecular weight of approximately 35 kDa. Conversely, conditioned medium prepared from PC-13 EC cells did not contain a factor immunologically related to IGF-II. Thus, differentiation of at least one EC cell line leads to increased production of a growth factor.

V. GROWTH FACTOR RECEPTORS EXPRESSED BY EC CELLS AND THEIR DIFFERENTIATED CELLS

The results described thus far indicate that a number of different growth factors are produced by EC cells and that differentiation modulates the production of these factors in a variety of ways. Consequently, one would expect a similar complexity for the regulation of growth factor receptors.

A. RECEPTORS FOR EGF

Rees et al. undertook the first study of growth factor receptors exhibited by EC cells and their differentiated cells.[86] These investigators determined that two EC cell lines, PC-13 and OC-15, do not bind radiolabeled EGF, whereas their differentiated cells exhibit significant numbers of high affinity receptors for EGF (approximately 30,000 receptors for OC-15-differentiated cells). The picture for F9 cells differs from that for PC-13 and OC-15 EC cells. Rizzino et al. demonstrated that F9 EC cells exhibit a small number of EGF receptors (a minimum of 700 per cell) and differentiation of F9 EC cells is accompanied by an increase in EGF binding of only threefold to fourfold.[22] Taken together, these studies not only established that differentiation is accompanied by an increase in EGF receptors, they also suggest that the regulation of cell proliferation might begin relatively early during mammalian development.

B. RECEPTORS FOR PDGF, TGF-β, AND FGF

More recently, the binding of other growth factors has been examined. Rizzino and co-workers have demonstrated that differentiation of both F9 and PC-13 EC cells is accompanied by a significant increase in the number of receptors for PDGF,[40] TGF-β,[90] and FGF.[60] In the cases of PDGF and TGF-β, both EC cell lines bind insignificant amounts of these growth factors, whereas their differentiated cells exhibit approximately 8,000 PDGF receptors and 6,000 TGF-β receptors per cell. A different picture has emerged for FGF receptors. Recent studies indicate that F9 and PC-13 EC cells exhibit a low number of FGF receptors (approximately 2,000 high affinity receptors).[60] However, as observed for the other receptors discussed thus far, differentiation leads to an increase in the number of FGF receptors (approximately tenfold).

Besides EC cells and their RA-induced differentiated cells, several cell lines derived from EC cells have been examined for growth factor receptors. PYS-2, a parietal endoderm-like cell line,[91] has been shown to exhibit receptors for PDGF,[40] TGF-β,[89] and FGF.[92] PSA-5E, a cell line originally reported to exhibit some properties of visceral extraembryonic endoderm,[93] has been shown to bind PDGF[40] and TGF-β,[90] but these cells bind less of each growth factor than PYS-2 cells. A notable exception is EGF. PSA-5E binds EGF, whereas PYS-2 does not.[86]

C. RECEPTORS FOR INSULIN AND INSULIN-LIKE GROWTH FACTORS

Receptors for insulin and the insulin-like growth factors (IGFs) have also been examined. PC-13 EC cells exhibit few, if any, receptors for insulin and differentiation of these EC cells leads to the appearance of receptors for insulin.[94] In contrast to PC-13 EC cells, F9 EC cells appear to express receptors for insulin,[95] and these cells respond to relatively low concentrations of insulin.[19] It should be noted that binding studies with insulin and insulin-like growth factors must be carefully performed, since some IGF receptors can also bind insulin. At least two different IGF receptors exist[96,97] (reviewed in detail in Chapters 4 to 6). Type 1 IGF receptors can bind IGF-I and IGF-II. In addition, Type 1 IGF receptors can bind insulin, although with a lower affinity. Type 2 IGF receptors only bind IGF-I and IGF-II. Studies by two groups have demonstrated that PC-13 EC cells and F9 EC cells bind both IGF-I and IGF-II.[89,98] Furthermore, Heath and Shi have demonstrated by cross-linking studies that PC-13 EC cells express both Type 1 and Type 2 IGF receptors.[89] Interestingly, the number of available IGF receptors decreases by

60% when PC-13 EC cells are induced to differentiate and this appears to be due to the production of IGF II, or a closely related factor, by the differentiated cells.[89]

D. INTRACELLULAR RECEPTORS

The failure of EC cells to bind certain growth factors raises an important question. Do EC cells fail to produce any receptors for these growth factors or do they express too few to be reliably detected? A recent attempt to answer this question involved a new approach. Weller et al. examined OC-15 EC cells for the presence of intracellular EGF receptors.[99] In this study, an antibody prepared against mouse EGF receptors was used to confirm the absence of detectable EGF receptors on the surface of OC-15 EC cells. Specifically, ^{125}I-EGF did not cross-link to an immunoprecipitable 170 kDa protein (the size of the mature EGF receptor) on the surface of OC-15 EC cells, whereas ^{125}I-EGF did cross-link to an immunoprecipitable 170 kDa protein on the surface of OC-15 differentiated cells. However, it appears that OC-15 EC cells contain fully matured EGF receptors that are not translocated to the cell surface, since addition of EGF to detergent-solubilized OC-15 EC cells induced phosphorylation of a 170 kDa band that could be immunoprecipitated by an antibody to the EGF receptor.

These findings raise a number of important questions. Do other EC cell lines that fail to bind EGF also possess intracellular EGF receptors? Are the putative intracellular EGF receptors uniformly distributed throughout the OC-15 EC cell population or are they expressed by a subset of OC-15 EC cells that have begun, but not completed, the process of differentiation? Are the putative intracellular EGF receptors derived from cell surface receptors that have been rapidly removed from the cell surface due to down regulation or receptor transmodulation? Lastly, do EC cells that fail to bind other growth factors, such as PDGF and TGF-β, possess intracellular receptors for these growth factors? Hopefully, these questions will be addressed in the near future because they hold some of the keys to understanding the ontogeny of growth factor receptors during development.

E. RECEPTORS EXPRESSED BY EMBRYONIC CELLS

Ultimately, intracellular receptors and their relevance must be addressed in the embryo itself. Currently, too little is known about growth factor receptors during the early stages of mammalian embryogenesis. The most notable exception is the ontogeny of EGF receptors. However, even for this receptor, much more work needs to be performed. To help situate the reader, a brief description of this work is provided. Further details concerning the ontogeny of EGF receptors during mammalian development are available in reviews by Adamson,[100,101] and in Chapter 7 of this volume.

The presence of EGF receptors during mouse development has been studied by several groups, including Adamson and co-workers. The earliest embryonic cells shown to exhibit EGF receptors are trophoblasts derived from cultured mouse blastocysts.[102] Unfortunately, it was not determined whether the inner cell mass and the early endoderm derived from inner cell mass are able to bind EGF. In other studies, extraembryonic tissues were examined at 2-d intervals starting at the 11th d of gestation and were also found to bind EGF *in vitro*.[103] This is especially true for the amnion. The visceral yolk sac also binds EGF, whereas parietal extraembryonic endoderm apparently binds very little EGF at the same stages of development. EGF binding to the mouse embryo proper has been examined during different stages of development, but these studies were restricted, for the most part, to measurements of EGF bound by the embryo as a whole. These studies demonstrated that EGF binding cannot be detected until the 9th d of gestation in the mouse.[104] As development proceeds, EGF binding remains relatively low until the later stages of mouse development, when it rises dramatically, during the 19th d of gestation, just prior to birth.[102] Although these studies with mouse embryonic tissues are an important start, additional studies with EGF and studies with other growth factors are needed to establish a clearer picture of growth factor receptors during early development.

VI. POSSIBLE ROLES OF GROWTH FACTORS PRODUCED BY EC CELLS AND THEIR DIFFERENTIATED CELLS

A. EVIDENCE FOR PARACRINE GROWTH CONTROL

The work discussed thus far indicates that production of growth factors and growth factor receptors are regulated differently in EC cells and their differentiated cells. The major questions posed by these findings are: what are the functions of these growth factors? are they involved in autocrine growth control, paracrine growth control, or both?

Nearly 10 years ago, Rizzino and co-workers established that EC cells do not respond to growth factors added to serum-free medium.[18,19] In contrast, differentiated cells derived from F9 and PC-13 EC cells respond to EGF,[86] PDGF,[40] FGF,[15,60] and TGF-β.[90] With the apparent exception of growth factors related to EGF, EC cells produce each of these growth factors. This supports early studies that suggested EC cells produce growth factors capable of influencing the growth of differentiated cells derived from the EC cells.[20] The most direct evidence for this hypothesis was reported by Heath and Isacke.[42] They demonstrated that a basic protein isolated from PC-13 EC cells (probably the FGFb-related growth factor described above) could stimulate growth of PC-13-differentiated cells.

Not only are EC cells able to regulate the growth of their differentiated cells by paracrine growth control, but the converse also appears to be true. Recent findings argue that differentiated cells derived from EC cells produce factors capable of influencing the growth of their parental EC cells. PC-13-differentiated cells produce an IGF II related growth factor for which PC-13 EC cells express receptors,[89] and PC-13 EC cells are known to respond to IGF II by increased cell proliferation.[105]

B. EVIDENCE FOR AUTOCRINE GROWTH CONTROL

In addition to paracrine growth control, there is good reason to suspect that autocrine growth control occurs as well. There are several reports of EC cells producing factors that are capable of regulating their own growth.[60,88] For example, F9 and PC-13 EC cells produce a factor closely related to FGFb and both EC cell lines exhibit a small number of high affinity receptors for FGF (approximately 2,000 per cell). However, direct proof that this factor stimulates growth of the EC cells is lacking. In this regard, exogenously added FGF has not been observed to stimulate growth of either EC cell line under any conditions tested, including culture in serum-free medium and culture in reduced serum at low cell density. If the factor produced by EC cells plays an autocrine role, why does exogenously added FGF fail to stimulate the growth of EC cells? There are several possible explanations. First, EC cells may produce as much of this factor as they require. Second, the effect of this factor may not be readily observed *in vitro*. Alternatively, this factor may not have a role in autocrine growth control. In addition, we do not know whether their FGF receptors are fully functional. Since intracellular EGF receptors can be detected in at least some EC cell lines,[99] it is possible that the FGF receptors expressed by EC cells are not able to generate a mitogenic signal. Lastly, FGF may affect the behavior of these cells in ways other than growth. For instance, nerve growth factor has recently been shown by Kahan and Kramp to stimulate the migration, but not the growth, of LT EC cells.[106]

Autocrine factors may also be produced by EC-derived differentiated cells. PC-13-differentiated cells appear to express receptors for the IGF-II-related growth factor that they produce and at least some of the PC-13-derived differentiated cells respond to exogenously added IGF-II. The same situation appears to exist for TGF-β. Recent studies appear to indicate that F9- and PC-13-differentiated cells produce TGF-β. This is of particular interest because F9- and PC-13-differentiated cells express receptors for TGF-β, and this factor inhibits their growth. If these differentiated cells do produce TGF-β then growth factors produced by EC-derived differentiated cells could play roles in both positive and negative autocrine growth control.

The finding that EC-derived differentiated cells appear to produce a factor able to inhibit their growth raises an obvious question. Why would a cell produce a factor that inhibits its own growth? One likely explanation is that TGF-β helps regulate important functions performed by the differentiated cell. In the case of parietal extraembryonic endoderm, two functions can be suggested. First, TGF-β may help control the production of Reichert's membrane. This possibility is suggested by the findings that TGF-β influences the production of extracellular matrices in other systems.[78-81] Second, TGF-β may be responsible for the differential growth rate exhibited by parietal and visceral extraembryonic endoderm. Although both cell types are equally numerous in embryos at the 5th d of gestation, by the 7th d of gestation visceral extraembryonic endoderm has increased approximately tenfold, whereas parietal endoderm has only doubled in cell number.[107]

Although an autocrine function for TGF-β is very attractive, the possibility still exists that it regulates growth by a paracrine mechanism, not an autocrine mechanism. In this regard, studies demonstrating production of TGF-β and expression of TGF-β receptors by EC-derived differentiated cells have not excluded the possibility that the differentiated cell population contains two cell types: one that produces TGF-β and one that expresses TGF-β receptors and responds to this factor by growth inhibition.

C. FACTORS THAT INFLUENCE THE DIFFERENTIATION OF EC CELLS

Lastly, there is another factor that appears to be directly relevant to the control of EC cell proliferation and differentiation. Several years ago, Smith and Hooper demonstrated that STO cells (the fibroblast cells used as a feeder layer to maintain certain multipotent EC cells as undifferentiated populations) condition their medium with a factor that interferes with the differentiation of EC cells.[108] In addition, the conditioned medium was used to maintain multipotent EC cells in the absence of a feeder layer. Recently, this factor, or a related one, has been identified in medium conditioned by buffalo rat liver (BRL) cells.[109] The factor has not been fully characterized, but it is reported to exhibit an apparent molecular weight of 20 to 30 kDa and it is stable at low pH.

Although little is known about the factor released by BRL cells, it appears to be very important and warrants further study. Besides the obvious need to fully characterize this factor, several important questions need to be addressed. First, how does this factor affect EC cells? Does it directly block differentiation of EC cells or it does it indirectly interfere with differentiation by promoting growth of the EC cells? Second, do EC cells produce a similar factor? In regard to the second question, Koopman and Cotton have reported that NG2, a multipotent EC cell line, conditions its medium with a factor that interferes with its own differentiation.[110] However, it is unclear whether the factor released by NG2 EC cells and the factor released by BRL cells are related and whether they act by a common mechanism. A third question is, does serum contain low concentrations of the factor produced by BRL cells? Again, the answer is unknown, but serum does contain factors that influence whether EC cells will differentiate. This is clearly illustrated by the differentiation of OC-15 EC cells into parietal endoderm-like cells when OC-15 EC cells are transferred from serum-containing medium to serum-free medium.[15]

This section has attempted to suggest likely functions for the growth factors that influence the growth of EC cells and their differentiated cells. Currently, there are more questions than answers. However, given the complexities of the growth factors produced by EC cells and their differentiated cells, these studies illustrate the utility of working with EC cells. Without understanding this simpler model system, it will be extremely difficult to understand the roles of growth factors during the early stages of mammalian development.

VII. PRODUCTION OF GROWTH FACTORS BY EARLY MOUSE EMBRYOS

It is clear from the work reviewed thus far that many different growth factors are produced by EC cells and their differentiated cells and that at least some of these growth factors play roles in paracrine growth control and perhaps autocrine growth control. Given the similarities between EC cells and early embryonic cells, one would expect early embryos to produce similar growth factors. Although this remains to be determined, recent studies have provided some evidence that EC cells and early embryos do produce similar growth factors. This work has been reviewed recently[6] and is only summarized and updated here.

The finding that EC cells produce growth factors with TGF activity (induce the soft agar growth of NRK cells) prompted a similar investigation of early mouse embryos. Rizzino demonstrated that mouse blastocysts cultured en masse also produce growth factors that induce the soft agar growth of NRK cells.[111] Furthermore, these factors are produced both by trophoblast monolayers and by cells derived from the inner cell mass. The exact identities of the cells that produce these factors have not been determined, but these factors appear to be produced by cells that form relatively early during development. This conclusion is based on the fact that some of these studies were performed in serum-free medium where the *in vitro* development of the mouse embryo is limited and probably does not progress beyond the equivalent of the 5th or 6th d of gestation *in vivo*.

Although these findings established that mouse embryos are capable of producing growth factors, the factors were not identified. Other studies by Rizzino et al. established that the soft agar growth of NRK-49F cells is stimulated by various combinations of four different growth factors: TGF-β, EGF, PDGF, and FGF.[37] More recently, a bioassay based on different cells was employed to help identify which growth factors are produced by early embryos. This bioassay is based on the finding that FGF and/or PDGF can induce the soft agar growth of NR-6-R cells.[53] Although TGF-β can potentiate the effects of FGF and PDGF, it cannot induce the soft agar growth of these cells on its own. Using this bioassay, Kelly and Rizzino have provided evidence that early mouse embryos produce an FGF-like growth factor that is recognized by antibodies prepared against bovine FGFb.[112] In addition, it appears that early mouse embryos produce TGF-β.[112] Studies are in progress to determine more definitively which factors are produced (see Note Added in Proof at end of chapter).

VIII. DISCUSSION AND CONCLUSIONS

It is clear from the work described in this review that production of growth factors is dramatically altered when EC cells differentiate. Three different patterns appear to exist, creating a complex picture. The production of at least two growth factors is shut off when EC cells differentiate, the production of a third growth factor appears to be unaffected when EC cells differentiate, and the production of a fourth is turned on when differentiation occurs. A summary of the results observed with F9 EC cells is given in Table 1. This level of complexity is unlikely to be due to chance alone given its occurrence in more than one EC cell line.

This type of careful orchestration of growth factors is what one would expect during the early stages of mammalian embryogenesis when the various embryonic cell types begin to appear. Without diffusible factors and a careful regulation of their production, coordination between the various early embryonic cells would depend solely on cell-cell interactions, which seems most unlikely. During the past decade, it has become abundantly clear that growth factors exert profound effects on the growth and differentiation of somatic cells. It is very likely that growth factors play similar roles during the early stages of development. The important question is when? Given the finding that early mouse embryos produce growth factors, it seems very likely that growth factors begin their regulatory roles as soon as the different cell types begin to appear.

TABLE 1
Summary of Growth Factors Produced by F9 EC Cells
and Their Differentiated Cells

Growth factor produced	RA EC cells ————> Differentiated cells		Assay employed
EGF-like	nd	nd	(RRA)[86]
PDGF-like	+	nd	(RRA, RIA)[40]
TGF-β-like	+	+	(RRA)[85]
FGFb-like	+	nd	(RRA, BIOASSAY)[60]

Note: nd, not detected; RRA, radioreceptor assay; RIA, radioimmunoassay.

TABLE 2
Relative Growth Factor Binding by F9 EC Cells and Their
Differentiated Cells

Cells	Relative binding			
	PDGF[40]	TGF-β[90]	FGF[60]	EGF[22]
F9 EC	1	1	1	1
F9 EC-diff Day 2	4	3	6	nd
F9 EC-diff Day 3	29	5	nd	nd
F9 EC-diff Day 4	nd	nd	7	nd
F9 EC-diff Day 5	nd	14	nd	4
F9 EC-diff Day 6	85	16	5	nd
Maximum receptor number	(8,300)	(6,000)	(17,000)	(~2,500)

Note: nd, not determined.

The regulation of growth factor receptors exhibited by EC cells and their differentiated cells is as complex as the regulation of growth factor production. A summary of the data for F9 EC cells and their differentiated cells is given in Table 2. Our current understanding suggests that regulation of growth factors and regulation of growth factor receptors are interdependent. For three different growth factors, related to PDGF, FGFb, and IGF-II, a lower level of production is correlated with a higher number of receptors. This pattern is consistent with the development of interdependence between the EC cells and their endoderm-like differentiated cells. More importantly, it predicts that extraembryonic endoderm, which begins to form during the 4th day of gestation, undergoes a similar transformation and becomes dependent on growth factors produced by the inner cell mass and/or embryonic ectoderm.

Thus far, the major focus of research in the area of embryonic growth factors has been to determine which growth factors and growth factor receptors are produced and when they appear. The work described in this review argues that growth factors and growth factor receptors are present very early during mammalian development. However, no cause-and-effect relationship has been established between the presence of growth factors and the presence of growth factor receptors. This is the next major challenge for investigators in this field. Now that some of the growth factors and the growth factor receptors involved have been identified, it will soon be possible to use a variety of tools, including antibodies and antisense RNA, to address the question of function. However, before work on growth factor functions can proceed, it will be necessary to clarify which members of each growth factor family are expressed by EC cells and early mouse embryos. One can anticipate that this issue will soon be resolved, since most of the genes for these growth factor families have been cloned and their gene sequences published.

Similar clarifications regarding the growth factor receptors are also needed. Genes for EGF and PDGF receptors have been identified and sequenced. However, it is not known whether there is more than one receptor for EGF. Evidence for PDGF and other growth factor receptors indicates that this is a legitimate concern. The most strongly documented evidence for multiple members of a growth factor receptor family is the existence of different receptors for the IGF family.[96,97] Recent evidence suggests a similar situation for EGF receptors. The gene for the EGF receptor is closely related to its oncogene counterpart, v-erb-B, and to the oncogene neu.[101] In addition, there is more than one receptor for TGF-β.[113]

Once the correct genes for each growth factor and growth factor receptor have been identified, it will be possible to begin unraveling the molecular mechanisms that regulate the expression of these genes. An understanding of how these genes are regulated during differentiation will be a major advance in cell and molecular biology.

At present, the roles of the growth factors described in this review are unknown. Speculations that these growth factors play important roles in autocrine and paracrine growth control are supported by little direct evidence. The most direct evidence is that EC cells and their differentiated cells appear to influence each other's growth when they are co-cultured.[24] However, this does not address the issue of whether the growth factors produced by EC cells play a significant role in autocrine growth control. When cells in culture are shown to produce a growth factor, the tendency is to argue for autocrine growth control if one can show that addition of the growth factor to the culture medium stimulates growth of the cells. The flaw in this reasoning is the cells under study may not be homogeneous. In fact, homogeneous populations in cell culture are the exception rather than the rule. Thus, direct proof of autocrine growth control requires a demonstration that the cells responding to exogeneously added growth factor do produce this factor. In nearly all studies where autocrine growth control has been proposed, this issue has not been carefully addressed.

Another important issue that needs to be addressed is the mechanisms by which growth factors exert their effects. Do they only regulate cell proliferation, or do they regulate determination and differentiation as well? In other systems, there are numerous examples of growth factors affecting differentiation. One of the first examples was the effects of EGF on keratinocytes. Twenty five years ago, Cohen and Elliott demonstrated that EGF could stimulate keratinization by stimulating the proliferation of keratinocytes.[114] More recently, FGF has been shown to inhibit myogenesis,[115] stimulate induction of mesoderm in explants of *Xenopus* embryos,[116] stimulate angiogenesis,[50-52] increase the response of primary anterior pituitary cells to thyrotropin-releasing factor, and increase their release of prolactin and thyrotropin.[117] EGF[118,119] and TGF-β[78-81] have been shown to influence the production of extracellular matrices by many different cell types. Moreover, TGF-β has been shown to affect the differentiation of many cell types (Section III.D) and several lines of evidence discussed in this review indicate that growth factors directly or indirectly interfere with the differentiation of multipotent EC cells (Section VI.C). Lastly, recent studies have shown that FGF and TGF-β can induce or promote mesoderm formation in amphibian embryos.[71,120,121] This is of particular interest given that early mouse embryos appear to produce both FGF and TGF-β.[112]

There are other reasons for suspecting that growth factors play a major role in the regulation of determination and differentiation. The decapentaplegic gene complex (DPP-C) of *Drosophila*, which has been implicated in the determination of the dorsal structures and in the morphogenesis of imaginal disks, exhibits 25 to 38% amino acid sequence identity with the C-termini of TGF-β, inhibin, and Mullerian-inhibiting substance.[122] A similar situation exists for EGF. In *Drosophila*, the notch (N) gene affects the early development of the central nervous system, and the sequence of this gene indicates that it could code for a transmembrane protein with an extracellular domain that could contain as many as 36 repeating units related to EGF.[123]

Thus, the question is not whether growth factors affect differentiation, but how do they affect differentiation? More specifically, do they affect differentiation by influencing cell proliferation, or do they regulate the expression of genes involved in differentiation and determination independently of growth? In some cases, growth factors appear to influence differentiation by stimulating cell proliferation. In other cases, growth factors appear to influence differentiation by blocking cell proliferation. TGF-β appears to be the best example of the latter, but it remains to be determined whether inhibition of cell proliferation is a cause or a consequence of differentiation. On the other hand, the relationships between growth factors and the *Drosophila* regulatory genes suggest that, at least in some cases, the effects of growth factors on differentiation have nothing to do with cell proliferation.

It is evident from the work reviewed in this chapter that many important questions concerning the roles of growth factors during early development remain to be addressed. Most questions raised in this review are likely to be answered in the near future through continued study of EC cells and early embryos. In addition, comparisons of EC cells and early embryos are likely to allow us to address other, more general questions. In particular, why do virtually all tumor cells produce growth factors? If EC cells accurately reflect the types of growth factors produced by their embryonic nontransformed counterparts, then this may also be true of other tumor cells and their nontransformed counterparts. The implications of such a finding would be far reaching. It has generally been assumed that the growth factors produced by tumor cells are due to inappropriate expression of growth factor genes. Moreover, it is widely believed that ectopically produced growth factors are responsible in large measure for the unregulated growth of the tumor cells. Evidence for both assumptions is seriously lacking. In nearly all cases, the nontransformed counterparts of tumor cells are not available for investigation. Given this situation, EC cells may not only help us define the roles of growth factors during early development, they may ultimately help us to better understand why tumor cells produce growth factors. Specifically, they may help us answer the question: are the growth factors produced by tumor cells a cause or a consequence of their unregulated growth?

ACKNOWLEDGMENTS

Heather Rizzino is thanked for critically reading this review and for making many helpful suggestions. She is also thanked for excellent editorial assistance. This work was supported by grants from the National Institute of Child Health and Human Development (HD 19837, HD 21568) and the National Cancer Institute (Laboratory Cancer Research Center Support Grant CA 36727).

NOTE ADDED IN PROOF

After this chapter was written, it was reported that mouse blastocysts express several growth factor genes, including TGF-β1 and PDGF A-chain.[125]

REFERENCES

1. **Pierce, G. B. and Cox, W. F.**, Neoplasms as caricatures of tissue renewal, in *Cell Differentiation and Neoplasia,* Saunders, G. F., Ed., Raven Press, New York, 1978, 57.
2. **Solter, D. and Damjanov, I.,** Teratocarcinoma and the expression of oncodevelopmental genes, in *Methods in Cancer Research,* Vol. 18, Fishman, W. H. and Busch, H., Eds., Academic Press, New York, 1979, 277.
3. **Hogan, B. L. M., Barlow, D. P., and Tilly, R.,** F9 Teratocarcinoma cells as a model for the differentiation of parietal and visceral endoderm in the mouse embryo, *Cancer Surveys,* 2, 115, 1983.

4.   **Martin, G. R.,** Teratocarcinomas and mammalian embryogenesis, *Science,* 209, 768, 1980.

5.   **Rizzino, A.,** Growth and differentiation of embryonal carcinoma cells in defined and serum-free media, in *Cell Culture Methods for Molecular and Cell Biology,* Vol. 3, Barnes, D. W., Sirbasku, D. A., and Sato, G. H., Eds., Alan R. Liss, New York, 1984, 107.

6.   **Rizzino, A.,** Defining the roles of growth factors during early mammalian development, in *The Mammalian Preimplantation Embryo: Regulation of Growth and Differentiation In Vitro,* Bavister, B. D., Ed., Plenum Press, New York, 1987, 151.

7.   **Bernstine, E. G., Hooper, M. L., Grandchamp, S., and Ephrussi, B.,** Alkaline phosphatase activity in mouse teratoma, *Proc. Natl. Acad. Sci. U.S.A.,* 70, 3899, 1973.

8.   **Strickland, S. and Mahdavi, V.,** The induction of differentiation in teratocarcinoma stem cells by retinoic acid, *Cell,* 15, 393, 1978.

9.   **Strickland, S., Smith, K. K., and Marotti, K. R.,** Hormonal induction of differentiation in teratocarcinoma stem cells: generation of parietal endoderm by retinoic acid and dibutyryl cAMP, *Cell,* 21, 347, 1980.

10.   **Jetten, A. M., Jetten, M. E. R., and Sherman, M. I.,** Stimulation of differentiation of several murine embryonal carcinoma cell lines by retinoic acid, *Exp. Cell Res.,* 124, 381, 1979.

11.   **Kuff, E. L. and Fewell, J. W.,** Induction of neural-like cells and acetylcholinesterase activity in cultures of F9 teratocarcinoma treated with retinoic acid and dibutyryl cyclic adenosine monophosphate, *Dev. Biol.,* 77, 103, 1980.

12.   **Edwards, M. K. S. and McBurney, M. W.,** The concentration of retinoic acid determines the differentiated cell types formed by a teratocarcinoma cell line, *Dev. Biol.,* 98, 187, 1983.

13.   **Adamson, E. D., Gaunt, S. J., and Graham, C. F.,** The differentiation of teratocarcinoma stem cells is marked by the types of collagen which are synthesized, *Cell,* 17, 469, 1979.

14.   **Rizzino, A.,** Growth and differentiation of embryonal carcinoma cells in defined media: the role of fibronectin, in *Cold Spring Harbor Conferences on Cell Proliferation: Growth of Cells in Hormonally Defined Media,* Vol. 9, Sato, G., Pardee, A., and Sirbasku, D., Eds., Cold Spring Harbor Laboratory, Cold Spring Harbor, NY, 1982, 209.

15.   **Rizzino, A.,** Two multipotent embryonal carcinoma cell lines irreversibly differentiate in defined media, *Dev. Biol.,* 95, 126, 1983.

16.   **Martin, G. R. and Evans, M. J.,** Differentiation of clonal lines of teratocarcinoma cells: formation of embryoid bodies *in vitro, Proc. Natl. Acad. Sci. U.S.A.,* 72, 1441, 1975.

17.   **Martin, G. R., Wiley, L. M., and Damjanov, I.,** The development of cystic embryoid bodies *in vitro* from clonal teratocarcinoma stem cells, *Dev. Biol.,* 61, 230, 1977.

18.   **Rizzino, A. and Sato, G.,** Growth of embryonal carcinoma cells in serum-free medium, *Proc. Natl. Acad. Sci. U.S.A.,* 75, 1844, 1978.

19.   **Rizzino, A. and Crowley, C.,** Growth and differentiation of embryonal carcinoma cell line F9 in defined media, *Proc. Natl. Acad. Sci. U.S.A.,* 77, 457, 1980.

20.   **Rizzino, A., Terranova, V., Rohrbach, D., Crowley, C., and Rizzino, H.,** The effects of laminin on the growth and differentiation of embryonal carcinoma cells in defined media, *J. Supramol. Struct.,* 13, 243, 1980.

21.   **Rizzino, A.,** Embryonal carcinoma cells release factors with transforming growth factor (TGF) activity, *J. Cell Biol.,* 95, 181a, 1982.

22.   **Rizzino, A., Orme, L. S., and De Larco, J. E.,** Embryonal carcinoma cell growth and differentiation: production of and response to molecules with transforming growth factor activity, *Exp. Cell Res.,* 143, 143, 1983.

23.   **Rizzino, A.,** Model systems for studying the differentiation of embryonal carcinoma cells, *Cell Biol. Int. Rep.,* 7, 559, 1983.

24.   **Isacke, C. M. and Deller, M. J.,** Teratocarcinoma cells exhibit growth cooperativity *in vitro, J. Cell. Physiol.,* 117, 407, 1983.

25.   **Raines, E. W. and Ross, R.,** Platelet-derived growth factor. I. High yield purification and evidence for multiple forms, *J. Biol. Chem.,* 257, 5154, 1982.

26.   **Chiu, I.-M., Reddy, E. P., Givol, D., Robbins, K. C., Tronick, S. R., and Aaronson, S. A.,** Nucleotide sequence analysis identifies the human *c*-sis proto-oncogene as a structural gene for platelet-derived growth factor, *Cell,* 37, 123, 1984.

27.   **Johnsson, A., Heldin, C.-H., Wasteson, A., Westermark, B., Deuel, T. F., Huang, J. S., Seeburg, P. H., Gray, A., Ullrich, A., Scrace, G., Stroobant, P., and Waterfield, M. D.,** The *c*-sis gene encodes a precursor of the B chain of platelet-derived growth factor, *EMBO J.,* 3, 921, 1984.

28.   **Betsholtz, C., Johnsson, A., Heldin, C.-H., Westermark, B., Lind, P., Urdea, M. S., Eddy, R., Shows, T. B., Philpott, K., Mellor, A. L., Knott, T. J., and Scott, J.,** cDNA sequence and chromosomal localization of human platelet-derived growth factor A-chain and its expression in tumor cell lines, *Nature (London),* 320, 695, 1986.

29. **Waterfield, M. D., Scrace, T., Whittle, N., Stroobant, P., Johnsson, A., Wasteson, A., Westermark, B., Heldin, C.-H., Huang, J. S., and Deuel, T. F.,** Platelet-derived growth factor is structurally related to the putative transforming protein p28[sis] of simian sarcoma virus, *Nature (London), 304,* 35, 1983.

30. **Doolittle, R. F., Hunkapiller, M. W., Hood, L. E., Devare, S. G., Robbins, K. C., Aaronson, S. A., and Antoniades, H. N.,** Simian sarcoma virus onc gene, *v*-sis, is derived from the gene (or genes) encoding a platelet-derived growth factor, *Science,* 221, 275, 1983.

31. **Gazit, A., Igarashi, H., Chiu, I.-M., Srinivasan, A., Yaniv, A., Tronick, S. R., Robbins, K. C., and Aaronson, S. A.,** Expression of the normal human sis/PDGF-2 coding sequence induces cellular transformation, *Cell,* 39, 89, 1984.

32. **Johnsson, A., Betsholtz, C., Heldin, C.-H., and Westermark, B.,** Antibodies against platelet-derived growth factor inhibit acute transformation by simian sarcoma virus, *Nature (London),* 317, 438, 1985.

33. **Robbins, K. C., Antoniades, H. N., Devare, S. G., Hunkapiller, M. W., and Aaronson, S. A.,** Structural and immunological similarities between simian sarcoma virus gene product(s) and human platelet-derived growth factor, *Nature (London),* 305, 605, 1983.

34. **Heldin, C.-H., Johnsson, A., Wennergren, S., Wernstedt, C., Betsholtz, C., and Westermark, B.,** A human osteosarcoma cell line secretes a growth factor structurally related to a homodimer of PDGF A-chains, *Nature (London),* 319, 511, 1986.

35. **Heldin, C.-H., Wasteson, A., and Westermark, B.,** Platelet-derived growth factor, *Mol. and Cell. Endocrinol.,* 39, 169, 1985.

36. **Rizzino, A. and Bowen-Pope, D.,** Production of and response to PDGF-like factors by early embryonic cells, *Fed. Proc.,* 43, 519, 1984.

37. **Rizzino, A., Ruff, E., and Rizzino, H.,** Induction and modulation of anchorage-independent growth by platelet-derived growth factor, fibroblast growth factor, and transforming growth factor-b, *Cancer Res.,* 46, 2816, 1986.

38. **Ross, R., Raines, E. W., and Bowen-Pope, D. F.,** The biology of platelet-derived growth factor, *Cell,* 46, 155, 1986.

39. **Gudas, L. J., Singh, J. P., and Stiles, C. D.,** Secretion of growth regulatory molecules by teratocarcinoma stem cells, in *Teratocarcinoma Stem Cells, Cold Spring Harbor Conferences on Cell Proliferation,* Vol. 10, Silver, L. M., Martin, G. R., and Strickland, S., Eds., Cold Spring Harbor Laboratory, Cold Spring Harbor, NY, 1983, 229.

40. **Rizzino, A. and Bowen-Pope, D. F.,** Production of PDGF-like growth factors by embryonal carcinoma cells and binding of PDGF to their endoderm-like differentiated cells, *Dev. Biol.,* 110, 15, 1985.

41. **Tiesman, J., Meyer, A., Hines, R. N., and Rizzino, A.,** Production of growth factors related to fibroblast growth factor and platelet-derived growth factor by human embryonal cells, *In Vitro Cell Dev. Biol.,* 24, 1209, 1988.

42. **Heath, J. K. and Isacke, C. M.,** PC13 embryonal carcinoma-derived growth factor, *EMBO J.,* 3, 2957, 1984.

43. **Abraham, J. A., Mergia, A., Whang, J. L., Tumolo, A., Friedman, J., Hjerrild, K. A., Gospodarowicz, D., and Fiddes, J. C.,** Nucleotide sequence of a bovine clone encoding the angiogenic protein, basic fibroblast growth factor, *Science,* 233, 545, 1986.

44. **Jaye, M., Howk, R., Burgess, W., Ricca, G. A., Chiu, M.-I., Ravera, M. W., O'Brien, S. J., Modi, W. S., Maciag, T., and Drohan, W. N.,** Human endothelial cell growth factor: cloning, nucleotide sequence, and chromosome localization, *Science,* 233, 541, 1986.

45. **Gospodarowicz, D., Baird, A., Cheng, J., Lui, G. M., Esch, F., and Böhlen, P.,** Isolation of fibroblast growth factor from bovine adrenal gland: physicochemical and biological characterization, *Endocrinology,* 118, 82, 1986.

46. **Gospodarowicz, D., Cheng, J., Lui, G. M., Baird, A., Esch, F., and Böhlen, P.,** Corpus luteum angiogenic factor is related to fibroblast growth factor, *Endocrinology,* 117, 2383, 1985.

47. **Sommer, A., Brewer, M. T., Thompson, R. C., Moscatelli, D., Presta, M., and Rifkin, D. B.,** A form of human basic fibroblast growth factor with an extended amino terminus, *Biochem. Biophys. Res. Commun.,* 144, 543, 1987.

48. **Neufeld, G. and Gospodarowicz, D.,** Basic and acidic fibroblast growth factors interact with the same cell surface receptors, *J. Biol. Chem.,* 261, 5631, 1987.

49. **Lobb, R. R. and Fett, J. W.,** Purification of two distinct growth factors from bovine neural tissue by heparin affinity chromatography, *Biochemistry,* 23, 6295, 1984.

50. **Esch, F., Baird, A., Ling, N., Ueno, N., Hill, F., Denoroy, L., Klepper, R., Gospodarowicz, D., Böhlen, P., and Guillemin, R.,** Primary structure of bovine pituitary basic fibroblast growth factor (FGF) and comparison with the amino-terminal sequence of bovine brain acidic FGF, *Proc. Natl. Acad. Sci. U.S.A.,* 82, 6507, 1985.

51. **Lobb, R. R., Alderman, E. M., and Fett, J. W.,** Induction of angiogenesis by bovine brain derived class 1 heparin-binding growth factor, *Biochem.,* 24, 4969, 1985.

52. **Thomas, K. A., Rios-Candelore, M., Gimenez-Gallego, G., DiSalvo, J., Bennett, C., Rodkey, J., and Fitzpatrick, S.,** Pure brain-derived acidic fibroblast growth factor is a potent angiogenic vascular endothelial cell mitogen with sequence homology to interleukin 1, *Proc. Natl. Acad. Sci. U.S.A.,* 82, 6409, 1985.

53. **Rizzino, A. and Ruff, E.,** Fibroblast growth factor induces the soft agar growth of two non-transformed cell lines, *In Vitro Cell. Dev. Biol.,* 22, 749, 1986.

54. **Moore, R., Casey, G., Brookes, S., Dixon, M., Peters, G., and Dickson, C.,** Sequence, topography and protein coding potential of mouse int-2: a putative oncogene activated by mouse mammary tumour virus, *EMBO J.,* 5, 919, 1986.

55. **Taira, M., Yoshida, T., Miyagawa, K., Sakamoto, H., Terada, M., and Sugimura, T.,** cDNA sequence of human transforming gene hst and identification of the coding sequence required for transforming activity, *Proc. Natl. Acad. Sci. U.S.A.,* 84, 2980, 1987.

56. **Bovi, P. D., Curatola, A. M., Kern, F. G., Greco, A., Ittmann, M., and Basilico, C.,** An oncogene isolated by transfection of Kaposi's sarcoma DNA encodes a growth factor that is a member of the FGF family, *Cell,* 50, 729, 1987.

57. **Thomas, K. A. and Gimenez-Gallego, G.,** Fibroblast growth factors: broad spectrum mitogens with potent angiogenic activity, *Trends Biochem. Sci.,* 11, 1, 1986.

58. **Folkman, J. and Klagsbrun, M.,** Angiogenic factors, *Science,* 235, 442, 1987.

59. **van Veggel, J. H., van Oostwaard, T. M. J., de Laat, S. W., and van Zoelen, E. J. J.,** PC13 embryonal carcinoma cells produce a heparin-binding growth factor, *Exp. Cell Res.,* 169, 280, 1987.

60. **Rizzino, A., Kuszynski, C., Ruff, E., and Tiesman, J.,** Production and utilization of growth factors related to fibroblast growth factor by embryonal carcinoma cells and their differentiated cells, *Dev. Biol.,* 129, 61, 1988.

61. **Jakobovits, A., Shackleford, G. M., Varmus, H. E., and Martin, G. R.,** Two proto-oncogenes implicated in mammary carcinogenesis, int-1 and int-2, are independently regulated during mouse development, *Proc. Natl. Acad. Sci. U.S.A.,* 83, 7806, 1986.

62. **Holley, R. W., Armour, R., and Baldwin, J. H.,** Density-dependent regulation of growth of BSC-1 cells in cell culture: growth inhibitors formed by the cells, *Proc. Natl. Acad. Sci. U.S.A.,* 75, 1864, 1978.

63. **Roberts, A. B., Anzano, M. A., Lamb, L. C., Smith, J. M., and Sporn, M. B.,** New class of transforming growth factors potentiated by epidermal growth factor: isolation from non-neoplastic tissues, *Proc. Natl. Acad. Sci. U.S.A.,* 78, 5339, 1981.

64. **Moses, H. L., Branum, E. L., Proper, J. A., and Robinson, R. A.,** Transforming growth factor production by chemically transformed cells, *Cancer Res.,* 41, 2842, 1981.

65. **Assoian, R. K., Komoriya, A., Meyer, C. A., Miller, D. M., and Sporn, M. B.,** Transforming growth factor-b in human platelets, *J. Biol. Chem.,* 258, 7155, 1983.

66. **Cheifetz, S., Weatherbee, J. A., Tsang, M. L.-S., Anderson, J. K., Mole, J. E., Lucas, R., and Massagué, J.,** The transforming growth factor-b system, a complex pattern of cross-reactive ligands and receptors, *Cell,* 48, 409, 1987.

67. **Seyedin, S. M., Segarini, P. R., Rosen, D. M., Thompson, A. Y., Bentz, H., and Graycar, J.,** Cartilage-inducing factor-b is a unique protein structurally and functionally related to transforming growth factor-b, *J. Biol. Chem.,* 262, 1946, 1987.

68. **Ikeda, T., Lioubin, M. N., and Marquardt, H.,** Human transforming growth factor type b2: production by a prostatic adenocarcinoma cell line, purification, and initial characterization, *Biochemistry,* 26, 2406, 1987.

69. **Wrann, M., Bodmer, S., de Martin, R., Siepl, C., Hofer-Warbinek, R., Frei, K., Hofer, E., and Fontana, A.,** T cell suppressor factor from human glioblastoma cells is a 12.5-kd protein closely related to transforming growth factor-b, *EMBO J.,* 6, 1633, 1987.

70. **Ohta, M., Greenberger, J. S., Anklesaria, P., Bassols, A., and Massagué, J.,** Two forms of transforming growth factor-b distinguished by multipotential haematopoietic progenitor cells, *Nature (London),* 329, 539, 1987.

71. **Rosa, F., Roberts, A. B., Danielpour, D., Dart, L. L., Sporn, M. B., and Dawid, I. B.,** Mesoderm induction in amphibians: the role of TGF-b2-like factors, *Science,* 239, 283, 1988.

72. **Tucker, R. F., Shipley, G. D., and Moses, H. L.,** Growth inhibitor from BSC-1 cells closely related to platelet type b transforming growth factor, *Science,* 226, 705, 1984.

73. **Ignotz R. A. and Massagué, J.,** Type b transforming growth factor controls the adipogenic differentiation of 3T3 fibroblasts, *Proc. Natl. Acad. Sci. U.S.A.,* 82, 8530, 1985.

74. **Seyedin, S. M., Thomas, T. C., Thompson, A. Y., Rosen, D. M., and Piez, K. A.,** Purification and characterization of two cartilage-inducing factors from bovine demineralized bone, *Proc. Natl. Acad. Sci. U.S.A.,* 82, 2267, 1985.

75. **Masui, T., Wakefield, L. M., Lechner, J. F., Laveck, M. A., Sporn, M. B., and Harris, C. C.**, Type b transforming growth factor is the primary differentiation-inducing serum factor for normal human bronchial epithelial cells, *Proc. Natl. Acad. Sci. U.S.A.*, 83, 2438, 1986.

76. **Massagué, J., Cheifetz, S., Endo, T., and Nadal-Ginard, B.**, Type b transforming growth factor is an inhibitor of myogenic differentiation, *Proc. Natl. Acad. Sci. U.S.A.*, 83, 8206, 1986.

77. **Sparks, R. L. and Scott, R. E.**, Transforming growth factor type b is a specific inhibitor of 3T3 T mesenchymal stem cell differentiation, *Exp. Cell Res.*, 165, 345, 1986.

78. **Ignotz, R. A. and Massagué, J.**, Transforming growth factor-b stimulates the expression of fibronectin and collagen and their incorporation into the extracellular matrix, *J. Biol. Chem.*, 261, 4337, 1986.

79. **Seyedin, S. M., Thompson, A. Y., Bentz, H., Rosen, D. M., Mcpherson, J. M., Conti, A., Siegel, N. R., Galluppi, G. R., and Piez, K. A.**, Cartilage-inducing factor-a, *J. Biol. Chem.*, 261, 5693, 1986.

80. **Roberts, A. B., Sporn, M. B., Assoian, R. K., Smith, J. M., Roche, N. S., Wakefield, L. M., Heine, U. I., Liotta, L. A., Falanga, V., Kehrl, J. H., and Fauci, A.**, Transforming growth factor type b: rapid induction of fibrosis and angiogenesis *in vivo* and stimulation of collagen formation *in vitro*, *Proc. Natl. Acad. Sci. U.S.A.*, 83, 4167, 1986.

81. **Varga, J. and Jimenez, S. A.**, Stimulation of normal human fibroblast collagen production and processing by transforming growth factor-β, *Biochem. Biophys. Res. Commun.*, 138, 974, 1986.

82. **Massagué, J.**, The TGF-b family of growth and differentiation factors, *Cell*, 49, 437, 1987.

83. **Rizzino, A.**, Transforming growth factor-b: multiple effects on cell differentiation and extracellular matrices, *Dev. Biol.*, 130, 411, 1988.

84. **Sporn, M. and Roberts, A. B.**, Autocrine growth factors and cancer, *Nature (London)*, 313, 745, 1985.

85. **Kelly, D. and Rizzino, A.**, Unpublished data.

86. **Rees, A. R., Adamson, E. D., and Graham, C. F.**, Epidermal growth factor receptors increase during the differentiation of embryonal carcinoma cells, *Nature (London)*, 281, 309, 1979.

87. **Rizzino, A.**, Unpublished data.

88. **Jakobovits, A., Banda, M. J., and Martin, G. R.**, Embryonal carcinoma-derived growth factors: specific growth-promoting and differentiation-inhibiting activities, in *Cancer Cells: Growth Factors and Transformation*, Vol. 3, Feramisco, J., Ozanne, B., and Stiles, C., Eds., Cold Spring Harbor Laboratory, Cold Spring Harbor, NY, 1985, 393.

89. **Heath, J. K. and Shi, W.-K.**, Developmentally regulated expression of insulin-like growth factors by differentiated murine teratocarcinomas and extraembryonic mesoderm, *J. Embryol. Exp. Morphol.*, 95, 193, 1986.

90. **Rizzino, A.**, Appearance of high affinity receptors for type b transforming growth factor during differentiation of murine embryonal carcinoma cells, *Cancer Res.*, 47, 4386, 1987.

91. **Lehman, J. M., Speers, W. C., Swartzendruber, D. E., and Pierce, G. B.**, Neoplastic differentiation: characteristics of cell lines derived from a murine teratocarcinoma, *J. Cell. Physiol.*, 84, 13, 1974.

92. **Kuszynski, C. and Rizzino, A.**, Unpublished data.

93. **Adamson, E. D., Evans, M. J., and Magrane, G. G.**, Biochemical markers of the progress of differentiation in cloned teratocarcinoma cell lines, *Eur. J. Biochem.*, 79, 607, 1977.

94. **Heath, J. K., Bell, S., and Rees, A. R.**, Appearance of functional insulin receptors during the differentiation of embryonal carcinoma cells, *J. Cell Biol.*, 91, 293, 1981.

95. **Nagarajan, L. and Anderson, W. B.**, Insulin promotes the growth of F9 embryonal carcinoma cells apparently by acting through its own receptor, *Biochem. Biophys. Res. Commun.*, 106, 974, 1982.

96. **Massagué, J. and Czech, M. P.**, The subunit structure of two distinct receptors for insulin like growth factors I and II and their relationship to the insulin receptor, *J. Biol. Chem.*, 257, 5038, 1982.

97. **Massagué, J., Guillette, B., and Czech, M. P.**, Affinity labeling of multiplication stimulating activity receptors in membranes from rat and human tissues, *J. Biol. Chem.*, 256, 2122, 1981.

98. **Nagarajan, L., Nissley, S. P., Rechler, M. M., and Anderson, W. B.**, Multiplication-stimulating activity stimulates the multiplication of F9 embryonal carcinoma cells, *Endocrinology*, 110, 1231, 1982.

99. **Weller, A., Meek, J., and Adamson, E. D.**, Preparation and properties of monoclonal and polyclonal antibodies to mouse epidermal growth factor (EGF) receptors: evidence for cryptic EGF receptors in embryonal carcinoma cells, *Development*, 100, 357, 1987.

100. **Adamson, E. D.**, Extraembryonic tissues as sources and sinks of humoral factors in development: teratocarcinoma model systems, in *Cellular Endocrinology: Hormonal Control of Embryonic and Cellular Differentiation*, Serrero, G. and Hayashi, J., Eds., Alan R. Liss, New York, 1986, 159.

101. **Adamson, E. D.**, Oncogenes in development, *Development*, 99, 449, 1987.

102. **Adamson, E. D. and Meek, J.**, The ontogeny of epidermal growth factor receptors during mouse development, *Dev. Biol.*, 103, 62, 1984.

103. **Adamson, E. D. and Warshaw, J. B.,** Down-regulation of epidermal growth factor receptors in mouse embryos, *Dev. Biol.,* 90, 430, 1982.
104. **Hortsch, M., Schlessinger, J., Gootwine E., and Webb, C. G.,** Appearance of functional EGF receptor kinase during rodent embryogenesis, *EMBO J.,* 2, 1937, 1983.
105. **Heath, J. K. and Deller, M. J.,** Serum-free culture of PC13 murine embryonal carcinoma cells, *J. Cell. Physiol.,* 115, 225, 1983.
106. **Kahan, B. W. and Kramp, D. C.,** NGF stimulation of mouse embryonal carcinoma cell migration, *Cancer Res.,* 47, 6324, 1987.
107. **Enders, A. C., Given, R. L., and Schlafke, S.,** Differentiation and migration of endoderm in the rat and mouse at implantation, *Anat. Rec.,* 190, 65, 1978.
108. **Smith, T. A. and Hooper, M. L.,** Medium conditioned by feeder cells inhibits the differentiation of embryonal carcinoma cultures, *Exp. Cell Res.,* 145, 458, 1983.
109. **Smith, A. G. and Hooper, M. L.,** Buffalo rat liver cells produce a diffusible activity which inhibits the differentiation of murine embryonal carcinoma and embryonic stem cells, *Dev. Biol.,* 121, 1, 1987.
110. **Koopman, P. and Cotton, R. G. H.,** A factor produced by feeder cells which inhibits embryonal carcinoma cell differentiation, *Exp. Cell Res.,* 154, 233, 1984.
111. **Rizzino, A.,** Early mouse embryos produce and release factors with transforming growth factor activity, *In Vitro Cell. Dev. Biol.,* 21, 531, 1985.
112. **Kelly, D. and Rizzino, A.,** Unpublished data.
113. **Cheifetz, S., Like, B., and Massagué, J.,** Cellular distribution of type I and type II receptors for transforming growth factor-b, *J. Biol. Chem.,* 261, 9972, 1986.
114. **Cohen, S. and Elliot, G. A.,** The stimulation of epidermal keratinization by a protein isolated from the submaxillary gland of the mouse, *J. Invest. Dermatol.,* 40, 1, 1963.
115. **Linkhart, T. A., Lim, R. W., and Hauschka, S. D.,** Regulation of normal and variant mouse myoblast proliferation and differentiation by specific growth factors, in *Growth of Cells in Hormonally Defined Media, Book B,* Sato, G. H., Pardee, A. B., and Sirbasku, D. A., Eds., Cold Spring Harbor Laboratory, Cold Spring Harbor, NY, 1982, 867.
116. **Slack, J. M. W., Darlington, B. G., Heath, J. K., and Godsave, S. F.,** Mesoderm induction in early *Xenopus* embryos by heparin-binding growth factors, *Nature (London),* 326, 197, 1987.
117. **Baird, A., Mormede, P., Ying, S.-Y., Wehrenberg, W. B., Veno, N., Ling, N., and Guillemin, R.,** A nonmitogenic pituitary function of fibroblast growth factor: regulation of thyrotropin and prolactin secretion, *Proc. Natl. Acad. Sci. U.S.A.,* 82, 5545, 1985.
118. **Chen, L. B., Gudor, R. C., Sun, T.-T., Chen, A. B., and Mosesson, M. W.,** Control of a cell surface major glycoprotein by epidermal growth factor, *Science,* 197, 776, 1977.
119. **Lembach, K. H.,** Enhanced synthesis and extracellular accumulation of hyaluronic acid during stimulation of quiescent human fibroblasts by mouse epidermal growth factor, *J. Cell Physiol.,* 89, 277, 1976.
120. **Slack, J. M. W., Darlington, B. G., Heath, J. K., and Godsave, S. F.,** Mesoderm induction in early *Xenopus* embryos by heparin-binding growth factors, *Nature (London),* 326, 197, 1987.
121. **Kimelman, D. and Kirshner, M.,** Synergistic induction of mesoderm by FGF and TGF-b and the identification of an mRNA coding for FGF in the early *Xenopus* embryo, *Cell,* 51, 869, 1987.
122. **Padgett, R. W., St. Johnston, R. D., and Gelbart, W. M.,** A transcript from a *Drosophila* pattern gene predicts a protein homologous to the transforming growth factor-b family, *Nature (London),* 325, 81, 1987.
123. **Knust, E., Dietrich, U., Tepass, U., Bremer, K. A., Weigel, D., Vässin, H., and Campos-Ortega, J. A.,** EGF homologous sequences encoded in the genome of *Drosophila melanogaster,* and their relation to neurogenic genes, *EMBO J.,* 6, 761, 1987.
124. **Tiesman, J. and Rizzino, A.,** Unpublished data.
125. **Rappolee, D. A., Brenner, C. A., Schultz, R., Mark, D., and Werb, Z.,** Developmental expression of PDGF, TGF-α, and TGF-β genes in preimplantation mouse embryos, *Science,* 241, 1823, 1988.

Chapter 8

GROWTH FACTOR SIGNALING IN EARLY MAMMALIAN DEVELOPMENT

Marit Nilsen-Hamilton

TABLE OF CONTENTS

I. THE ROLE OF GROWTH FACTORS IN DEVELOPMENT

For normal embryonic and fetal development, the cellular processes of proliferation, movement, and differentiation must be strictly coordinated. In addition to regulating these three cellular processes in individual cells, growth factors coordinate development by modulating intercellular communication. In this chapter, growth factors, other than those of the insulin family, will be discussed, with specific reference to their roles in the regulation of development. I shall discuss the role of growth factors in development with special consideration given to: (1) the particular developmental events in which growth factors are believed to participate, (2) the period in development when each of the growth factors functions, (3) how proliferation and differentation are coordinated between cells, and (4) how the developing embryo integrates and controls the network of developmental signals. Other review articles related to this topic are listed.[1-12] Insulin and insulin-like growth factors and their significance in development are discussed in Chapters 4 to 6 of this volume.

Embryonic and extraembryonic fetal tissues are rich sources of growth factors.[1,13] The mature growth factors are small proteins with molecular weights ranging from 6,000 to about 30,000 and are generally fully active at concentrations in the picomolar to nanomolar range. Rather than acting on distant cells in an endocrine or exocrine mode, growth factors probably more often act locally to stimulate proliferation, either in a paracrine or autocrine fashion.

A. INTEGRATION OF CELLULAR PROLIFERATION AND DIFFERENTIATION IN DEVELOPMENT

To achieve fine developmental control, the messages carried by signaling molecules should be clear and delivered to the appropriate targets. This is accomplished by the specific interaction of each polypeptide growth factor with a specific cell-surface receptor. Within the developing organism, there also must be an intricate communication network. Growth factors are an important part of this network. In addition to delivering a message to the target cell, growth factors induce cells to produce secreted proteins that may act also as intercellular messengers.[2] A good example of an intercellular messenger cascade initiated by mitogenic stimuli is the network of soluble protein lymphokines that coordinates the immune response.

An important aspect of development is that many different proliferative and differentiation events occur simultaneously in different localities in the embryo. Proliferative and differentiation signals and responses must be similarly localized. A proliferative response will occur only when there is the coincident localization of active growth factor and receptor. Thus, there are a number of factors that determine the temporal, spatial, and cell-specific proliferation and differentiation that occurs during development. These factors include: (1) the developmental time period during which the growth factor or other intercellular message is produced, (2) the spatial distribution of the cells producing the growth factor, (3) the spatial distribution of the cells with receptors for the growth factor relative to the distribution of the cells producing the growth factor, (4) the effective radius of growth factor activity within the tissue (determined by the rate of production and the rate of degradation as a function of time and space), (5) the time during which specific receptors are exposed on the target cell surface, (6) the type of cells that express the particular receptor, and (7) the ability of the cell that expresses the receptor to respond to the delivered message.

B. GROWTH FACTORS OF IMPORTANCE IN MAMMALIAN DEVELOPMENT

The requirement for growth factors is probably present at the earliest stages of development. Preimplantation embryos from several mammalian species are blocked in their development at different stages, according to the species, during *in vitro* culture. This developmental block can be removed if the embryos are transferred back to the oviduct[14] or cultured in explanted organ cultures of the ampullary region of the oviduct.[15] This suggests that the culture media lack essential requirements needed for growth.

TABLE 1

Growth factor	Related proteins and genes	Ref. to reviews
Epidermal growth factor (EGF)	Notch *(Drosophila melanogaster)*	1,6,12,16—20
Transforming growth factor type alpha (TGF-alpha)	Delta *(D. melanogaster)* *lin-12 (Caenorhabditis elegans)* Vaccinia Virus Growth factor Sea-urchin embryonic mRNA	
Fibroblast growth factor (FGF)	*int-2* Mesoderm-inducing factor *(Xenopus laevis)*	6—8,16,18,21
Platelet-derived growth factor (PDGF)	*sis*	6,16,18,22,23—25
Transforming growth factor type beta (TGF-beta) (also called cartilage-inducing factors A and B, BSC-1 growth inhibitor, polyergin)	Decapentaplegic *(D. melanogaster)* Mesoderm-inducing factor *(Xenopus laevis)* Inhibins β chain Activin Mullerian inhibiting substance	6,18,26—30

Although it is clear that growth factors play an important role in mammalian development, our understanding of the nature of their involvement is still rudimentary. Many facts about the cellular responses to growth factors have been revealed from studies of cells in culture. However, there is much to learn about how these cellular responses are integrated in the developing embryo. A list of well-characterized growth factors is presented in Table 1. Many of these growth factors appear to act early in mammalian development. Other growth factors such as nerve growth factor (NGF), interleukin 2 (IL-2), and granulocyte-macrophage colony stimulating factor (GM-CSF), which are important for the development of specific differentiated cell types, generally appear to act later in development or in postnatal life and are not discussed here in detail.

The specific growth factor(s) required for proliferation and differentiation vary as a function of the developmental stage, as seen in neuronal development. Neurons dissected from the superior ganglion of 14-d mouse embryos do not require NGF for survival or neurite extension *in vitro* but do require a neurite extension factor found in conditioned medium.[31] In contrast, most neurons from 16- to 17-d old embryos required NGF for survival. This change in growth factor requirements may reflect a developmental change in the cell surface receptors expressed by these cells.

An understanding of the molecular and cellular basis of development, requires an understanding of the mechanisms whereby mitogenic growth factors regulate diverse cellular and tissue responses. Several growth factors, including EGF, TGF-α, PDGF, aFGF, and bFGF, have been shown to stimulate large subsets of embryonic cells to proliferate. Most nonhematopoietic cells respond to EGF and TGF-α. TGF-α has approximately 40% sequence identity with EGF[32] and modulates cellular responses by binding to the EGF receptor; TGF-α appears to be the fetal form of EGF. FGF stimulates proliferation of mesoderm-derived cells and neuronal cell types. Acidic FGF and bFGF have 55% amino acid sequence identity.[33,34] PDGF stimulates most cells of mesenchymal origin to proliferate. TGF-β is unusual in that it has the ability to both stimulate and inhibit proliferation depending upon the cell type and its environment. Two TGF-betas have been identified. Amino acid sequences reported for TGF-β1[36-38] and TGF-β2[35,39,40] isolated from several species show that the molecules are homologous. The complete amino acid sequences of human TGF-β1 and TGF-β2 show 71.4% identity.[39] The mechanisms whereby these growth factors stimulate specific cells in the early embryo remain to be clarified.

There is a genetically inherited factor, called the Ped (preimplantation embryo development) gene, that determines the rate of cell division in preimplantation embryos.[41] The Ped gene appears to be located in the Qa-2 region of the mouse major histocompatibility complex.[42] Its location in this chromosomal region which contains other genes that code for cell surface and secreted proteins suggests that the Ped gene may code for a membrane or secreted protein that regulates proliferation in the early mouse embryo.

C. STAGE AND TISSUE-SPECIFIC DEVELOPMENTAL REGULATION

Different growth factors may act at various stages of development. For example, FGF or a related growth factor probably acts very early in development. The sequence of the protooncogene *int-2* is related to bFGF gene[43] and thus may encode an FGF-like molecule. *Int-2* is expressed in the early embryo 7 d after fertilization.[44] Four species of *int-2* mRNA were found in periimplantation mouse embryos and in embryonal carcinoma (EC) cells.[44] EC cells are teratocarcinoma stem cells that share many characteristics with normal early embryonic cells. Some EC cell lines can be induced to differentiate into a variety of cell types with characteristics of one of the three primitive germ layers.[45] There are a number of EC cell lines including the PSA, PC13, and OC15 cells that will be discussed later in this section. The *int-2* mRNAs were particularly abundant in EC derivatives of the primitive endodermal lineage.[44] Other evidence that FGF is important in early development comes from studies of amphibian development in which an FGF-like molecule appears to induce mesoderm development in the early amphibian embryo.[46-49]

Further evidence that growth factors act in early development is indicated by the fact that pluripotent EC cells secrete growth factors.[4] EC cells have been used by many investigators as a model of mouse embryo cells at 3.5 to 7 d of gestation.[50] Embryonal carcinoma-derived growth factor (ECDGF; M_r = 17,000) has been isolated from conditioned medium of PC13 cells[51] and several growth-promoting activities were identified in PSA-1-cell-derived conditioned medium.[52] Analysis of their specificities and chemical properties show that none of these EC cell-derived growth factors is the same as EGF or PDGF. However, it is not yet clear whether these EC-cell-derived growth factors are related to FGF.

Whereas FGF may act in the preimplantation embryo, TGF-α and PDGF probably act after implantation. In studies of the expression of the EGF and PDGF receptors on EC cells and their differentiated derivatives, it was found that EC cells do not possess EGF or PDGF receptors while receptor expression could be detected following differentiation. EC cells do not bind EGF,[4] but EGF receptors are detected on the PSA-5E cell line (related to visceral endoderm), on PC13-END cells (related to extraembryonic mesoderm), and on endodermal derivatives of OC15 cells.[53] The developmental period of expression of EGF receptors and TGF-α in the embryo suggests that this growth factor probably acts initially on extraembryonic trophoblast cells, the first differentiated cell type to form during embryogenesis.

D. GROWTH FACTOR HOMOLOGUES IN NONMAMMALIAN
DEVELOPMENT

The participation of growth factors in the control of development is phylogenetically ancient. Homologues to mammalian growth factors have been found in invertebrates. The notch and delta loci of *Drosophila melanogaster* encode polypeptides that show substantial sequence identity to mammalian EGF. The notch polypeptide chain of 2703 amino acids contains 36 repeated EGF-like coding sequences,[54] while the extracellular domain of the delta gene product contains a tandem array of 9 EGF-like repeats.[55] The notch and delta proteins, like the EGF and TGF-α precursors, are probably transmembrane proteins. The developmental function of these two neurogenic genes is to commit ventral ectoderm cells to differentiate into dermoblasts rather than neuroblasts. A third *Drosophila* gene product that is expressed in all ectodermal derivatives, except those of the central nervous system, also has a sequence related to that of EGF.[56]

EGF-related sequences have also been found in developmentally regulated genes in nematodes and sea urchins. The gene, *lin-12*, of the nematode *Caenorhabditis elegans*, also contains EGF-like sequences.[57] Similarly, a cDNA complementary to an embryonic sea urchin mRNA sequence has been found to have a high percentage of identity with the sequence of EGF.[58] These findings have opened the way for systematic analyses of the function of EGF-like molecules in development by taking advantage of the extensive knowledge of the genetics of these invertebrate species.

FGF and TGF-β have also been implicated as regulators of embryonic development in nonmammalian species. The *Drosophila* homeotic locus, decapentaplegic, contains regions of identity with the TGF-b gene.[59] TGF-β and FGF-like molecules are probably the natural inducers of mesoderm in *Xenopus* embryos. FGF induces animal caps of the *Xenopus* embryo to form mesoderm.[47,48] TGF-β and FGF act synergistically to induce muscle actin expression in animal hemisphere cells of the *Xenopus* embryo.[46] Of the two forms of mammalian TGF-β, only TGF-β2 induces mesodermal development in *Xenopus* animal region explants.[49] Vg1, a maternal mRNA localized in the vegetal hemisphere of frog eggs, encodes a protein with 38% identity to human TGF-β1. Following fertilization of *Xenopus* eggs, the Vg1 mRNA is distributed to endodermal cells.[60] A peptide with 89% identity to the C-terminus of bovine bFGF was found also in the *Xenopus* embryo during the time period from oogenesis through gastrulation.[46] Other evidence that FGF is involved in regulating early embryonic development has come from experiments showing that low concentrations of purified bFGF mimic the mesoderm-induction activity of the ventrovegetal signal in *Xenopus* embryos[47] and that the mesoderm-inducing activity in chick embryos had a high affinity for heparin,[61] which is a property that distinguishes FGF.[62]

E. GROWTH FACTORS AND ONCOGENES AS PROLIFERATIVE REGULATORS

As befits their oncogenic activity, some oncogene products are related to growth factors and their receptors. The protooncogene, *sis,* encodes the B-chain of PDGF.[63-65] *Int-2* encodes a polypeptide related to bFGF and aFGF.[43] Several proto-oncogenes, including *fms, erb B,* and *ros,* encode growth factor receptors or are homologous with growth factor receptor genes.[66,67] Some proto-oncogenes, such as *myc* and *fos,* encode proteins whose expression is regulated by growth factors.[68,69] Others encode proteins such as tyrosine protein kinases *(src, yes, syn, kit, abl, met)* or GTP-binding proteins *(ras)* that may be involved in transmission of growth factor signals.[66,67]

The expression of many proto-oncogenes is tissue-specific and developmentally regulated[70-81] (reviewed in Chapter 9 of this volume). The expression of proto-oncogenes, such as *myc,* has been clearly associated with the proliferative effects of growth factors.[68] The involvement of proto-oncogene-like proteins in development is also evident in *Drosophila melanogaster* in which certain genetic loci have been found to have sequences related to the mammalian proto-oncogenes, *myc,*[82] *int-*1,[83] and *rel.*[84] The significance of these oncogenes in pattern formation during development is discussed in a recent review.[85]

II. INTERCELLULAR COMMUNICATION IN EMBRYONIC DEVELOPMENT

A. GAP JUNCTIONAL COMMUNICATION

Communication among cells is essential for the coordinated development of a multicellular organism. One example of communication requiring cell to cell contact is via gap junctions. Small molecules move from one cell to another through well-defined, intermembrane channels.[86,87] Gap junctions provide a means of transferring nutritional precursors as well as informational molecules. Although the participation of gap junctional communication in normal growth and development is not well defined, it is known that loss of the ability to communicate in this way is a characteristic of oncogenically transformed cells.[88] The experimental evidence for the role of gap junctions in development has been reviewed recently.[89]

Gap junctional communication appears during the first 6 h of the eight-cell stage preceding compaction in the murine embryo.[90] As the trophoblast and inner-cell-mass cells differentiate, junctional communication between these two cell types decreases.[91] The importance of junctional communication for normal development is indicated by the results of studies in which microinjection of antibodies raised against the major protein extracted from rat liver gap junctions inhibited gap-junctional communication and resulted in the failure of eight-cell stage mouse embryos to compact.[92] Similarly, early embryos of the DDK inbred mouse strain were found to have reduced gap-junctional communication.[93] The DDK locus is expressed as an early acting lethal condition when the DDK eggs are fertilized by spermatozoa of another mouse strain. DDK/C3H embryos could be rescued to the blastocyst stage when incubated in methylamine, which increased gap-junctional communication. However, the methylamine treatment was unable to reverse the spontaneous decompaction that occurred in these embryos from the 16-cell stage onward.[92] Thus, it seems that gap-junctional communication is important in regulating compaction in the mammalian embryo.

There is also evidence for the participation of gap junctional communication in the development of nonmammalian embryos. Injection of antibodies directed against a rat liver gap junction protein into *Xenopus* blastomeres resulted in developmental defects that appeared to be due to a failure of neural induction.[94] Development of the neural tube is normally induced by the underlying mesoderm, and inhibition of communication between the two tissues could explain the developmental failure. In the mollusk *Lymnaea stagnalis,* cellular compartments, defined by gap junctional connections, changed progressively during development. The boundaries of these compartments corresponded with the developmental boundaries delineating groups of cells with restricted developmental fates.[95] Thus, in some cases, junctional communication may mediate specific developmental signals. In other cases, communication through gap junctions may distinguish groups of cells with similar developmental fates.

Tumor promoters inhibit communication via gap junctions.[96,97] This effect is believed to be mediated by the increased activity of protein kinase C brought about by the tumor promoters. Increased cytoplasmic calcium levels also inhibit junctional communication.[86-88] Some growth factors stimulate the turnover of phosphatidyl inositol with the resulting production of inositol triphosphate and diacylglycerol, which is the natural stimulator of protein kinase C.[98,99] Inositol triphosphate releases calcium into the cytoplasm from an intracellular store. Both of these actions of growth factors are expected to inhibit junctional communication.

B. INTERCELLULAR COMMUNICATION MEDIATED BY CELL-SURFACE PROTEINS

Intercellular communication may involve cell-to-cell contact or may occur via specific cell surface proteins. Membrane-associated growth stimulators and inhibitors have been identified.[100-103] In some cases, cell-to-cell interactions require the simultaneous recognition of more than one cell surface protein, as in the recognition of foreign antigens by T-cells.

The precursors of EGF and TGF-α are transmembrane proteins.[104-107] The active soluble forms of these growth factors are released by proteolytic cleavage. The sequence of the membrane-bound portion of the precursor is highly conserved between human and rat, suggesting the possible importance of this segment. For example, the remaining membrane-bound fragment may perform a function after the EGF sequence is cleaved off. It is also possible that the membrane-bound precursor might, in some cases, regulate proliferation through direct cell-to-cell contact. Evidence to support this comes from studies of *Drosophila,* which indicate that the EGF-like protein encoded in the notch and delta genes may act at the cell surface or in a very restricted region around the cells that produce the notch protein, rather than acting on a distant target.[54,108-110]

C. INTERCELLULAR COMMUNICATION INVOLVING SOLUBLE MEDIATORS

A second means of intercellular communication in embryogenesis occurs via soluble secreted signals. Mediators that are likely to be important in embryonic development include steroids, prostaglandins, cyclic nucleotides, secreted proteins, and hormones. Probably, all types of intercellular signals — autocrine, paracrine, and endocrine — regulate embryonic development. Many of these intercellular signals are likely to be growth factors or growth-factor induced proteins. Endocrine hormones, which reach their destination via the bloodstream, are important components of the fetal-maternal communication system. Paracrine regulators, which affect cells in the immediate vicinity of the producer cells, probably constitute one of the most important regulatory systems for growth and development in the early embryo and in the feto-placental unit.

Autocrine regulatory systems may also be part of normal embryonic development. Unlike the paracrine and endocrine modes, autocrine control involves autoregulation of proliferation by factors produced by the cell itself.[111] The autocrine regulatory mode has the potential of allowing a particular cell type to proliferate independently of external signals. Further, there is evidence that autocrine growth regulation can mediate tumorigenesis.[112] Because coordination of cellular proliferation and differentiation is of paramount importance to normal embryonic and fetal development, any autocrine regulatory loop that exists *in vivo* is almost certainly regulated via communication with other cell types in the developing embryo.

Autocrine regulatory systems might amplify a proliferative signal in a particular region of a developing tissue or convert an endocrine or paracrine signal into a local proliferative signal. For example, placental lactogen stimulates rat embryonic fibroblasts to synthesize and secrete IGF-II,[113] and TGF-β stimulates AKR-2B mouse embryo cells to express the *sis* gene that encodes the PDGF B chain.[114]

Normal development depends upon specific localization of proliferative signals. The capacity of many different embryonic cells to respond to a particular growth factor with increased proliferation is restricted to those cells within the effective radius of the cells producing the growth factor. The varying rate of growth-factor inactivation provides one means of local control over its activity. Polypeptide growth factors are inactivated by endocytosis and degradation of the receptor-growth factor complex in the lysosomes as exemplified by the EGF system.[20,115,116] The rate at which cells internalize and degrade EGF is influenced by the cytoplasmic calcium concentration[117,118] and by phosphorylation of the EGF receptor (discussed later in this review). The balance between the rate at which the growth factor is produced and diffuses through the tissues and the rate at which it is degraded by the surrounding target cells determines the region in the embryo in which the growth factor will be effective.

The effective distance that the growth factor moves from the secreting cell can also be determined by the affinity of the growth factor for the components of the extracellular matrix. For example, aFGF and bFGF have very high affinities for heparin.[21,62,119] Therefore, all of the FGF released by cells in the developing embryo may rapidly become associated with extracellular matrix and so remain in the original location of the cells that produced it.

Secretion of an inactive form of growth factor by some cells can result in localization of growth-factor activity. The growth factor is functional only following contact with an activator that is localized in specific areas of the developing organism. In some instances the activating molecule may be a protease. For example, TGF-β is secreted as a larger precursor that may be inactive until it is processed to the mature form.[120,36] The TGF-α precursor, which may itself be active, is a transmembrane protein from which the growth factor is probably released by extracellular proteolytic cleavage.[106] In other instances, activation may be achieved by dissociating the mature growth factor from an inhibitor. For example, TGF-β and PDGF bind alpha$_2$-macroglobulin with a very high affinity and specificity.[121-123] Thus, some cells might regulate the local activity of these two growth factors by secreting alpha$_2$-macroglobulin.

D. EPITHELIAL-MESENCHYMAL INTERACTIONS

Epithelial-mesenchymal interactions are probably implicated in the response of developing embryonic tissues to growth factors. For example, in the developing tooth of the mouse embryo, the EGF receptors are located on the mesenchymal and not on the epithelial cells on Day 14. Yet, on Days 14 through 17, EGF stimulates proliferation of dental epithelium and inhibits proliferation of the dental mesenchyme.[124] Thus, proliferation of the epithelium may be mediated through a paracrine signal released by the mesenchymal cells in response to EGF or TGF-α. It was also found that whereas EGF inhibited proliferation of the mesenchyme in intact tissue, disaggregated dental mesenchymal cells proliferated *in vitro* when cultured with EGF.[124] Similarly, mesenchymal-epithelial interactions are important in early lung development. On Day 13, EGF receptors were localized to lung mesenchymal cells of the developing mouse, yet EGF stimulated epithelial branching rather than mesenchymal proliferation.[125] Observations of embryonic hair follicle growth also provide evidence for paracrine growth regulation. The nuclear labeling index, a measure of mitogenic activity, has been shown to decline with distance from the hair follicle anlagen. From these results, it was suggested that a diffusible ligand, such as EGF, which is produced by the dermal papilla, acts as a mitogenic factor in hair growth.[126]

Specific interactions between cells during development often involve the secretion of proteins that form part of the extracellular matrix. This topic has been discussed in several recent reviews.[7,8,127,128] One way in which the extracellular matrix can interact with growth factors to provide a developmental signal is by serving as an insoluble network of binding sites for growth factors, such as aFGF and bFGF, that bind heparin tightly.[62,129] The embryo might define local regions of high- or low-growth factor concentration by controlling the composition and deposition of the extracellular matrix.

III. DEVELOPMENTAL EXPRESSION OF GROWTH FACTORS AND THEIR RECEPTORS

Each growth factor probably acts at specific stages of development on particular cells and tissues. The developmental period during which the growth factor acts and the site of action is determined by the location of the growth factor and its receptor and the time at which they are expressed.

A. TRANSFORMING GROWTH FACTOR TYPE ALPHA

The developmental expression of TGF-α has been studied at the level of protein and mRNA expression. The maximum level of TGF-α mRNA measured by Northern blots and *in situ* hybridization has been found in the placental decidua on Day 8 (plug date = d 0) of gestation in the rat.[130] Maximum TGF-α activity was detected in the mouse embryo by radioreceptor assay on Day 7. The level rapidly declined, then a second peak of TGF-α appeared on Day 13 that again declined to a minimum by Day 17.[131] However, no TGF-α mRNA was detected in the embryo.[130] The TGF-α activity previously reported to be in the embryo may have come from extraembryonic tissue contaminating the preparation. Thus, the TGF-α produced in the decidua may be a paracrine regulator of cell proliferation in the embryo and extraembryonic tissue from Day 7 onwards. These and other studies of the expression of EGF receptors in embryonic tissues and on EC cells and their differentiated derivatives are consistent with the hypothesis that TGF-α stimulates the proliferation of mesenchymal and epithelial cells during early development. Experimental evidence suggests that TGF-α acts in mid-gestation and that the primary target organs are the placenta, lungs, kidney, palate, and skin.

B. THE EGF/TGF-ALPHA RECEPTOR

The extent to which a given cell responds to a particular growth factor is determined by the concentration of the growth factor and by the density and affinity of the receptors. The organism can exert a fine control over localized cell growth and differentiation by controlling these variables in specific regions of the embryo as a function of development.

Studies of the cellular proliferative responses in various embryonic tissues suggest that EGF-like molecules (perhaps TGF-α) regulate proliferation and maturation in the developing lung,[132] palate,[133,134] teeth,[135,136] vascular elements,[137] skin,[138,139] kidney, and salivary gland,[136] and the gastrointestinal tract.[140] Embryonal expression of the EGF receptor is tissue-specific and developmentally regulated. EGF-binding activity has been detected in trophoblast cells 6.5 d after fertilization[141] and in the placenta and the embryo from Day 11 through Day 18.[142,143] The maximum amount of EGF binding occurred in embryos on the 13th d of gestation, with the amnion containing the most EGF-binding activity. There was less EGF binding in the lungs, trophoblast, limbs, visceral yolk sac, liver, brain, parietal endoderm, and heart, in that order.[143] EGF receptors have been found also in human extraembryonic tissues.[144] Both the affinity of the EGF receptor and the number of EGF-binding sites change with murine development. In the mouse placenta and embryonic tissue, the affinity of the EGF receptor for EGF slowly declines by about 50% between Days 11 and 18 of gestation but remains unchanged over this period in brain and liver.[141] The epidermis is also a target of EGF-like growth factors.[135] The distribution of EGF receptors in embryonic rat skin varies as a function of development and is generally correlated spatially with relatively undifferentiated or rapidly proliferating epidermal cells.[126]

From a comparison of the patterns of EGF binding in a variety of tissues, it has been suggested that an EGF-like molecule may play a role in the formation of epitheliomesenchymal organs. It has been proposed that, in each of these tissues, EGF or TGF-α stimulates epithelial proliferation during initial epithelial bud formation and branching morphogenesis.[136] These studies, and studies of the ontogeny of the EGF receptor kinase activity, suggest that EGF, or TGF-α, stimulates proliferation in certain developing tissues in the latter half to two thirds of gestation.

The cellular distribution of the EGF receptor in specific organs varies with the developmental stage. This was clearly demonstrated in tissue sections of mouse embryonic submandibular jaw at 17 d of gestation. At this age, the developing jaw contains the germs of all three molar teeth at different stages of development. In sections of the developing jaw, it was found that the EGF-binding pattern varied in a stage-specific manner rather than as a function of developmental age.[136] These results suggest that proliferation of cells in the developing jaw is regulated by EGF only during specific stages of differentiation.

The distribution of EGF receptors in the *Drosophila* embryo corresponds in general to the distribution of proliferating cells in the embryo. Surveys of the cell-specific expression of the EGF receptor gene (DER) mRNA in *Drosophila* by *in situ* hybridization showed a uniform distribution of gene expression in the cellular blastoderm embryo. Larger concentrations of mRNA were localized in the proliferating tissues of the imaginal disk and the brain cortex of the larvae.[145,146] Certain nondividing cell populations did not express DER. Thus, it appears that a major function of EGF-like growth factors in the developing embryo is to stimulate proliferation of certain cell lineages.

A number of polypeptide growth factor receptors are tyrosine protein kinases, including the receptors for EGF,[147] IGF-I,[148] and PDGF.[149] The FGF receptor is also probably included in this group,[150] although one initial investigation failed to reveal tyrosine kinase activity.[151] In contrast, tyrosine protein kinase activity has not been found in association with the TGF-β receptors.[152] For those that have the activity, the tyrosine kinase activity of the receptor is believed to be instrumental in transducing the extracellular growth factor signal into an intracellular regulatory response (see Chapter 3). A number of intracellular protein substrates of receptor kinases have been identified, including the receptors themselves.[66]

EGF receptor kinase activity has been found in the mouse embryo from Day 10 of gestation onwards.[143] Activity was also found in the uterus, placenta, decidua, and amnion. No EGF receptor kinase activity was found in the visceral and parietal yolk sac. In all tissues but the placenta, receptor activation resulted in autophosphorylation of tyrosine residues; in the placenta, phosphorylation was mainly on serine. On Day 13, the fetal skin, skeletal muscle, head, trunk, legs, lungs, and intestines were shown to express the EGF receptor kinase, whereas the fetal liver, heart, and brain did not. EGF receptor kinase activity in the liver and brain were reported to rise rapidly between Days 17 and 18.[153] The expression of EGF receptor kinase activity correlates in general with the results of other studies on EGF-binding activity in embryonic tissues.[143] However, not all studies have reported similar findings.[153]

The factors that regulate the level of expression of the EGF receptor in the early embryo are unknown. Regulation of the rate of transcription of the EGF receptor gene is one means of varying the density of EGF receptors on the cell surface. EGF regulates the expression of the gene coding for its own receptor.[154-156] Other growth factors which regulate the cell-surface expression of the EGF receptor include PDGF,[157-160] FGF,[157] and TGF-β.[161,162] Several steroid hormones have also been found to regulate the expression of the EGF receptor. For example, 17-b-estradiol administered to immature female rats tripled the EGF-binding capacity of uterine membranes by a mechanism that was inhibited by cycloheximide and actinimycin D.[163] The administration of glucocorticoids also has been demonstrated to result in increased EGF binding by human fibroblasts,[164] cultured C3H10T1/2 mouse embryo cells,[165] and cells of the fetal rabbit lung.[166] The ability of glucocorticoids to increase EGF receptor expression may explain the synergistic action of growth factors and glucocorticoids in stimulating cellular proliferation.[167,168]

The extent of growth factor binding to a tissue is a function of receptor density and the affinity of the receptors for the ligand. Some well-defined posttranslational mechanisms may act to regulate the number of growth factor receptors. For example, cell-surface exposure of receptors can be decreased by receptor endocytosis, a process known as "down-regulation." However, specific features of the cytoplasmic domain of the EGF receptor are known to be necessary for endocytosis of the receptor, and mutant EGF receptors with altered cytoplasmic domains, produced through recombinant DNA techniques, are not subject to ligand-induced endocytosis. Point mutations at position 721 to replace Lys with Met[169,170] or Ala[171] in the putative active site of the tyrosine protein kinase have resulted in receptors that were not down-regulated. The Met[721] receptor was not internalized; however, the Ala[721] receptor was internalized normally. Consequently, it has been proposed that the altered Ala[721] receptor was recycled to the cell surface without being degraded. Both of these mutants lacked tyrosine protein kinase activity. Other mutant receptors subjected to C-terminal truncation also lacked tyrosine kinase activity and displayed reduced ligand-induced internalization.[172,173] However, the association of tyrosine kinase activity and ligand-induced internalization did not appear to be absolute. A mutant receptor with an insertion of four additional amino acid residues after residue 708 that lacked tyrosine kinase activity was down-regulated.[174] One correlation consistent to all of these studies is that the kinase-negative mutants were unable to transduce the EGF signal into a response initiating DNA synthesis or other metabolic alterations associated with normal EGF receptor function.[169-174] These and other data strongly suggest that the tyrosine protein kinase action initiates the events by which growth factors regulate DNA synthesis and proliferation.

Heterologous growth factors are able to regulate the ability of cells to bind EGF by altering the rate of internalization[159] or externalization of the receptor[175] or by altering the ligand binding affinity.[160] The cell-surface expression of high affinity EGF binding sites has been shown to be decreased by PDGF[157-160] and FGF[157] and increased by TGF-β.[161,162]

Differentiation of OC15 embryonal carcinoma cells into endoderm-like cells is associated with responsiveness to EGF. Although undifferentiated OC15 cells do not bind or respond to EGF, solubilized OC15 cells contain readily detectible EGF-activated receptor kinase activity.[175] These results suggest that embryonal stem cells may contain an intracellular pool of active EGF receptors that can be rapidly translocated to the cell surface following the appropriate stimulus.

Internalization of EGF receptor can be regulated directly by protein kinase C, a calcium- and phospholipid-dependent kinase.[176] Protein kinase C probably initiates internalization of the receptor by phosphorylating the EGF receptor on thr-654, which is located in the cytoplasmic domain near the trans-membrane domain.[177-180] Some growth factors activate protein kinase C in responsive cells by stimulating the turnover of phosphatidyl inositol, which results in the production of diacylglycerol, a natural activator of protein kinase C.[181-184] The EGF receptor is also phosphorylated on thr-654 in response to a PDGF stimulus.[179]

IV. NONPROLIFERATIVE FUNCTIONS OF GROWTH FACTORS IN DEVELOPMENT

A. GROWTH FACTORS AND CELLULAR DIFFERENTIATION

In addition to their importance in regulating cell proliferation in development, growth factors and their receptors play a major role in cellular differentiation. One of the earliest differentiation events in the developing embryo is the regional specification of the three embryonic cell groups; ectoderm, mesoderm, and endoderm. Specific morphology-inducing factors (morphogens) have been identified from studies of amphibian embryonic development. The ventrovegetal morphogens that induce mesodermal development are thought to be peptides related to bFGF and TGF-β.[46-49]

Although there is not a general concensus, a number of studies have shown that the effects of growth factors on cell differentiation are independent of those on proliferation. For example, TGF-β inhibits the differentiation of adipocytes[185] and skeletal and smooth muscle cells,[186-189] without altering the growth rates of the affected cells. Although there is not always a reciprocal relationship between proliferation and differentiation, growth factors can affect both processes simultaneously. For example, TGF-β stimulates the differentiation of a cell line derived from intestinal crypt epithelium concomitant with the inhibition of proliferation.[190] FGF also stimulates the differentiation of many cell types, including preadipocytes, neurons, endothelial cells, and chondrocytes.[7,8] In contrast, FGF inhibits the differentiation of muscle cells and stimulates their proliferation.[191]

Growth factors affect differentiation and maturation of fetal tissues *in vivo*. EGF promotes the maturation of the fetal lung without increasing lung size,[132] accelerates tooth eruption and eyelid opening in newborn mice,[135] and promotes development of the epithelial lining of the fetal gut.[140] Nerve growth factor[192,193] promotes the survival and differentiation of neuronal cells *in vitro* and is essential for development of sympathetic and sensory neurons *in vivo*.

B. GROWTH FACTORS AS INHIBITORS OF CELLULAR PROLIFERATION

Cellular proliferation in the embryo is temporally restricted, and current thinking is that it is initiated by growth factor(s). Cessation of proliferation, equally important in orderly developmental processes, could be caused by depletion of the required growth factor or by the appearance of a specific inhibitor of proliferation. The latter is more likely because it provides the developing organism with an additional control over cellular multiplication. TGF-β, interferon, and tumor necrosis factor alpha (TNF-α, also known as cachectin) inhibit cell proliferation and may have a role in regulating development. Specific receptors have been identified for each of these growth regulators, but mechanisms underlying the inhibitory signals generated by these receptors remain unclear. Unlike the receptors for EGF, PDGF, and IGF-I, which are tyrosine protein kinases, the receptors for TNF-α, TGF-β, and interferon have not yet been shown to possess enzymatic activity.

Cell growth can be arrested in several ways. TGF-β and interferon arrest cells in G_0 of the cell cycle, thereby inhibiting both DNA replication and subsequent mitosis. The existence of cells, such as the placental giant cells, in which the genome has multiplied many times without cell division,[194] suggests the presence of other growth factors that may inhibit cytokinesis without inhibiting DNA synthesis. Proliferation (hyperplasia) can be dissociated from growth in size (hypertrophy). For example, TGF-β inhibits DNA synthesis and proliferation without inhibiting many of the early metabolic responses to insulin in proximal kidney epithelial cells.[195] Thus, TGF-β uncouples hypertrophy from hyperplasia in these cells and allows them to grow in size without DNA duplication or mitosis.

Other growth inhibitors of significance in development are the interferons, for which three classes have been described: alpha, beta, and gamma. Although they were recognized first as antiviral agents secreted in response to viral infection, interferons also inhibit the proliferation of many cell types.[196] Interferons have two well-known effects on animal cells. One effect is to induce cells to synthesize (2′ 5′) oligo-adenylate synthetase whose polyadenylated product stimulates an oligo-A-dependent endoribonuclease (RNAase F). The ribonuclease cleaves single-stranded RNA at UpUp, UpGp, or UpAp. The second effect of interferons is to activate a dsRNA-dependent protein kinase that phosphorylates and thereby inactivates, the a subunit of the translational initiation factor, EIF-2. Progression from G_1 to S in the cell cycle requires protein synthesis;[197] thus, these inhibitory actions of interferon on translation may be responsible for its ability to inhibit proliferation. Several growth factors induce cells to produce interferons,[198] suggesting a negative feedback mode for interferons in growth-factor-stimulated cellular proliferation.[198]

Selective cell death, an important aspect of normal embryogenesis and fetal development,[199] may be mediated in part by growth factors. For example, although EGF stimulates cellular proliferation in most cells, A 431 epidermal carcinoma cells that express excessive numbers of EGF receptors die in response to EGF.[200] This observation suggests that a developmental "switch" that initiates selective cell death could involve increased expression of the EGF receptor.

V. GROWTH-FACTOR-REGULATED PROTEINS IN DEVELOPMENT

A. GENES REGULATED BY GROWTH FACTORS

There is evidence to show that expression of a number of cellular genes is enhanced when proliferation is stimulated by growth factors or serum. Further, expression of some of these genes is cell-cycle dependent.[201] It has been estimated that the subset of genes regulated by growth factors in Balb/c 3T3 cells is fewer than 50,[202-204] although this number may vary in other cell types. These genes, which code for intracellular and secreted proteins, can be divided into several groups. One group of growth-factor-regulated genes encodes proteins that bind DNA. These proteins are the most likely candidates to be directly involved in the regulation of transcription. Growth factors, e.g., PDGF, stimulate the expression of *myc* and *fos* that encode putative DNA-binding proteins.[71,205-207] Another growth-factor-induced gene has been found with a sequence containing three zinc fingers.[208] The zinc finger motif is indicative of a DNA-binding protein.

A second group of growth-factor-induced genes encode proteins responsible for the general increase in cellular metabolism required for cell proliferation. Examples of genes in this category include the gene encoding a calcium-binding protein (clone 2A9) and the gene encoding the ATP/ADP translocase (clone 2F1).[209] Still another group of growth-factor-regulated genes encode cytoskeletal proteins such as actin[210,211] and vimentin (clone 4F1).[209] Genes encoding secreted proteins are another important subset of growth-factor-regulated genes. Examples include the genes encoding fibronectin,[212-214] collagen,[215,216] and MRP/PLF (alternatively called mitogen-regulated protein or proliferin).[217,218] Many of the growth-factor-regulated genes encode secreted proteases or their inhibitors. Genes in this category include the gene for cathepsin L (also called major excreted protein or MEP),[217,219-221] collagenase,[222-224] transin (also known as stromolysin),[225] the collagenase inhibitor, TIMP,[224] and the inhibitor of plasminogen activator, PAI-1[226-230] (also called IIP48).

A number of growth-factor-regulated genes have been shown to be expressed at specific times and in specific cell types during development. These genes include those that encode *fos* and *myc*,[73,74,77,79-81] vimentin, a calcium-binding protein, the ATP/ADP translocase,[209] and MRP/PLF.[231,232] Each gene is expressed in an individual developmental pattern both with respect to tissue type and developmental age. The protein products of these growth-factor-induced genes are probably central participants in the regulation of the rate of initiation of DNA synthesis and proliferation. However, many of the growth-factor-induced proteins may be involved in functions other than autoregulation. For example, these secreted proteins may serve as paracrine regulators of growth or differentiation signals that coordinate embryonic development. They may also participate in forming or reforming an extracellular matrix or basement membrane. These structures are important in tissue cohesion and may determine the developmental path of adjacent cells and delineate developing organs.

Fos and *myc*, two proto-oncogenes regulated by growth factors, have been found in specific regions of the early embryo. The distribution of *myc* expression in the placenta is consistent with a role for this proto-oncogene in proliferation.[80] The proto-oncogene *fos* is expressed in early development in cells that later become the placenta.[79] In subsequent developmental stages, the highest levels of *fos* are expressed in cells of mesenchymal origin.[79] Consistent with these findings are two reports of high levels of *fos* mRNA in developing bone and mesenchymal tissues of early mouse and human fetuses.[80,81]

B. GROWTH-FACTOR-INDUCED SECRETED PROTEINS

PDGF, EGF, and FGF are relatively nonspecific growth factors with respect to target cell types. However, the ability of cells to coordinate their behavior is greatly enhanced if the signal sent by a particular cell conveys a message that includes information about the sender. For example, although a single mitogenic agent stimulates the proliferation of many diverse types of lymphoid cell, various cell types respond differently by secreting different lymphokines. In a similar way, the "broad action" growth factors, such as PDGF, EGF, and FGF, may initiate the production of different signals in disparate cell types. This has been shown for the induction of secreted proteins by EGF and FGF in a number of cell lines; the pattern of secreted proteins produced in response to growth factors was shown to be cell line specific.[2,23] Cells of a similar histiotype but derived from different species have similar secretion profiles and produce proteins of the same molecular weight in response to a given growth factor stimulus. Some cells respond differently depending on their density or their degree of quiescence. Each cell type produces the same secreted proteins in response to EGF, FGF, or PDGF. In contrast, TGF-β has different effects from these growth factors on protein secretion patterns. Insulin and the IGFs generally do not change the cellular secretion pattern. Thus, proteins secreted in response to a mitogenic signal may reflect the cell source, but not necessarily the precise nature of the original signal.

Increases in the secreted levels of proteins in response to growth factors are accompanied by increases in the levels of the corresponding mRNAs. Many growth-factor-induced secreted proteins are components of the extracellular matrix, proteases, or protease inhibitors that act on the extracellular matrix. However, some growth-factor-induced secreted proteins appear to be signals that coordinate development. An example is mitogen-regulated protein (MRP), also known as proliferin (PLF) or MRP/PLF.

MRP/PLF is a heavily glycosylated protein with a median M_r of 34,000. It is synthesized and secreted by Swiss 3T3 cells in culture when they are stimulated by growth factors.[217] After the discovery of MRP in Swiss 3T3 cells, the mRNA for this protein was independently cloned from Balb/c 3T3 cells and named proliferin (PLF) in recognition of its 32% amino acid identity with prolactin.[218] PLF has been shown to be the same protein as MRP.[232,236] MRP/PLF is produced in response to growth factors by Swiss 3T3 cells[217] and by one clone of BALB/c 3T3 cells[203] but not by another.[221] It is also secreted constitutively by BNL cells, a BALB/c normal liver-derived cell line.[233,235]

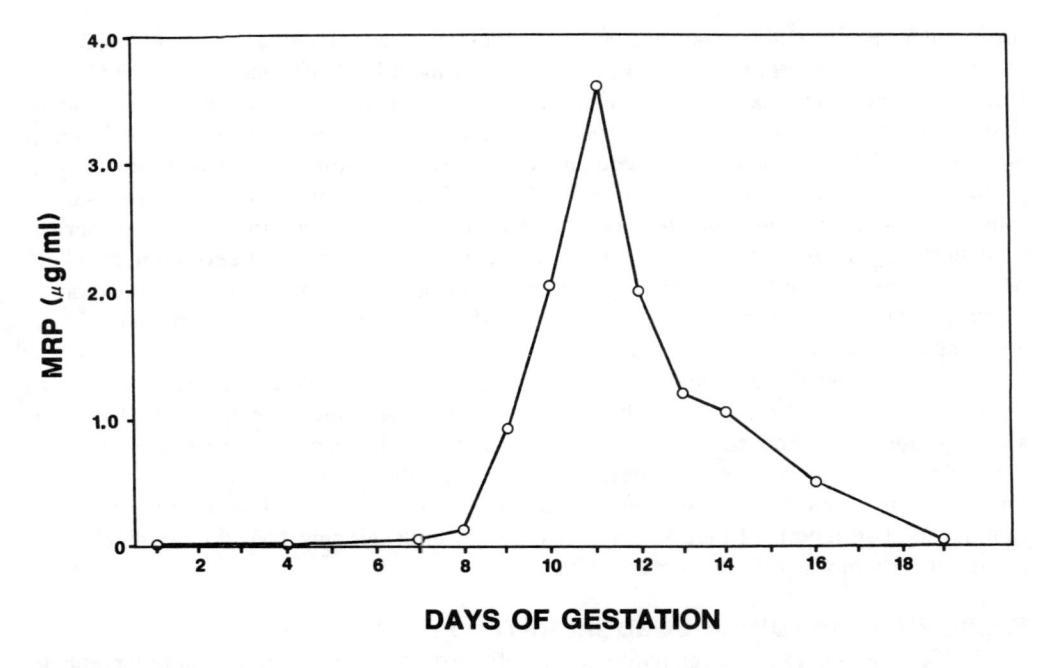

FIGURE 1. MRP levels in the maternal circulation were measured in blood plasma of mice at various gestational stages. The day of the plug was designated Day 0. MRP levels were measured by radioimmunoassay. These data were provided by A. M. D. Mubaidin.

Of the tissues surveyed, MRP/PLF mRNA was found only in the placenta.[231] In this study, the mRNA level in the placenta was found to peak on Day 10 of murine gestation and then slowly decline over the next 6 d, at which time it reached a stable minimum.

Using immunofluorescent staining, MRP/PLF was localized with trophoblast cells of the fetal placenta between Days 9 and 13 of development.[232,236] Between Days 13 and 14, there was a sudden drop in the amount of MRP/PLF in the placenta. When paraffin-embedded placental tissue sections were stained using an avidin-biotin-peroxidase stain, the MRP/PLF was found specifically in the Golgi region of the placental trophoblastic giant cells,[232] suggesting that the protein was synthesized and secreted by the giant cells. The fraction of placental giant cells that stained for MRP/PLF was compared and analyzed as a function of gestational age, and consistent with previous results, the percent of placental giant cells containing MRP/PLF declined between Days 13 and 14 of gestation. MRP/PLF is also secreted into the maternal circulation in mid-gestation (Figure 1). The level of MRP/PLF in the plasma rises rapidly between Days 8 and 11, peaks on Day 10 to 11, and drops rapidly thereafter.[232]

It is well documented that growth factors induce MRP/PLF in 3T3 cells by increasing the amount of MRP/PLF mRNA available for translation.[217,218,235-237] This induction is at least partly due to an increase in the rate of transcription of MRP/PLF mRNA.[237] Either another protein must be synthesized before the MRP/PLF mRNA level increases, or a protein that turns over rapidly is required for induction of the MRP/PLF gene. This is evident because the production of MRP/PLF and its mRNA[237] is inhibited by pretreatment with cycloheximide.

TGF-α is believed to be the fetal form of EGF and to play an important role in fetal development. Although it has only 32% sequence identity with EGF, TGF-α interacts with the EGF receptor and is functionally equivalent to EGF. EGF stimulates the production of MRP/PLF by 3T3 cells.[217] In the rat, TGF-α mRNA has been found concentrated in the maternal decidua.[130] The giant cells that contain the most MRP/PLF are located in the basal zone, adjacent to the maternal decidua basalis. Both MRP/PLF and TGF-α appear in the placenta in midgestation.[130,232,233,243] It is likely that MRP/PLF gene expression is regulated in the giant cells by the TGF-α produced in the decidua.

The rate of decline in the amount of MRP/PLF in the placenta and in the maternal circulation is very rapid and appears to be regulated by three mechanisms. The first mechanism is at the level of MRP/PLF mRNA which begins to decline after Day 10 of gestation[231] and may be regulated by TGF-β.[244]

A second means of decreasing MRP/PLF is at the posttranslational level. The secreted level of MRP is modulated by crinophagy or a related phenomenon.[221,233] Crinophagy is the fusion of secretory vesicles and lysosomes, with subsequent degradation of the vesicle contents. The level of the related protein, prolactin, in the pituitary is also modulated by crinophagy.[238] Degradation of newly synthesized MRP/PLF may represent a developmentally regulated process that rapidly reduces the amount of MRP/PLF secreted by the giant cells of the placenta.

The decline of MRP/PLF in the fetus and the maternal circulation also may be achieved by its removal from the circulation via binding to the mannose-6-phosphate receptor.[239] The level of expression of the mannose-6-phosphate receptor in the fetal and maternal liver increases in late development. During the same developmental period, the level of expression of MRP/PLF in the placenta and in the maternal circulation rapidly declines. The peak level of expression of the receptor in maternal and fetal liver occurs on Day 16.[239] By this time, the amount of MRP/PLF that can be detected in the placenta or in the mother is at a minimum (Figure 1). Thus, it appears that the distinct developmental profile of MRP/PLF is a consequence of multiple levels of regulation, including both positive and negative control at the translational level and negative control at the posttranslational and postsecretional levels.

TGF-β decreases the level of MRP/PLF[244] and MRP/PLF mRNA[297] in 3T3 cells in culture. TGF-β is found in the placenta[245] and therefore may be a developmental regulator of MRP/PLF production. It will be of interest to determine whether the level of TGF-β in the placenta increases as a function of gestational age. The sensitivity of 3T3 cells to inhibition of MRP/PLF synthesis by TGF-β increases when the cells are quiescent and at high density. The MRP/PLF mRNA level in the placenta may also be regulated negatively by TGF-β. This could occur by means of changes in the level of TGF-β with fetal age, changes in the sensitivity of the cells to inhibition by TGF-β, or both.

A model for the regulation of MRP/PLF mRNA levels in the placental giant cells throughout development is shown in Figure 2. The mRNA is believed to increase in response to TGF-α secreted by the decidua basalis. The decrease in MRP mRNA levels may be in response to TGF-β, which is found in the fetal placenta. The MRP/PLF secreted by the giant cells is believed to bind to as yet unidentified cells in fetal and maternal tissue. The function of MRP/PLF may be to stimulate proliferation of cells that express MRP/PLF receptors. In analogy with the demonstrated function of other members of the protein family (growth hormone, prolactin, and placental lactogen) which have been shown to induce IGF production in adult or fetal cells,[113,246,247] MRP/PLF is proposed to be an inducer of IGF in specific fetal and maternal cells. Thus, MRP/PLF may coordinate the growth of the fetus with that of the placenta.

MRP/PLF production and proliferation are closely coupled both *in vitro* and *in vivo*. First, MRP/PLF is a major protein secreted by proliferating, growth-factor-stimulated 3T3 cells.[217] Second, the period of maximum production of MRP/PLF *in vivo* corresponds with the period of placental growth (Figure 3). Third, the expression of MRP/PLF in 3T3 cells is associated with the property of immortality.[235] Fourth, the ability of tumor promoters to stimulate the expression of MRP/PLF in quiescent cells is lost in mutants that are resistant to the mitogenic activity of these tumor promoters.[242] The results of these studies suggest that some of the factors necessary for the expression of MRP/PLF are the same as the factors necessary for the initiation of DNA synthesis.

In light of the results discussed above, MRP/PLF has been suggested to function as a

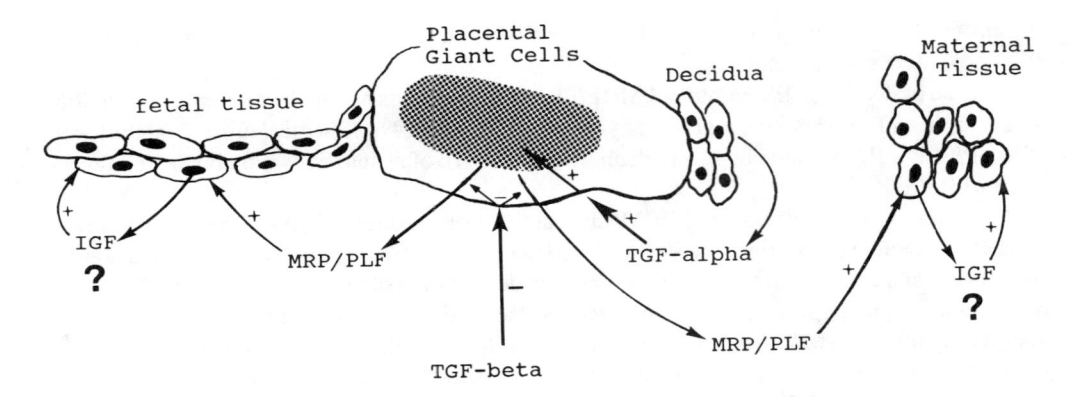

FIGURE 2. A model proposing the means whereby MRP levels in the placenta and in the maternal circulation are regulated. Evidence from studies of cells in culture supports the proposal that TGF-α and TGF-β regulate the level of MRP mRNA in the giant cells. MRP is proposed to act on other fetal and maternal cells to induce IGFs.

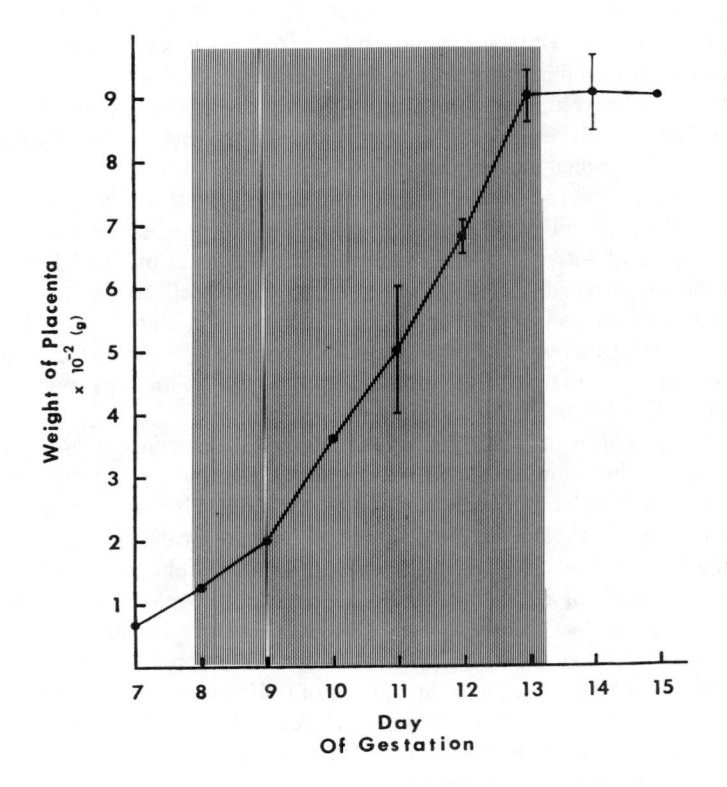

FIGURE 3. The relationship between presence of MRP in the placental giant cells and placental growth. The gray area defines the period during gestation when MRP is detected in the placental basal zone giant cells using an avidin-biotin-peroxidase staining procedure. The day of the plug was designated Day 0. The filled circles represent an average value for the weights of two or three placentas at each stage of gestation. The error bars represent the standard deviations of the averages. For the points with no obvious error, the length of the bar was shorter than the diameter of the circle.

proliferation factor,[217,218] probably in the paracrine or endocrine mode, rather than as an autocrine growth factor. The evidence against an autocrine role for MRP/PLF includes: (1) anti-MRP/PLF antibody does not inhibit growth-factor-induced DNA synthesis in Swiss 3T3 cells, and (2) a Balb/c 3T3 cell line that does not make MRP/PLF[221] responds to growth factors with increased DNA synthesis to the same extent as a Balb/c 3T3 cell line that produces MRP/PLF.[235] Although a growth factor role for MRP/PLF was suggested upon its original discovery[217] and after its sequence identity with prolactin was recognized,[218,235] no evidence has been produced so far to support or refute this hypothesis. A recent report showed that when MRP/PLF was expressed in C3H10T^{1}/2-derived myoblasts then their differentiation was inhibited.[299] Thus, although MRP/PLF could be a regulator of placental proliferation, another explanation for the observations just discussed is that the production of MRP/PLF is regulated by either proliferation of the placenta or by the growth factor(s) that regulate placental or embryonic cell proliferation.

As well as MRP/PLF, the placenta makes several other prolactin-like proteins, including the placental lactogens, which have been discussed in a recent review,[248] and proliferin-related protein (PRP).[249] Two placental lactogens that have been identified in the mouse are placental lactogen I (PL-I) and placental lactogen II (PL-II). PL-I appears in midgestation, whereas PL-II appears later in gestation.[250] Proliferin-related protein mRNA also appears in midgestation, peaking 2 d later (Day 12) than MRP/PLF mRNA.[249] Although the developmental profiles of the appearance of MRP/PLF and PL-I in the maternal blood are similar, the two molecules are distinguishable by their molecular size,[217,251] antigenic determinants,[252] and activities.[239,251] It is not yet known whether placental lactogen or PRP genes are regulated by growth factors as is the MRP/PLF gene. However, the production of human placental lactogen by primary human placental cell cultures was reported to be stimulated by EGF.[253] Similarly, expression of the prolactin gene has been shown to be stimulated by both EGF and FGF.[254,255]

Many functions have been attributed to the placental lactogens, including the regulation of mammary gland secretory activity, maternal intermediary metabolism, fetal IGF production, and luteal progesterone production.[248] The effects of the placental lactogens are presumed to be mediated by their interactions with growth hormone and prolactin receptors. In support of this mechanism, specific receptors for ovine placental lactogen have been found in fetal ovine liver.[256] Preliminary evidence shows that at very high concentrations, MRP/PLF stimulates the proliferation of NB2 cells[298] this indicates that MRP/PLF interacts very weakly with prolactin receptors. A strong possibility is that MRP and the placental lactogens are a set of related proteins that coordinate the proliferation and differentiation of the fetoplacental unit with the proliferation and differentiation of particular cells in maternal organs and tissues.

VI. TRANSFORMING GROWTH FACTOR TYPE BETA

A. MULTIFUNCTIONAL ROLE IN PROLIFERATION AND DIFFERENTIATION

TGF-β is a 25 kDa homodimer that was independently discovered at least three times: as an inhibitor of anchorage-dependent growth cell in monolayer culture,[257] as a stimulator of anchorage-independent cell growth in agar,[258,259] and as a cartilage-inducing factor.[260] There are two species of TGF-β.[35] Type 1 TGF-β is identical to cartilage-inducing factor A.[261] The BSC-1 growth inhibitor[40] and cartilage-inducing factor B[262] are Type 2 TGF-β's. The cDNAs for both TGF-β Types 1 and 2 have been cloned and sequenced.[36-40] From the deduced amino acid sequences, it appears that TGF-β is produced as a larger precursor from which the 112 amino acid monomer is cleaved.[36-40] There is also evidence for three different TGF-β receptors on the surfaces of cells in culture.[263]

TGF-β stimulates the proliferation of cells of mesenchymal origin and inhibits the proliferation of other cells. The inhibitory effects of TGF-β are more apparent in cells of epithelial origin. However, the proliferation of fibroblasts,[264] endothelial cells,[265-267] and T-lymphocytes[268] is inhibited by TGF-β. The ability to both inhibit and stimulate proliferation is not unique to TGF-β. EGF[200] and interferon[269] inhibit the proliferation of some cells while stimulating the proliferation of others. EGF and interferon initiate cellular responses that are predominantly stimulatory or inhibitory. For TGF-β, there are many examples of both types of cellular response.

Although the ability of TGF-β to act on various cell types in opposite ways is puzzling, it may not reflect fundamental differences of the initial cellular responses to the growth factor. Different types of responses may result from the initiation of a number of metabolic changes in all responsive cells, some of which lead to proliferation, while others lead to proliferative arrest. Depending on the individual responses in specific cell types and the environment of the cells, the balance of these effects could lead to cellular proliferation or inhibition of proliferation. This hypothesis is discussed in more detail below with specific reference to the effects of TGF-β on the expression of genes encoding proteases, protease inhibitors, extracellular matrix components, and *sis*.

B. REGULATION OF EXTRACELLULAR MATRIX ACCUMULATION

A major function of TGF-β is to regulate the expression of genes whose products comprise the extracellular matrix. Genes coding for proteases and protease inhibitors that affect turnover of the extracellular matrix are regulated by TGF-β and by other growth factors. The ultimate effect of TGF-β is usually to conserve the extracellular matrix. For example, TGF-β decreases the expression of genes encoding proteases that degrade protein components of the extracellular matrix. The proteases regulated in this way by TGF-β include plasminogen activator,[227,270] cathepsin L[244,271] (also known as MEP), transin,[225] and collagenase.[224] By contrast, in adult human skin fibroblasts, TGF-β increases the activity of plasminogen activator.[228] However, it is not clear whether the increase in plasminogen activator activity in these TGF-β-treated cells is a direct effect of the growth factor or a secondary effect of the induction of the *sis* gene (and thus production of the B-chain of PDGF) by TGF-β. TGF-β also increases the expression of protease inhibitors, including the gene for the inhibitor of plasminogen activator, PAI-1[226,227,229,230] (also known as IIP48), and the gene coding for the tissue inhibitor of metalloproteases (TIMP), a collagenase inhibitor.[224] Intradermal treatment of rats with TGF-β resulted in fibrosis *in vivo*,[215] increased accumulation of total protein, collagen, and DNA in wound chambers implanted *in vivo*,[272] and accelerated the rate of healing of excisional wounds.[273] TGF-β also stimulated the accumulation of collagen and fibronectin by cells in culture.[213,215] In addition, TGF-β increased the expression of the fibronectin gene in several types of normal and transformed cells,[213] increased the production of collagen by rat lung fibroblasts,[274] and stimulated the synthesis of proteoglycans in human arterial smooth muscle cells.[275] Thus, it has been proposed that by increasing the expression of genes that code for protease inhibitors and extracellular matrix components and by decreasing the expression of genes coding for proteases, TGF-b acts as an extracellular matrix-sparing agent that stimulates the formation of extracellular matrix.[30,229] This concept is consistent with its occurrence in high concentrations in bone,[260] where the extracellular matrix is a major component.

The ability of TGF-β to shift the balance towards increased accumulation of the extracellular matrix has far reaching developmental significance. In addition to proliferating, cells in a growing tissue must produce sufficient extracellular matrix to form a cohesive tissue with the appropriate morphology. One function of the proteases induced by growth factors, such as EGF, FGF, and PDGF, is to degrade the extracellular matrix to allow cellular migration or to provide space for the daughter cells. However, to maintain the integrity of the developing embryo, destruction of the extracellular matrix must be restricted. TGF-β may be the major embryonic factor that limits the destructive effects of growth-factor-induced proteases in developing tissues.

It has been proposed that a major function of TGF-β in embryonic development is to modulate extracellular matrix accumulation for growth and to limit proteolytic digestion during rapid cell proliferation. In this model, the role of growth factors such as PDGF, EGF, or FGF is to stimulate cell proliferation and migration, while the role of TGF-β is to spare the extracellular matrix and to allow proliferating cells to form a solid tissue. The model is consistent with the observation that in promoting growth in soft agar, TGF-β often acts synergistically with EGF, FGF, TGF-α, or PDGF to stimulate anchorage-independent growth.[258,276] These data fit the hypothesis that TGF-β stimulates the accumulation of the extracellular matrix, while EGF, TGF-α, PDGF, or FGF stimulate cellular proliferation.

C. OPPOSITE ACTIONS OF TGF-BETA

TGF-β stimulates DNA synthesis and proliferation in AKR-2B cells. The increased rate of DNA synthesis is preceded by inhibition of DNA synthesis that is delayed for several hours compared with the proliferative response of these cells to EGF.[264] The delayed effect of TGF-β on DNA synthesis is probably due to the formation of the *sis* gene product, produced in response to TGF-β and which acts as the immediate autocrine stimulator of AKR-2B cell proliferation.[114] Although TGF-β may induce all cells to express *sis*, the *sis* product can only act as an autocrine growth factor in those cells that possess the PDGF receptor. PDGF receptors are expressed only on cells of mesenchymal origin. Thus, the hypothesis that TGF-β stimulates cell proliferation indirectly by inducing the expression of *sis* is consistent with the observation that TGF-β generally stimulates proliferation in cells of mesenchymal origin, whereas it inhibits anchorage-dependent proliferation of many cells of epithelial origin. These results suggest that, in the embryo, selective proliferative responses to TGF-β may be determined by local expression of the PDGF receptor.

The ability of TGF-β to inhibit proliferation could also result from effects on the expression of genes encoding proteases and protease inhibitors. Proliferation of many cells is stimulated by proteases.[280-284] Cells can dampen the mitogenic effect of proteases by secreting protease inhibitors.[285] The negative effects of TGF-β on the proliferation of certain cells in monolayer culture may occur because of the decreased extracellular proteolytic activity induced by the action of the growth factor.

The opposite actions of TGF-β on proliferation could be explained by its ability to simultaneously induce *sis* and to regulate genes encoding proteases and protease inhibitors. The action of TGF-β on cellular proliferation may derive from the balance of positive (*sis* expression) and negative (decreased proteolysis) signals for DNA synthesis. The accumulation of extracellular matrix without cell proliferation might occur during developmental periods when specific membranes separating cells and tissues are deposited or during the development of tissues, such as cartilage and bone, in which large amounts of extracellular matrix accumulate. During the formation of new tissues, TGF-β and a proliferative growth factor (PDGF, EGF, or FGF) may be required to coordinate cellular responses. Cells would respond with either increased or decreased proliferation, depending on their state of differentiation and their sensitivities to the action of extracellular proteases and to PDGF. Thus, by regulating the expression of a small set of genes, TGF-β can either stimulate or inhibit proliferation and development. A model to explain the actions of TGF-β is shown schematically in Figure 4.

VII. GROWTH FACTORS, CELL MIGRATION, AND INVASION

Directed movements of cells within the embryo are essential in early mammalian development. At least two types of secreted stimulus could direct the movement of cells during embryogenesis. The first is the secretion of a "track" of a particular extracellular matrix by cells in the specified path. TGF-β may be important in the development of these directional "tracks". In this context, it will be of great interest to determine whether the effect of each form of TGF-β is to produce a different composition of extracellular matrix and thus a different type of "track".

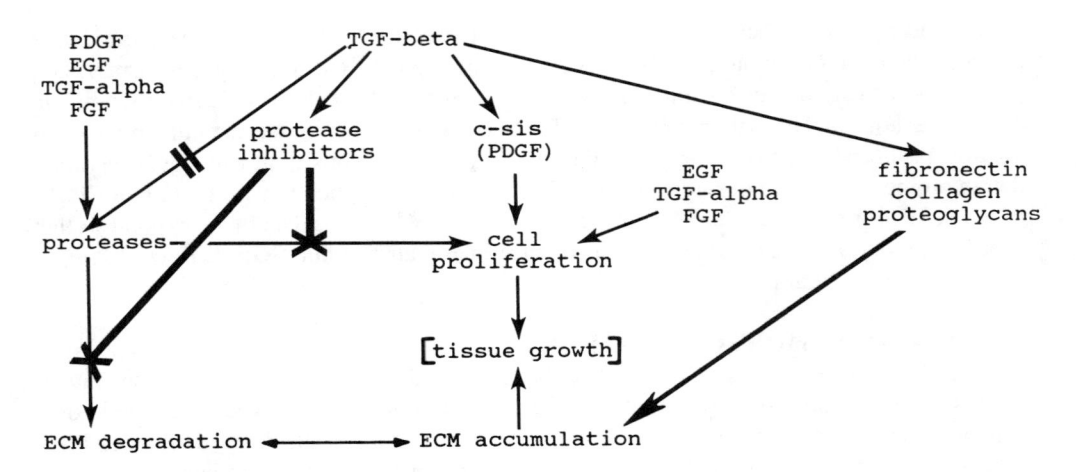

FIGURE 4.   A model to explain the bifunctional action of TGF-β in development. TGF-β has been shown to decrease the levels of secreted proteases and to increase the levels of protease inhibitors, thereby altering the balance of turnover of the extracellular matrix. TGF-β can also induce the PDGF-B chain (c-*sis*). PDGF, EGF, and FGF provide proliferative signals that can act synergistically with the TGF-β-induced accumulation of the extracellular matrix to stimulate tissue growth. TGF-β can also reduce proliferation by lowering extracellular proteolytic action. Whether the cell responds by initiating or ceasing proliferation will depend upon the relative strengths of these signals and the ability of the cell to respond to each signal. Thus, the bifunctional action of TGF-β may derive from the products of a single set of genes that are coordinately regulated by TGF-β.

Cells may also respond differently at the level of induction and repression of particular genes within the set of regulated genes to TGF-β Types 1, 2, and 3. A difference in the actions of different TGF-βs is indicated by the observation that TGF-β Type 2, but not Type 1, stimulates mesodermal morphogenesis in *Xenopus*.

The second type of signal for directed cell movement is a soluble chemotactic factor. A common property of growth factors is their ability to stimulate chemotaxis.[286-291] The signal for directed cell movement may also arise from a combination of soluble and deposited signals. For example, FGF binds heparin very tightly and probably immediately attaches to the extracellular matrix after it is released.

Tissue invasion is another aspect of development that is probably regulated by growth factors. The first major invasive action of the embryo is initiated by the extra-embryonal tissue when the trophoblast cells invade the decidua. In addition to migration, invasion is facilitated by secreted proteases. For example, mouse trophoblast cells secrete plasminogen activator which allows them to invade the uterine wall and allows embryonic implantation.[292-293] A number of growth factors, including EGF, FGF, and PDGF, stimulate the production of secreted proteases. Proteases induced by these growth factors include plasminogen activator,[294-296] cathepsin L.[221,271] (also known as MEP), transin,[225] and collagenase.[222,223] In early to midgestation of the rodent (starting at Day 7 in the mouse and Day 8 in the rat), the deciduomata produces TGF-α mRNA,[130] and TGF-α can be detected in the embryo.[131] It is possible that TGF-α induces the synthesis and secretion of proteases by trophoblast cells. TGF-α may also act as a chemotactic factor for trophoblast invasiveness.

VIII. SUMMARY

The importance of growth factors in early mammalian development is evident. The results of studies of the effects of FGF, PDGF, TGF-α, and TGF-β on cells in culture and in embryonic and fetal development, suggest that these growth factors regulate proliferation, differentiation, migration, and invasion. Individual growth factors probably act in development on particular cell types and in a specific window of time and space. The developmental activity of each growth factor is determined by: (1) the expression of the specific receptor on the target cell, (2) production of the growth factor, (3) degradation and inactivation of the growth factor, and (4) the state of cell differentiation and, therefore, the nature and extent of the response to the growth factor. In addition to stimulating proliferation, differentiation, and migration in specific cells, growth factors also induce a number of secreted proteins. Some of these proteins contribute to or modify the extracellular matrix. Others are probably paracrine messengers that coordinate the developmental program among specific embryonic and fetal cells and between the mother and the fetus.

ACKNOWLEDGMENT

The author thanks the following people for their contributions to this chapter: Richard T. Hamilton, for reviewing the manuscript and for valuable advice in its preparation; Adnan Mubaidin, for providing the data in Figure 1; and Yng-Ju Jang, for providing the data shown in Figure 3. This work was funded in part by grants GM33528 and CA39256 from the National Institutes of Health and by a grant from the Iowa State Biotechnology Council.

REFERENCES

1. **Gospodarowicz, D.,** Epidermal and nerve growth factors in mammalian development, *Annu. Rev. Physiol.,* 43, 251, 1981.
2. **Nilsen-Hamilton, M. and Hamilton, R. T.,** Secreted proteins, intercellular communication and the mitogenic response, *Cell Biol. Int. Rep.,* 6, 815, 1982.
3. **Adamson, E. D.,** Growth factors in development, in *The Biological Basis of Reproductive and Developmental Medicine,* Warshaw, J. B., Ed., Elsevier, New York, 1983, 307.
4. **Jakobovits, A.,** The expression of growth factors and growth factor receptors during mouse embryogenesis, in *Oncogenes and Growth Control,* Kahn, P. and Graf, T., Eds., Springer-Verlag, Berlin, Heidelberg, 1986, 8.
5. **Adamson, E. D.,** Extraembryonic tissues as sources and sinks of humoral factors in development, teratocarcinoma model systems, in *Cellular Endocrinology, Hormonal Control of Embryonic and Cellular Differentiation,* Alan R. Liss, New York, 1986, 159.
6. **Deuel, T. F.,** Polypeptide growth factors, roles in normal and abnormal cell growth, *Annu. Rev. Cell Biol.,* 3, 443, 1987.
7. **Gospodarowicz, D., Neufeld, G., and Schweigerer, L.,** Molecular and biological characterization of fibroblast growth factor, an angiogenic factor which also controls the proliferation and differentiation of mesoderm and neuroectoderm derived cells, *Cell Differ.,* 19, 1, 1986.
8. **Gospodarowicz, D., Ferrara, S., Neufield, G., and Schweigerer, L.,** Structural characterization and biological function of fibroblast growth factor, *Endocrinol. Rev.,* 8, 95, 1987.
9. **Kidder, G. M.,** Intercellular communication during mouse embryogenesis, in *The Mammalian Preimplantation Embryo: Regulation of Growth and Differentiation In Vitro,* Bavister, B. D., Ed., Plenum Press, New York, 1987, 43.
10. **Ohlsson, R. I. and Pfeifer-Ohlsson, S. B.,** Cancer genes, proto-oncogenes, and development, *Exp. Cell Res.,* 173, 1, 1987.
11. **Rizzino, A.,** Defining the roles of growth factors during early mammalian development, in *The Mammalian Preimplantation Embryo,* Bavister, B. D., Ed., Plenum Press, New York, 1987, 151.
12. **Mercola, M. and Stiles, C. D,** Growth factor superfamilies and mammalian embryogenesis, *Development,* 102, 451, 1988.
13. **Gospodarowicz, D., Cheng, J., Lui, G.-M., Fujii, D. K., Baird, A., and Böhlen, P.,** Fibroblast growth factor in the human placenta, *Biochem. Biophys. Res. Commun.,* 128, 554, 1985.
14. **Whittingham, D. G. and Biggers, J. D.,** Fallopian tube and early cleavage in the mouse, *Nature (London),* 213, 942, 1967.

15. **Whittingham, D. G. and Biggers, J. D.,** Development of zygotes in cultured mouse oviducts. I. The effect of varying oviductal conditions, *J. Exp. Zool.,* 169, 391, 1968.
16. **James, R. and Bradshaw, R. A.,** Polypeptide growth factors, *Annu. Rev. Biochem.,* 53, 259, 1984.
17. **Derynck, R.,** Transforming growth factor-alpha, in *Oncogenes and Growth Control,* Kahn, P. and Graf, T., Eds., Springer-Verlag, Berlin, 1986. 58.
18. **Goustin, A. S., Leof, E. B., Shipley, G. D., and Moses, H. L.,** Growth factors and cancer, *Cancer Res.,* 46, 1015, 1986.
19. **Carpenter, G., Goodman, L., and Shaver, L.,** The physiology of epidermal growth factor, in *Oncogenes and Growth Control,* Kahn, P. and Graf, T., Eds., Springer-Verlag, Berlin, Heidelberg, 1986, 65.
20. **Carpenter, G.,** Receptors for epidermal growth factor and other polypeptide mitogens, *Annu. Rev. Biochem.,* 56, 881, 1987.
21. **Lobb, R. R., Harper, J. W., and Fett, J. W.,** Purification of heparin-binding growth factors, *Anal. Biochem.,* 154, 1, 1986.
22. **Deuel, T. F. and Huang, J. S.,** Platelet-derived growth factor: structure, function and roles in normal and transformed cells, *J. Clin. Invest.,* 74, 669, 1984.
23. **Stiles, C. D.,** The biological role of oncogenes — insights from platelet-derived growth factor: Rhoads memorial award lecture, *Cancer Res.,* 45, 5215, 1985.
24. **Heldin, C.-H. and Westermark, B.,** Role of PDGF-like growth factors in autocrine stimulation of growth of normal and transformed cells, in *Oncogenes and Growth Control,* Kahn, P. and Graf, T., Eds., Springer-Verlag, Berlin, 1986, 43.
25. **Heldin, C.-H., Betsholz, C., Claesson-Welsh, L., and Westermark, B.,** Subversion of growth regulatory pathways in malignant transformation, *Biochim. Biophys. Acta,* 907, 219, 1987.
26. **Massagué, J.,** The transforming growth factors, *Trends in Biochem. Sci.,* 10, 237, 1985.
27. **Moses, H. L. and Leof, E. B.,** Transforming growth factor-beta, in *Oncogenes and Growth Control,* Kahn, P. and Graf, T., Eds., Springer-Verlag, Berlin, 1986, 51.
28. **Sporn, M. B., Roberts, A. B., Wakefield, L. M., and Assoian, K.,** Transforming growth factor-beta: biological function and chemical structure, *Science,* 233, 532, 1986.
29. **Massagué, J.,** The TGF-beta family of growth and differentiation factors, *Cell,* 49, 437, 1987.
30. **Sporn, M. B., Roberts, A. B., Wakefield, L. M., and de Crombugghe, B.,** Mini-review: some recent advances in the chemistry and biology of transforming growth factor-beta, *J. Cell Biol.,* 105, 1039, 1987.
31. **Coughlin, M. D. and Collins, M. B.,** Nerve growth factor-independent development of embryonic mouse sympathetic neurons in dissociated cell culture, *Dev. Biol.,* 110, 392, 1985.
32. **Marquardt, H., Hunkapillar, M. W., Hood, L. E., and Todaro, G. J.,** Rat transforming growth factor type 1: structure and relation to epidermal growth factor, *Science,* 223, 1079, 1984.
33. **Gimenez-Gallego, G., Rodkey, C., Bennett, C., Rios-Candelore, M., Disalvo, J., and Thomas, K.,** Brain-derived fibroblast growth factor: complete amino acid sequence and homologies, *Science,* 230, 1385, 1985.
34. **Esch, F., Baird, A., Ling, N., Ueno, N., Hill, F., Denoroy, L., Klepper, R., Gospodarowicz, D., Böhlen, P., and Guilleman, R.,** Primary structure of bovine pituitary basic fibroblast growth factor (FGF) and comparison with amino terminal sequence of bovine brain acidic FGF, *Proc. Natl. Acad. Sci., U.S.A.,* 82, 6507, 1985.
35. **Cheifetz, S., Weatherbee, J. A., Tsang, M. L.-S., Anderson, J. K., Mole, J. E., Lucas, R., and Massagué, J.,** The TGF-beta system, a complex of cross-reactive ligands and receptors, *Cell,* 48, 409, 1987.
36. **Derynck, R., Jarrett, J. A., Chen, E. Y., Eaton, D. H., Bell, J. R., Assoian, R. K., Roberts, A. B., Sporn, M. B., and Goeddel, D. V.,** Human transforming growth factor-beta cDNA sequence and expression in normal and transformed cells, *Nature (London),* 316, 701, 1985.
37. **Derynck, R., Jarrett, J. A., Chen, E. Y., and Goeddel, D. V.,** The murine transforming growth factor-beta precursor, *J. Biol. Chem.,* 261, 4377, 1986.
38. **Sharples, K., Plowman, G. D., Roe, T. M., Twardzik, D. R., and Purchio, A. F.,** Cloning and sequence analysis of simian transforming growth factor-beta cDNA, *DNA,* 6, 239, 1987.
39. **Marquardt, H., Lioubin, M. N., and Ikeda, T.,** Complete amino acid sequence of human transforming growth factor type beta 2, *J. Biol. Chem.,* 262, 12127, 1987.
40. **Hanks, S. K., Armour, R., Baldwin, J. H., Maldonado, F., Spiess, J., Holley, R. W.,** Amino acid sequence of the BSC-1 cell growth inhibitor (polyergin) deduced from the nucleotide sequence of the cDNA, *Proc. Natl Acad. Sci. U.S.A.,* 85, 79, 1988.
41. **Goldbard, S. B. and Warner, C. W.,** Genes affect the timing of early mouse embryo development, *Biol. Reprod.,* 27, 419, 1982.
42. **Warner, C. M., Gollnick, S. O., and Goldbard, S. B.,** Linkage of the preimplantation-embryo-development (Ped) gene to the mouse major histocompatibility complex (MHC), *Biol. Reprod.,* 36, 606, 1987.
43. **Dickson, C. and Peters, G.,** Potential oncogene product related to growth factors, *Nature (London),* 326, 833, 1987.

44. **Jakobovits, A., Shackleford, G. M., Varmus, H. E., and Martin, G. R.,** Two proto-oncogenes implicated in mammary carcinogenesis, int-1 and int-2, are independently regulated during mouse development, *Proc. Natl. Acad. Sci. U.S.A.,* 83, 7806, 1986.

45. **Martin, G. R.,** Teratocarcinomas and mammalian embryogenesis, *Science,* 209, 768, 1980.

46. **Kimelman, D. and Kirschner, M.,** Synergistic induction of mesoderm by FGF and TGF-beta and the identification of an mRNA coding for FGF in the early *Xenopus* embryo, *Cell,* 51, 869, 1987.

47. **Slack, J. M. W., Darlington, B. G., Heath, J. K., and Godsave, S. F.,** Mesoderm induction in early *Xenopus* embryos by heparin binding growth factors, *Nature (London),* 326, 197, 1987.

48. **Smith, J. C.,** A mesoderm-inducing factor is produced by a *Xenopus* cell line, *Development,* 99, 3, 1987.

49. **Rosa, F., Roberts, A. B., Danielpour, D., Dart, L. L., Sporn, M. B., and Dawid, I. B.,** Mesoderm induction in amphibians: the role of TGF-beta2-like factors, *Science,* 239, 1988.

50. **Martin, G. R.,** Isolation of a pluripotent cell line from early mouse embryos cultured in medium conditioned by teratocarcinoma stem cells, *Proc. Natl. Acad. Sci. U.S.A.,* 78, 7634, 1981.

51. **Heath, J. K. and Isacke, C. M.,** PC13 Embryonal carcinoma derived growth factor, *EMBO J.,* 3, 2957, 1984.

52. **Jakobovits, A., Schwab, M., Bishop, J. M., and Martin, G. R.,** Expression of N-*myc* in teratocarcinoma stem cells and mouse embryos, *Nature (London),* 318, 188, 1985.

53. **Rees, A. R., Adamson, E. D., and Graham, C. F.,** Epidermal growth factor receptors increase during the differentiation of embryonal carcinoma cells, *Nature (London),* 281, 309, 1979.

54. **Wharton, K. A., Johansen, K. M., Xu, T., and Artavanis-Tsakonas, S.,** Nucleotide sequence from the neurogenic locus Notch implies a gene product that shares homology with proteins containing EGF-like repeats, *Cell,* 43, 567, 1985.

55. **Vässin, H., Bremer, K. A., Knust, E., and Campos-Ortega, J. A.,** The neurogenic gene *Delta* of *Drosophila melanogaster* is expressed in neurogenic territories and encodes a putative transmembrane protein with EGF-like repeats, *EMBO J.,* 6, 3431, 1987.

56. **Knust, E., Dietrich, U., Tepass, U., Bremer, K. A., Weigel, D., Vässin, H., and Campos-Ortega, J. A.,** EGF homologous sequences encoded in the genome of *Drosophila melanogaster,* and their relation to neurogenic genes, *EMBO J.,* 6, 761, 1987.

57. **Greenwald, I.,** *Lin-12,* a nematode homeotic gene, is homologous to a set of mammalian proteins that includes epidermal growth factor, *Cell,* 43, 583, 1985.

58. **Hursh , D. A., Andrews, M. E., and Raff, R. A.,** A sea urchin gene encodes a polypeptide homologous to epidermal growth factor, *Science,* 237, 1487, 1987.

59. **Padgett, R. W., St. Johnston, R. D., and Gelbart, W. M.,** A transcript from a *Drosophila* pattern gene predicts a protein homologous to the transforming growth factor-beta family, *Nature (London),* 325, 81, 1987.

60. **Weeks, D. L. and Melton, D. A.,** A maternal mRNA localized in the vegetal hemisphere in *Xenopus* eggs codes for a growth factor related to TGF-beta, *Cell,* 51, 861, 1987.

61. **Born, J., Davids, M., and Tiedemann, H.,** Affinity chromatography of embryonic inducing factors on heparin-Sepharose, *Cell. Diff.,* 21, 131, 1987.

62. **Gospodarowicz, D., Cheng, J., Lui, G. M., Baird, A., and Böhlen, P.,** Isolation of brain fibroblast growth factor by heparin-Sepharose affinity chromatography: identity with pituitary fibroblast growth factor, *Proc. Natl. Acad. Sci. U.S.A.,* 81, 6963, 1984.

63. **Devare, S. G., Reddy, E. P., Law, J. D., Robbins, K. C., and Aaronson, S. A.,** Nucleotide sequence of the simian sarcoma virus genome: demonstration that its acquired cellular sequences encode the transforming gene product p28^{sis}, *Proc. Natl. Acad. Sci. U.S.A.,* 80, 731, 1983.

64. **Doolittle, R. F., Hunkapillar, M. W., Hood, L. E., Devare, S. G., Robbins, K. C., Aaronson, S. A., and Antoniades, H. N.,** Simian sarcoma virus *onc* gene, v-*sis,* is derived from the gene (or genes) encoding a PDGF, *Science,* 221, 275, 1983.

65. **Waterfield, M. D., Scrace, G. T., Whittle, N., Stroobant, P., Johnson, A., Wasteson, A., Westermark, B., Heldin, C. H., Huang, J. S., and Deuel, T.,** Platelet-derived growth factor is structurally related to the putative transforming protein p28^{sis} of simian sarcoma virus, *Nature (London),* 304, 35, 1983.

66. **Hunter, T. and Cooper, J. A.,** Protein tyrosine kinases, *Annu. Rev. Biochem.,* 54, 897, 1985.

67. **Garrett, C. T.,** Critical review — oncogenes, *Clin. Chim. Acta,* 156, 1, 1986.

68. **Bravo, R. and Müller, R.,** Involvement of proto-oncogenes in growth control, the induction of c-fos and c-myc by growth factors, in *Oncogenes and Growth Control,* Kahn, P. and Graf, T., Eds., Springer-Verlag, Berlin, 1986, 252.

69. **Ferrari, S. and Baserga, R.,** Oncogenes and cell cycle genes, *BioEssays,* 7, 9, 1987.

70. **Müller, R., Slamon, D. J., Adamson, E. D., Tremblay, J. M., Muller, D., Cline, M. J., and Verma, I. M.,** Transcription of c-onc genes c-ras^{Ki} and c-fms during mouse development, *Mol. Cell. Biol.* 3, 1062, 1983.

71. **Müller, R., Bravo, R., Burckhardt, J., and Curran, T.,** Induction of the c-*fos* gene and protein by growth factors precedes activation of c-*myc, Nature (London),* 312, 716, 1984.

72. **Deschamps, J., Mitchell, R. L., Meijlink, F., Kruijer, W., Schubert, D., and Verma, I. M.,** Proto-oncogene fos is expressed during development, differentiation, and growth, in *Molecular Biology of Development: Cold Spring Harbor Symposium in Quantitiative Biology*, Vol. 50, Cuddihy, J. and Brown, D., Eds., Cold Spring Harbor Laboratories, Cold Spring Harbor, NY, 1985, 733.

73. **Jakobovits, A., Banda, M. J., and Martin, G. R.,** Embryonal carcinoma-derived growth factors: specific growth promoting and differentiation inhibiting activities, in *Growth Factors and Transformation, Cancer Cells*, Vol. 3, Feramisco, J., Ozanne, B., and Stiles, C., Eds., Cold Spring Harbor Laboratories, Cold Spring Harbor, NY, 1985, 393.

74. **Kirsch, I. R., Bertness, V., Silver, J., and Hollis, G. F.,** Regulated expression of the c-myb and c-myc oncogenes during erythroid differentiation, *J. Cell. Biochem.*, 32, 11, 1986.

75. **Zimmerman, K. A., Yancopoulos, G. D., Collum, R. G., Smith, R. K., Kohl, N. E., Denis, K. A., Nau, M. M., Witte, O. N., Toran-Allerand, D., Gee, C. E., Minna, J. D., and Alt, F. W.,** Differential expression of myc family genes during murine development, *Nature (London)*, 319, 780, 1986.

76. **Goldman, D. S., Kiessling, A. A., Millette, C. F., and Cooper, G. M.,** Expression of c-mos RNA in germ cells of male and female mice, *Proc. Natl. Acad. Sci. U.S.A.*, 84, 4509, 1987.

77. **Leibovitch, M.-P., Leibovitch, S. A., Hillion, J., Guillier, M., Schmitz, A., and Harel, J.,** Possible role of c-fos, c-N-ras and c-mos proto-oncogenes in muscular development, *Exp. Cell Res.*, 170, 80, 1987.

78. **Leon, J., Guerrero, I., and Pellicer, A.,** Differential expression of the ras gene family in mice, *Mol. Cell. Biol.*, 7, 1535, 1987.

79. **Müller, R., Slamon, D. J., Tremblay, J. M., Cline, M. J., and Verma, I. M.,** Differential expression of cellular oncogenes during pre- and postnatal development of the mouse, *Nature (London)*, 299, 640, 1982.

80. **Pfeifer-Ohlsson, S., Goustin, A. S., Rydnect, J., Wahlstrom, T., Bjersing, L., Stehelin, D., and Ohlsson, R.,** Spatial and temporal pattern of cellular myc oncogene expression in developing human placenta: implications for embryonic cell proliferation, *Cell*, 38, 585, 1984.

81. **Dony, C. and Gruss, P.,** Proto-oncogene c-fos expression in growth regions of fetal bone and mesodermal web tissue, *Nature (London)*, 328, 711, 1987.

82. **Villares, R. and Cabrera, C. V.,** The *achaete-scute* gene complex of *D. melanogaster:* conserved domains in the subset of genes required for neurogenesis and their homology to *myc*, *Cell*, 50, 415, 1987.

83. **Rijsewijk, F., Schuermann, M., Wagenaar, E., Parren, P., Weigel, D., and Nusse, R.,** The *Drosophila* homologue of the mouse mammary oncogene *int-1* is identical to the segment polarity gene *wingless*, *Cell*, 50, 649, 1987.

84. **Steward, R.,** *Dorsal*, an embryonic polarity gene in *Drosophila*, is homologous to the vertebrate proto-oncogene, c-*rel*, *Science*, 238, 692, 1987.

85. **Tsonis, P. A.,** Oncogenes take a place in pattern formation, *Trends Biochem. Sci.*, 13, 4, 1988.

86. **Loewenstein, W. R.,** Junctional intercellular communication: the cell-to-cell membrane channel, *Physiol. Rev.*, 61, 829, 1981.

87. **Evans, W. H.,** Gap junctions: towards a molecular structure, *BioEssays*, 8, 3, 1988.

88. **Loewenstein, W. R.,** Junctional intercellular communication and the control of growth, *Biochim. Biophys. Acta*, 560, 1, 1979.

89. **Green, C. R.,** Evidence mounts for the role of gap junctions during development, *BioEssays*, 8, 7, 1988.

90. **Goodall, H. and Johnson, M. H.,** Use of carboxyfluorescein diacetate to study formation of permeable channels between mouse blastomeres, *Nature (London)*, 295, 524, 1982.

91. **Lo, C. W. and Gilula, N. B.,** Gap junctional communication in the preimplantation mouse embryo, *Cell*, 18, 399, 1979.

92. **Lee, S., Gilula, N. B., and Warner, A. E.,** Gap junctional communication and compaction during preimplantation stages of mouse development, *Cell*, 51, 851, 1987.

93. **Buehr, M., McLaren, A., and Warner, A.,** Reduced gap junctional communication is associated with the lethal condition characteristic of DDK mouse eggs fertilized by foreign sperm, *Development*, 101, 449, 1987.

94. **Warner, A. E., Guthrie, S. C., and Gilula, N. B.,** Antibodies to gap-junctional protein selectively disrupt junctional communication in the early amphibian embryo, *Nature (London)*, 311, 127, 1984.

95. **Serras, F. and van den Biggelaar, J. A. M.,** Is the mosaic embryo also a mosaic of communication compartments?, *Dev. Biol.*, 120, 132, 1987.

96. **Murray, A. W. and Fitzgerald, D. J.,** Tumor promoters inhibit metabolic cooperation in cocultures of epidermal and 3T3 cells, *Biochem. Biophys. Res. Commun.*, 91, 395, 1979.

97. **Yotti, L. P., Chang, C. C., and Trotsko, J. E.,** Elimination of metabolic cooperation in chinese hamster cells by a tumor promoter, *Science*, 206, 1089, 1979.

98. **Nishizuka, Y.,** The role of protein kinase C in cell surface signal transduction and tumour promotion, *Nature (London)*, 308, 693, 1984.

99. **Bell, R. M.,** Protein kinase C activation by diacylglycerol second messengers, *Cell*, 45, 631, 1986.

100. **Whittenberger, B., Raben, D., Lieberman, M. A., and Glaser, L.,** Inhibition of growth of 3T3 cells by extract of surface membranes, *Proc. Natl. Acad. Sci. U.S.A.,* 75, 5457, 1978.

101. **Cassel, D., Wood, P. M., Bunge, R. P., and Glaser, L.,** Mitogenicity of brain axolemma membranes and soluble factors for dorsal root ganglion schwann cells, *J. Cell. Biochem.,* 18, 433, 1982.

102. **Lieberman, M. A.,** Mitogenic proteins of the 3T3 plasma membrane, *J. Cell. Physiol.,* 114, 73, 1983.

103. **Vale, R. D., Peterson, S. W., Matiuck, N. V., and Fox, C. F.,** Purified plasma membranes inhibit polypeptide growth factor-induced DNA synthesis in subconfluent 3T3 cells, *J. Cell Biol.,* 98, 1129, 1984.

104. **Gray, A., Dull, T. J., and Ullrich, A.,** Nucleotide sequence of epidermal growth factor cDNA predicts a 128,000 molecular weight protein precursor, *Nature (London),* 303, 722, 1983.

105. **Scott, J., Urdea, M., Quiroga, M., Sanchez-Pascador, R., Fong, N., Selby, M., Rutter, W. J., and Bell, G. I.,** Structure of mouse submaxillary messenger RNA encoding epidermal growth factor and seven related proteins, *Science,* 221, 236, 1983.

106. **Derynck, R., Roberts, A. B., Winkler, M. E., Chen, E. Y., and Goeddel, D. V.,** Human transforming growth factor-alpha: precursor structure and expression in *E. coli*, *Cell*, 38, 287, 1984.

107. **Frey, P., Forand, R., Maciag, T., and Shooter, E. M.,** The biosynthetic precursor of epidermal growth factor and the mechanism of its processing, *Proc. Natl Acad. Sci. U.S.A.,* 76, 6294, 1979.

108. **Hoppe, P. and Greenspan, R.,** Local function of the Notch gene for embryonic ectodermal pathway choice in *Drosophila*, *Cell*, 46, 773, 1986.

109. **Technau, G. M. and Campos-Ortega, J. A.,** Cell autonomy of expression of neurogenic genes of *Drosophila melanogaster*, *Proc. Natl. Acad. Sci. U.S.A.,* 84, 4500, 1987.

110. **Hartley, D. A., Xu, T., and Artavanis-Tsakonas, S.,** The embryonic expression of the Notch locus of *Drosophila melanogaster* and the implications of point mutations in the extracellular domain of the predicted protein, *EMBO J.,* 6, 3407, 1987.

111. **Sporn, M. B. and Todaro, G. J.,** Autocrine secretion and malignant transformation of cells, *N. Engl. J. Med.,* 303, 878, 1980.

112. **Kaplan, P. L., Anderson, M., and Ozanne, B.,** Transforming growth factor(s) production enables cells to grow in the absence of serum: an autocrine system, *Proc. Natl. Acad. Sci. U.S.A.,* 79, 485, 1982.

113. **Adams, S. O., Nissley, S. P., Handwerger, S., and Rechler, M. M.,** Developmental patterns of insulin-like growth factor-I and -II synthesis and regulation in rat fibroblasts, *Nature (London),* 302, 150, 1983.

114. **Leof, E. B., Proper, J. A., Goustin, A. S., Shipley, G. D., DiCorleto, P. E., and Moses, H. L.,** Induction of c-sis mRNA and activity similar to platelet-derived growth factor by transforming growth factor beta: a proposed model for indirect mitogenesis involving autocrine activity, *Proc. Natl. Acad. Sci. U.S.A.,* 83, 2453, 1986.

115. **Carpenter, G. and Cohen, S.,** ^{125}I-labeled human epidermal growth factor, *J. Cell Biol.,* 71, 159, 1976.

116. **Fox, C. F., Linsley, P. S., and Wrann, M.,** Receptor remodeling and regulation in the action of epidermal growth factor, *Fed. Proc. Fed. Am. Soc. Exp. Biol.,* 41, 2988, 1982.

117. **Tupper, J. T. and Bodine, P. V.,** Calcium effects on epidermal growth factor receptor-mediated endocytosis in normal and SV40-transformed human fibroblasts, *J. Cell. Physiol.,* 115, 159, 1983.

118. **Korc, M., Matrisian, L. M., and Magun, B. E.,** Cytosolic calcium regulates epidermal growth factor endocytosis in rat pancreas and cultured fibroblasts, *Proc. Natl. Acad. Sci. U.S.A.,* 81, 461, 1984.

119. **Lobb, R. R. and Fett, J. W.,** Purification of two distinct growth factors from bovine neural tissue by heparin affinity chromatography, *Biochemistry,* 23, 6295, 1984.

120. **Lawrence, D. A., Pircher, R., Kryceve-Martinerie, C., and Julien, P.,** Normal embryo fibroblasts release transforming growth factors in a latent form, *J. Cell. Physiol.,* 121, 184, 1984.

121. **Huang, J. S., Huang, S. S., and Deuel, T. F.,** Specific covalent binding of platelet-derived growth factor to human plasma alpha$_2$-macroglobulin, *Proc. Natl. Acad. Sci. U.S.A.,* 81, 342, 1984.

122. **Raines, E. W., Bowen-Pope, D. F., and Ross, R.,** Plasma binding proteins for platelet-derived growth factor that inhibit its binding to cell surface receptors, *Proc. Natl. Acad. Sci. U.S.A.,* 81, 3424, 1984.

123. **O'Conner-McCourt, M. D. and Wakefield, L. M.,** Latent TGF-beta in serum: a specific complex with alpha$_2$-macroglobulin, *J. Biol. Chem.* 262, 14090, 1987.

124. **Partanen, A.-M., Ekblom, P., and Thesleff, I.,** Epidermal growth factor inhibits morphogenesis and cell differentiation in cultured mouse embryonic teeth, *Dev. Biol.,* 111, 84, 1985.

125. **Goldin, G. V. and Opperman, L. A.,** Induction of supernumerary tracheal buds and the stimulation of DNA synthesis in the embryonic chick lung and trachea by epidermal growth factor, *J. Embryol. Exp. Morphol.,* 60, 235, 1980.

126. **Green, M. R., Phil, D., and Couchman, J. R.,** Distribution of epidermal growth factor receptors in rat tissues during embryonic skin development, hair formation, and the adult hair growth cycle, *J. Invest. Derm.,* 83, 118, 1984.

127. **Kemp, R. B. and Hinchliffe, J. R., Eds.,** *Progress in Clinical and Biological Research: Matrices and Cell Differentiation,* Vol. 151, Alan R. Liss, New York, 1984.

128. **Caplan, A. I.,** The extracellular matrix is instructive, *BioEssays,* 5, 129, 1986.

129. **Shing, Y., Folkman, J., Sullivan, R., Butterfield, C., Murray, J., and Klagsbrun, M.,** Heparin affinity purification of a tumor-derived capillary endothelial cell growth factor, *Science,* 223, 1296, 1984.

130. **Han, V. K. M., Hunter, E. S., III, Pratt, R. M., Zendegui, J. G., and Lee, D. C.,** Expression of rat transforming growth factor alpha mRNA during development occurs predominantly in the maternal decidua, *Mol. Cell. Biol.* 7, 2335, 1987.

131. **Twardzik, D. R.,** Differential expression of transforming growth factor-alpha during prenatal development of the mouse, *Cancer Res.,* 45, 5413, 1985.

132. **Catterton, W. Z., Escobedo, M. B., Sexson, W. R., Gray, M. E., Sundell, H. W., and Stahlman, M. T.,** Effect of epidermal growth factor on lung maturation in fetal rabbits, *Pediatr. Res.,* 13, 104, 1979.

133. **Hassell, J. R.,** The development of rat palatal shelves in vitro, *Dev. Biol.,* 45, 90, 1975.

134. **Grove, R. I. and Pratt, R. M.,** Growth and differentiation of embryonic mouse palatal epithelial cells in primary culture, *Exp. Cell Res.,* 148, 195, 1983.

135. **Cohen, S.,** Isolation of a mouse submaxillary gland protein accelerating incisor eruption and eyelid opening in the new-born animal, *J. Biol. Chem.,* 237, 1555, 1962.

136. **Partanen, A.-M. and Thesleff, I.,** Localization and quantitation of ^{125}I-epidermal growth factor binding in mouse embryonic tooth and other embryonic tissues at different developmental stages, *Dev. Biol.* 120, 186, 1987.

137. **Thesleff, I., Ekblom, P., and Keski-Oja, J.,** Inhibition of morphogenesis and stimulation of vascular proliferation in embryonic tooth cultures by a sarcoma growth factor preparation, *Cancer Res.,* 43, 5902, 1983.

138. **Thorburn, G. D., Waters, M. J., Young, I. R., Dolling, M., Buntine, D., and Hopkins, P. S.,** Epidermal growth factor: a critical factor in fetal maturation? The fetus and independent life (Ciba Foundation Symposium 86), Pitman, London, 1981, 172.

139. **Tsao, M. C., Walthall, B. J., and Ham, R. G.,** Clonal growth of normal human epidermal keratinocytes in a defined medium, *J. Cell. Physiol.,* 110, 219, 1982.

140. **Mulvihill. S. J., Stone, M. M., Fonkalsrud, E. W., and Debas, H. T.,** Trophic effect of amniotic fluid on fetal gastrointestinal development, *J. Surg. Res.* 40, 291, 1986.

141. **Adamson, E. D. and Meek, J.,** The ontogeny of epidermal growth factor receptors during mouse development, *Dev. Biol.,* 103, 62, 1984.

142. **Nexø, E., Hollenberg, M. D., Figueroa, A., and Pratt, R. M.,** Detection of epidermal growth factor-urogastrone and its receptor during fetal mouse development, *Proc. Natl. Acad. Sci. U.S.A.,* 77, 2782, 1980.

143. **Adamson, E. D., Deller, M. J., and Warshaw, J. B.,** Functional EGF receptors are present on mouse embryo tissues, *Nature (London),* 291, 656, 1981.

144. **Chegini, N. and Rao, Ch. V.,** Epidermal growth factor binding to human amnion, chorion, decidua and placenta from mid- and term pregnancy. Quantitative light microscopic autoradiographic studies, *J. Clin. Endocrinol. Metab.,* 61, 529, 1984.

145. **Schejter, E. D., Segal, D., Glazer, L., and Shilo, B.-Z.,** Alternative 5¢ exons and tissue specific expression of the *Drosophila* EGF receptor homolog transcripts, *Cell,* 46, 1091, 1986.

146. **Kammermeyer, K. L. and Wadsworth, S. C.,** Expression of *Drosophila* epidermal growth factor receptor homologue in mitotic cell populations, *Development,* 100, 201, 1987.

147. **Cohen, S., Carpenter, G., and King, L.,** EGF-receptor prokin kinase interactions, *J. Biol. Chem.,* 255, 4834, 1980.

148. **Sasaki, N., Rees-Jones, R. W., Zick, Y., Nissley, S. P., and Rechler, M. M.,** Characterization of insulin-like growth factor I-stimulated tyrosine kinase activity associated with the beta-subunit of type I insulin-like growth factor receptors in rat liver cells, *J. Biol. Chem.,* 260, 9793, 1985.

149. **Daniel, T. O., Tremble, P. M., Frackelton, A. R., Jr., and Williams, L. T.,** Purification of the platelet-derived growth factor receptor by using an anti-phosphotyrosine antibody, *Proc. Natl Acad. Sci. U.S.A.,* 82, 2684, 1985.

150. **Coughlin, S. R., Barr, P. J., Cousens, L. S., Fretto, L. J., and Williams, L. T.,** Acidic and basic fibroblast growth factors stimulate tyrosine kinase activity *in vivo, J. Biol. Chem.,* 263, 988, 1988.

151. **Neufeld, G. and Gospodarowicz, D.,** The identification and partial characterization of the fibroblast growth factor receptor of baby hamster kidney cells, *J. Biol. Chem.,* 260, 13860, 1985.

152. **Fanger, B. O., Wakefield, L. M., and Sporn, M. B.,** Structure and properties of the cellular receptor for TGF-beta, *Biochemistry,* 25, 3083, 1986.

153. **Hortsch, M., Schlessinger, J., Gootwine, E., and Webb, C. G.,** Appearance of functional EGF receptor kinase during rodent embryogenesis, *EMBO J.,* 2, 1937, 1983.

154. **Clark, A. J. L., Ishii, S., Richert, N., Merlino, G., and Pastan, I.,** Epidermal growth factor regulates the expression of its own receptor, *Proc. Natl. Acad. Sci. U.S.A.,* 82, 8374, 1985.

155. **Kudlow, J. E., Cheung, C.-Y. M., and Bjorge, J. D.,** Epidermal growth factor stimulates the synthesis of its own receptor in a human breast cancer cell line, *J. Biol. Chem.*, 261, 4134, 1986.

156. **Earp, H. S., Austin, K. S., Blaisdell, J., Rubin, R. A., Nelson, K. G., Lee, L. W., and Grisham, J. W.,** Epidermal growth factor (EGF) stimulates EGF receptor synthesis, *J. Biol. Chem.*, 261, 4777, 1986.

157. **Fox, C. F., Wrann, M., Linsley, P., and Vale, R.,** Hormone-induced modification of EGF receptor proteolysis in the induction of EGF action, *J. Supramolec. Struct.*, 12, 517, 1979.

158. **Wrann, M., Fox, C. F., and Ross, R.,** Modulation of epidermal growth factor receptors on 3T3 cells by platelet-derived growth factor, *Science*, 210, 1363, 1980.

159. **Wharton, W., Leof, E., Pledger, W. J., and O'Keefe, E. J.,** Modulation of the epidermal growth factor receptor by platelet-derived growth factor and choleragen: effects on mitogenesis, *Proc. Natl. Acad. Sci. U.S.A.*, 79, 5567, 1982.

160. **Bowen-Pope, D. E., Dicorleto, P. E., and Ross, R.,** Interactions between the receptors for platelet-derived growth factor and epidermal growth factor, *J. Cell Biol.*, 96, 679, 1983.

161. **Assoian, R. K., Frolik, C. A., Roberts, A. B., Miller, D. M., and Sporn, M. B.,** Transforming growth factor-beta controls receptor levels for epidermal growth factor in NRK fibroblasts, *Cell*, 36, 35, 1984.

162. **Massagué, J.,** Transforming growth factor-beta modulates the high affinity receptors for epidermal growth factor and transforming growth factor-alpha, *J. Cell Biol.*, 100, 1508, 1985.

163. **Mukku, V. R. and Stancel, G. M.,** Regulation of epidermal growth factor receptor by estrogen, *J. Biol. Chem.*, 260, 9820, 1985.

164. **Baker, J. B. and Cunningham, D. D.,** Glucocorticoid-mediated alteration in growth factor binding and action, analysis of the binding change, *J. Supramolec. Struct.*, 9, 69, 1978.

165. **Ivanovic, V. and Weinstein, I. B.,** Glucocorticoids and benzo(a)pyrene have opposing effects on EGF receptor binding, *Nature (London)*, 293, 404, 1981.

166. **Sadiq, H. F. and Devaskar, U. P.,** Glucocorticoids increase pulmonary epidermal growth factor receptors in female and male fetal rabbit, *Biochem. Biophys. Res. Commun.*, 119, 408, 1984.

167. **Armelin, H.,** Pituitary extracts and steroid hormones in the control of 3T3 cell growth, *Proc. Natl. Acad. Sci. U.S.A.*, 70, 2702, 1973.

168. **Gospodarowicz, D.,** Localization of a fibroblast growth factor and its effect alone, and with hydrocortisone on 3T3 cell growth, *Nature (London)*, 249, 123, 1974.

169. **Chen, W. S., Lazar, C. S., Poenie, M., Tsien, R. Y., Gill, G. N., and Rosenfeld, M. G.,** Requirement for intrinsic protein tyrosine kinase in the immediate and late actions of the EGF receptor, *Nature (London)*, 328, 820, 1987.

170. **Glenney, J. R., Chen, W. S., Lazar, S. S., Walton, G. M., Zokas, L. M., Rosenfeld, M. G., and Gill, G. M.,** Ligand induced endocytosis of the EGF receptor i s blocked by mutational inactivation and by microinjection of anti-phosphotyrosine antibodies, *Cell*, 52, 675, 1988.

171. **Honeggar, A. M., Dull, T. J., Felder, S., Van Obberghen, E., Bellot, F., Szapary, D., Schmidt, A., Ullrich, A., and Schlessinger, J.,** Point mutation at the ATP binding site of EGF receptor abolishes protein-tyrosine kinase activity and alters cellular routing, *Cell*, 51, 199, 1987.

172. **Prywes, R., Livneh, E., Ullrich, A., and Schlessinger, J.,** Mutations in the cytoplasmic domain of the EGF receptor affect EGF binding and receptor internalization, *EMBO J.*, 5, 2179, 1986.

173. **Livneh, E., Prywes, R., Kashles, O., Reiss, N., Sasson, I., Mory, Y., Ullrich, A., and Schlessinger, J.,** Reconstitution of human epidermal growth factor receptors and its deletion mutants in cultured hamster cells, *J. Biol. Chem.*, 261, 12490, 1986.

174. **Livneh, E., Reiss, N., Berebt, E., Ullrich, A., and Schlessinger, J.,** An insertional mutant of the EGF growth factor receptor allows dissection of diverse receptor functions, *EMBO J.*, 6, 2669, 1987.

175. **Weller, A., Meek, J., and Adamson, E. D.,** Preparation and properties of monoclonal and polyclonal antibodies to mouse epidermal growth factor (EGF) receptors: evidence for cryptic EGF receptors in embryonal carcinoma cells, *Development*, 100, 351, 1987.

176. **Logsdon, C. D. and Williams, J. A.,** Intracellular Ca^{2+} and phorbol esters synergistically inhibit internalization of epidermal growth factor in pancreatic acini, *Biochem. J.*, 223, 893, 1984.

177. **Hunter, T., Ling, N., and Cooper, J. A.,** Protein kinase C phosphorylation of the EGF receptor at a threonine residue close to the cytoplasmic face of the plasma membrane, *Nature (London)*, 311, 480, 1984.

178. **Lin, C. R., Chen, W. S., Lazar, C. S., Carpenter, C. D., Gill, G. N., Evans, R. M., and Rosenfeld, M. G.,** Protein kinase C phosphorylation at thr 654 of the unoccupied EGF receptor and EGF binding regulate functional receptor loss by independent mechanisms, *Cell*, 44, 839, 1986.

179. **Davis, R. J. and Czech, M. P.,** Platelet-derived growth factor mimics phorbol diester action on epidermal growth factor receptor phosphorylation at threonine-654, *Proc. Natl. Acad. Sci. U.S.A.*, 82, 4080, 1985.

180. **Davis, R. J. and Czech, M. P.,** Inhibition of the apparent affinity of the epidermal growth factor receptor caused by phorbol diesters correlates with phosphorylation of threonine-654 but not other sites on the receptor, *Biochem. J.*, 233, 435, 1986.

181. **Habenicht, A. J. R., Glomset, J. A., King, W. C., Nist, C., Mitchell, C. D., and Ross, R.,** Early changes in phosphatidylinositol and arachadonic acid metabolism in quiescent Swiss 3T3 cells stimulated to divide by platelet-derived growth factor, *J. Biol. Chem.,* 256, 12329, 1981.

182. **Rozengurt, E., Rodriquez-Pena, M., and Smith, K. A.,** Phorbol esters, phospholipase C and growth factors rapidly stimulate the phosphorylation of a M_r 80,000 protein in intact Swiss 3T3 cells, *Proc. Natl. Acad. Sci. U.S.A.,* 80, 7244, 1983.

183. **Tsuda, T., Kaibuchi, K., Kawahara, Y., Fukuzaki, H., and Takai, Y.,** Induction of protein kinase C and calcium ion mobilization by fibroblast growth factor in Swiss 3T3 cells, *FEBS Lett.,* 187, 43, 1985.

184. **Kaibuchi, K., Tsuda, T., Kikuchi, A., Tanimoto, T., Yamashita, T., and Takai, Y.,** Possible involvement of protein kinase C and calcium ion in growth factor-induced expression of c-myc oncogene in swiss 3T3 fibroblasts, *J. Biol. Chem.,* 261, 1187, 1986.

185. **Ignotz, R. A. and Massagué, J.,** Type beta transforming growth factor controls the adipogenic differentiation of 3T3 fibroblasts, *Proc. Natl. Acad. Sci. U.S.A.,* 82, 8530, 1985.

186. **Florini, J. R., Roberts, A. B., Ewton, D. Z., Falen, S. L., Flanders, K. C., and Sporn, M. B.,** Transforming growth factor-beta, a very potent inhibitor of myoblast differentiation, identical to the differentiation inhibitor secreted by buffalo rat liver cells, *J. Biol. Chem.,* 261, 16509, 1986.

187. **Olson, E. N., Sternberg, E., Hu, J. S., Spizz, G., and Wilcox, C.,** Regulation of myogenic differentiation by type beta transforming growth factor, *J. Cell Biol.* 103, 1799, 1986.

188. **Massagué, J., Cheifetz, S., Endo, T., and Nadal-Ginard, B.,** Type beta transforming growth factor is an inhibitor of myogenic differentiation, *Proc. Natl. Acad. Sci. U.S.A.,* 83, 8206, 1986.

189. **Majack, R. A.,** Beta-type transforming growth factor specifies organizational behavior in vascular smooth muscle cell cultures, *J. Cell Biol.,* 105, 465, 1987.

190. **Kurokowa, M., Lynch, K., and Podolsky, D. K.,** Effects of growth factors on an intestinal epithelial cell line: transforming growth factor inhibits proliferation and stimulates differentiation, *Biochem. Biophys. Res. Commun.,* 142, 775, 1987.

191. **Gospodarowicz, D., Weseman, J., Moran, J., and Lindstrom, J. S.,** Effect of fibroblast growth factor on the division and fusion of bovine myoblasts, *J. Cell Biol.,* 70, 395, 1976.

192. **Bradshaw, R. A.,** Nerve growth factor, *Annu. Rev. Biochem.,* 47, 191, 1978.

193. **Calissano, P., Cattaneo, A., Biocca, S., Aloe, L., Mercanti, D., and Levi-Montalcini, R.,** The nerve growth factor, established findings and controversial aspects, *Exp. Cell Res.,* 154, 1, 1984.

194. **Barlow, P., Owen, D. A. J., and Graham, C.,** DNA synthesis in the preimplantation mouse embryo, *J. Embryol. Exp. Morphol.,* 27, 431, 1972.

195. **Fine, L. G., Holley, R. W., Nasri, H., and Badie-Dezfooly, B.,** BSC-1 growth inhibitor transforms a mitogenic stimulus into a hypertrophic stimulus for renal proximal tubular cells; relationship to sodium ion/hydrogen ion antiport activity, *Proc. Natl. Acad. Sci. U.S.A.,* 82, 6163, 1985.

196. **Clemens, M. J. and McNurlan, M. A.,** Regulation of cell proliferation and differentiation by interferons, *Biochem. J.,* 226, 345, 1985.

197. **Brooks, R. F.,** Continuous protein synthesis is required to maintain the probability of entry into S phase, *Cell,* 12, 311, 1977.

198. **Zullo, J., Hall, D., Rollins, B., and Stiles, C. D.,** Oncogenes and interferons: genetic targets for animal cell growth factors, in *Oncogenes and Growth Control,* Kahn, P. and Graf, T., Eds., Springer-Verlag, Berlin, 1986, 259.

199. **Saunders, J. W., Jr.,** Death in embryonic systems, *Science,* 154, 604, 1966.

200. **Kawamoto, T., Mendelsohn, J., Le, A., Sato, G. H., Lazar, C. S., and Gill, G. N.,** Relation of epidermal growth factor receptor concentration to the growth of human epidermoid carcinoma A431 cells, *J. Biol. Chem.,* 259, 7761, 1984.

201. **Denhardt, D. T., Edwards, D. R., and Parfett, C. L. J.,** Gene expression during the mammalian cell cycle, *Biochim. Biophys. Acta,* 865, 83, 1986.

202. **Cochran, B. H., Reffel, A. C., and Stiles, C. D.,** Molecular cloning of gene sequences regulated by PDGF, *Cell,* 33, 939, 1983.

203. **Linzer, D. I. H. and Nathans, D.,** Growth-related changes in specific mRNAs of cultured mouse cells, *Proc. Natl. Acad. Sci. U.S.A.,* 80, 4271, 1983.

204. **Lau, L. F. and Nathans, D.,** Identification of a set of genes expressed during the G_0/G_1 transition of cultured mouse cells, *EMBO J.,* 4, 3145, 1985.

205. **Kelly, K., Cochran, B. H., Stiles, C. D., and Leder, P.,** Cell-specific regulation of the c-*myc* gene by lymphocyte mitogens and PDGF, *Cell,* 35, 603, 1983.

206. **Greenberg, M. E. and Ziff, E. B.,** Stimulation of 3T3 cells induces transcription of the c-*fos* proto-oncogene, *Nature (London),* 311, 433, 1984.

207. **Kruijer, W., Cooper, J. A., Hunter, T., and Verma, I. M.,** PDGF induces rapid but transient expression of the c-*fos* gene and protein, *Nature (London),* 312, 711, 1984.

208. **Chavrier, P., Zerial, M., Lemaire, P., Almendral, J., Bravo, R., and Charnay, P.,** A gene ecoding a protein with zinc fingers is activated during G_0G_1 transition in cultured cells, *EMBO J.*, 7, 29, 1988.

209. **Soprano, K. J., Soprano, D. R., Cosenza, S., and Owen, T.,** Expression of growth-associated genes in various tissues of the fetal and adult rat, *Mol. Cell. Biochem.*, 75, 61, 1987.

210. **Elder, P. K., Schmidt, L. J., Ono, T., and Getz, M. J.,** Specific stimulation of actin gene transcription by epidermal growth factor and cycloheximide, *Proc. Natl. Acad. Sci., U.S.A.*, 81, 7476, 1984.

211. **Leof, E. B., Proper, J. A., Getz, M. J., and Moses, H. L.,** Transforming growth factor type beta regulation of actin mRNA, *J. Cell. Physiol.*, 127, 83, 1986.

212. **Chen, L. B., Gudor, R. C., Sun, T.-T., Chen, A. B., and Mosesson, M. W.,** Control of a cell surface major glycoprotein by epidermal growth factor, *Science,* 197, 776, 1977.

213. **Ignotz, R. A. and Massagué, J.,** Transforming growth factor-beta stimulates the expression of fibronectin and collagen and their incorporation into the extracellular matrix, *J. Biol. Chem.*, 261, 4337, 1986.

214. **Blatti, S. P., Foster, D. N., Ranganathan, G., Moses, H. L., and Getz, M. J.,** Induction of fibronectin gene transcription and mRNA is a primary response to growth-factor stimulation of AKR-2B cells, *Proc. Natl. Acad. Sci. U.S.A.*, 85, 1119, 1988.

215. **Roberts, A. B., Sporn, M. B., Assoian, R. K., Smith, J. M., Roche, N. S., Wakefield, L. M., Heine, U. I., Liotta, L. A., Falanga, V., Kehrl, J. H., and Fauci, A. S.,** TGF type beta: rapid induction of fibrosis and angiogenesis in vivo and stimulation of collagen formation *in vitro, Proc. Natl. Acad. Sci. U.S.A.*, 83, 4167, 1986.

216. **Penttinen, R. P., Kobayashi, S., and Bornstein, P.,** Transforming growth factor beta increases mRNA for matrix proteins both in the presence and absence of changes in mRNA stability, *Proc. Natl. Acad. Sci. U.S.A.*, 85, 1105, 1988.

217. **Nilsen-Hamilton, M., Shapiro, J. M., Massoglia S. L., and Hamilton, R. T.,** Selective stimulation by mitogens of incorporation of [^{35}S]methionine into a family of proteins released into the medium by 3T3 cells, *Cell*, 20, 19, 1980.

218. **Linzer, D. I. H. and Nathans, D.,** Nucleotide sequence of a growth-related mRNA encoding a member of the prolactin-growth hormone family, *Proc. Natl. Acad. Sci. U.S.A.*, 81, 4255, 1984.

219. **Gottesman, M. M.,** Transformation-dependent secretion of a low molecular weight protein by murine fibroblasts, *Proc. Natl. Acad. Sci. U.S.A.*, 75, 2767, 1978.

220. **Troen, B. R., Ascherman, D., Atlas, D., and Gottesman, M. M.,** Cloning and expression of the gene for the major excreted protein of transformed mouse fibroblasts: a secreted lysosomal protease regulated by transformation, *J. Biol. Chem.*, 263, 254, 1988.

221. **Nilsen-Hamilton, M., Hamilton, R. T., Allen, W. R., and Massoglia, S. L.,** Stimulation of the release of two glycoproteins from mouse 3T3 cells by growth factors and by agents that increase intralysosomal pH, *Biochem. Biophys. Res. Commun.*, 101, 411, 1981.

222. **Chua, C. C., Geiman, D. E., Keller, G. H., and Ladda, R. L.,** Induction of collagenase secretion in human fibroblast cultures by growth promoting factors, *J. Biol. Chem.*, 260, 5213, 1985.

223. **Bauer, E. A., Cooper, T. W., Huang, J. S., Altman, J., and Deuel, T. F.,** Stimulation of *in vitro* human skin collagenase expression by platelet-derived growth factor, *Proc. Natl. Acad. Sci. U.S.A.*, 82, 4132, 1985.

224. **Edwards, D. R., Murphy, G., Reynolds, J. J., Whitham, S. E., Docherty, A. J. P., Angel, P., and Heath, J. K.,** Transforming growth factor beta modulates the expression of collagenase and metalloproteinase inhibitor, *EMBO J.*, 6, 1899, 1987.

225. **Matrisian, L. M., Leroy, P., Ruhlman, C., Gesnel, M.-C., and Breathnach, R.,** Isolation of the oncogene and epidermal growth factor-induced transin gene: complex control in rat fibroblasts, *Mol. Cell Biol.*, 6, 1679, 1986.

226. **Nilsen-Hamilton, M. and Holley, R. W.,** Rapid selective effects by a growth inhibitor and epidermal growth factor on the incorporation of [^{35}S]-methionine into proteins secreted by African green monkey (BSC-1) cells, *Proc. Natl. Acad. Sci. U.S.A.*, 80, 5636, 1983.

227. **Laiho, M., Saksela, O., Andreasen, P. A., and Keski-Oja, J.,** Enhanced production and extracellular deposition of the endothelial-type plasminogen activator in cultured human lung fibroblasts by transforming growth factor-beta, *J. Cell Biol.*, 103, 2403, 1986.

228. **Laiho, M., Saksela, O., and Keski-Oja, J.,** Transforming growth factor beta alters plasminogen activator activity in human skin fibroblasts, *Exp. Cell Res.*, 164, 399, 1986.

229. **Thalacker, F. and Nilsen-Hamilton, M.,** Specific induction of secreted proteins by transforming growth factor-beta and 12-O-tetradecanoylphorbol-13-acetate, *J. Biol. Chem.*, 262, 2283, 1987.

230. **Lund, L. R., Riccio, A., Andreassen, P. A., Nielsen, L. S., Kristensen, P., Laiho, M., Saksela, O., Blasi, F., and Danø, K.,** Transforming growth factor-beta is a strong and fast acting positive regulator of the level of type-1 plasminogen activator inhibitor mRNA in WI-38 human lung fibroblasts, *EMBO J.*, 6, 1281, 1987.

231. **Linzer, D. I. H., Lee, S.-J., Ogren, L., Talamantes, F., and Nathans, D.,** Identification of proliferin mRNA and protein in mouse placenta, *Proc. Natl. Acad. Sci. U.S.A.*, 82, 4356, 1985.

232. **Jang, Y. J., Mubaidin, A. M. D., and Nilsen-Hamilton, M.,** Studies on mitogen-regulated protein, *J. Cell Biol.*, 105, 256a, 1987.

233. **Nilsen-Hamilton, M., Jang, Y.-J., Alvarez-Azaustre, E., and Hamilton, R. T.,** Regulation of the production of a prolactin-like protein (MRP/PLF) in 3T3 cells and in the mouse placenta, *Mol. Cell. Endo.*, 56, 179, 1988.

234. **Nilsen-Hamilton, M. and Hamilton, R. T.,** Detection of proteins induced by growth regulators, *Methods in Enzymol.*, 147, 427, 1987.

235. **Parfett, C. L. J., Hamilton, R. T., Howell, B. W., Edwards, D. R., Nilsen-Hamilton, M., and Denhardt, D. T.,** Characterization of a cDNA clone encoding murine mitogen-regulated protein: regulation of mRNA levels in mortal and immortal cell lines, *Mol. Cell Biol.*, 5, 3289, 1985.

236. **Nilsen-Hamilton, M., Hamilton, R. T., and Alvarez-Azaustre, E. A.,** Relationship between mitogen-regulated protein (MRP) and proliferin (PFL), a member of the prolactin/growth hormone family, *Gene*, 51, 163, 1987.

237. **Linzer, D. I. H. and Mordacq, J. C.,** Transcriptional regulation of proliferin gene expression in response to serum in transfected mouse cells, *EMBO J.*, 6, 2281, 1987.

238. **Smith, R. E. and Farquhar, M. G.,** Lysosome function in the regulation of the secretory process in cells of the anterior pituitary gland, *J. Cell Biol.*, 31, 319, 1966.

239. **Lee, S.-J. and Nathans, D.,** Proliferin secreted by cultured cells binds to mannose-6-phosphate receptors, *J. Biol. Chem.*, 263, 3521, 1988.

240. **Sen-Majumdar, A., Murthy, U., and Das, M.,** A new trophoblast-derived growth factor from human placenta: purification and receptor identification, *Biochemistry*, 25, 627, 1986.

241. **Sen-Majumdar, A., Murthy, U., Chianese, D., and Das, M.,** A specific antibody to a new peptide growth factor from human placenta: immunocytochemical studies on its location and biosynthesis, *Biochemistry*, 25, 634, 1986.

242. **Fienup, V. K., Jeng, M. H., Hamilton, R. T., and Nilsen-Hamilton, M.,** Relation between the regulation of DNA synthesis and the production of two secreted glycoproteins by 12-O-tetradecanoyl-phorbol-13-acetate in 3T3 cells and in phorbol ester nonresponsive 3T3 variants, *J. Cell. Physiol.*, 129, 151, 1986.

243. **Lee, D. C., Rochford, R., Todaro, G. J., and Villarreal, L. P.,** Developmental expression of rat transforming growth factor-alpha mRNA, *Mol. Cell. Biol.*, 5, 3644, 1985.

244. **Chiang, C.-P. and Nilsen-Hamilton, M.,** Opposite and selective effects of epidermal growth factor and human platelet transforming growth factor-beta on the production of secreted proteins by murine 3T3 cells and human fibroblasts, *J. Biol. Chem.*, 261, 10478, 1986.

245. **Frolik, C. A., Dart, L. L., Meyers, C. A., Smith, D. M., and Sporn, M. B.,** Purification and initial characterization of a type beta transforming growth factor from human placenta, *Proc. Natl. Acad. Sci. U.S.A.*, 80, 3676, 1983.

246. **Atkison, P. R., Weidman, E. R., Bhaumick, B., and Bala, R. M.,** Release of somatomedin-like activity by cultured WI-38 human fibroblasts, *Endocrinology*, 106, 2006, 1980.

247. **Clemmons, D. R., Underwood, L. E., and Van Wyk, J. J.,** Hormonal control of immunoreactive somatomedin production by cultured human fibroblasts, *J. Clin. Invest.*, 67, 10, 1981.

248. **Ogren, L. and Talamantes, F.,** Prolactins of pregnancy and their cellular source, *Int. Rev. Cytol.*, 112, 1, 1988.

249. **Linzer, D. I. H. and Nathans, D.,** A new member of the prolactin growth hormone gene family expressed in the mouse placenta, *EMBO J.*, 4, 1419, 1985.

250. **Soares, M. J. and Talamantes, F.,** Gestational effects on placental and serum androgen, progesterone and prolactin-like activity in the mouse, *J. Endocrinol.*, 95, 29, 1982.

251. **Colosi, P., Ogren, L., Thordarson, G., and Talamantes, F.,** Purification and partial characterization of two prolactin-like glycoprotein hormone complexes from the mid-pregnant mouse conceptus, *Endocrinology*, 120, 2500, 1987.

252. **Lee, S.-J. and Nathans, D.,** Secretion of proliferin, *Endocrinology*, 120, 208, 1987.

253. **Lai, W. H. and Guyda, H. J.,** Characterization and regulation of epidermal growth factor receptors in human placental cell cultures, *J. Clin. Endocrinol. Metab.*, 58, 344, 1984.

254. **Evans, G. A., David, D. N., and Rosenfeld, M. G.,** Regulation of prolactin and somatotropin mRNAs by thyroliberin, *Proc. Natl. Acad. Sci. U.S.A.*, 75, 1294, 1978

255. **Murdoch, G. H., Potter, E., Nicolaisen, A. K., Evans, R. M., and Rosenfeld, M. G.,** Epidermal growth factor rapidly stimulates prolactin gene transcription, *Nature (London)*, 300, 192, 1982.

256. **Freemark, M. and Handwerger, S.,** The glycogenic effects of placental lactogen and growth hormone in the ovine fetal liver are mediated through binding to specific fetal ovine placental lactogen receptors, *Endocrinology*, 118, 613, 1986.

257. **Holley, R. W., Armour, R., Baldwin, J. H.,** Density-dependent regulation of growth of BSC-1 cells in cell culture: growth inhibitors formed by the cells, *Proc. Natl. Acad. Sci. U.S.A.*, 75, 1864, 1978.

258. **Roberts, A. B., Anzano, M. A., Lamb, L. C., Smith, J. M., and Sporn, M. B.,** New class of transforming growth factors potentiated by epidermal growth factor; isolation from nonneoplastic tissues, *Proc. Natl. Acad. Sci. U.S.A.,* 78, 5339, 1981.

259. **Moses, H. L., Branum, E. L., Proper, J. A., and Robinson, R. A.,** Transforming growth factor production by chemically transformed cells, *Cancer Res.,* 41, 2842, 1981.

260. **Seyedin, S. M., Thomas, T. C., Thompson, A. Y., Rosen, D. M., and Piez, K. A.,** Purification and characterization of two cartilage-inducing factors from bovine demineralized bone, *Proc. Natl. Acad. Sci. U.S.A.,* 82, 2267, 1985.

261. **Seyedin, S. M., Thompson, A. Y., Bentz, H., Rosen, D. M., McPherson, J. M., Conti, A., Siegel, N. R., Galluppi, G. R., and Piez, K. A.,** Cartilage-inducing factor-A, *J. Biol. Chem.,* 261, 5693, 1986.

262. **Seyedin, S. M., Segarini, P. R., Rosen, D. M., Thompson, A. Y., Bentz, H., Graycar, J.,** Cartilage-inducing factor-B is a unique protein structurally and functionally related to transforming growth factor-beta, *J. Biol. Chem.,* 262, 1946, 1987.

263. **Cheifetz, S., Like, B., and Massagué, J.,** Cellular distribution of type I and type II receptors for transforming growth factor-beta, *J. Biol. Chem.,* 261, 9972, 1986.

264. **Shipley, G. D., Tucker, R. F., and Moses, H. L.,** Type beta transforming growth factor/growth inhibitor stimulates entry of monolayer cultures of AKR-2B cells in S phase after a prolonged prereplicative interval, *Proc. Natl. Acad. Sci. U.S.A.,* 82, 4147, 1985.

265. **Baird, A. and Durkin, T.,** Inhibition of endothelial cell proliferation by type-beta transforming growth factor: interactions with acidic and basic fibroblast growth factors, *Biochem. Biophys. Res. Commun.,* 138, 476, 1986.

266. **Fràter-Schöder, M., Müller, G., Birchmeier, W., and Böhlen, P.,** Transforming growth factor beta inhibits endothelial cell proliferation, *Biochem. Biophys. Res. Commun.,* 137, 295, 1986.

267. **Heimark, R. L., Twardzik, D. R., and Schwartz, S. M.,** Inhibition of endothelial regeneration by type-beta transforming growth factor from platelets, *Science,* 233, 1078, 1986.

268. **Kehrl, J. H., Roberts, A. B., Wakefield, L. M., Jakowlew, S., Sporn, M. B., and Fauci, A. S.,** Transforming growth factor beta is an important immunomodulatory protein for human B-lymphocytes, *J. Immunol.,* 137, 3855, 1986.

269. **Matheson, D. S., Green, B., and Tan, Y. H.,** Human interferons alpha and beta inhibit T-cell-dependent and stimulate T-cell-independent mitogenesis and natural cytotoxicity: relationship to chromosome 21, *Cell. Immunol.,* 65, 366, 1981.

270. **Saksela, O., Moscatelli, D., and Rifkin, D. B.,** The opposing effects of basic fibroblast growth factor and transforming growth factor beta on the regulation of plasminogen activator activity in capillary endothelial cells, *J. Cell Biol.,* 105, 957, 1987.

271. **Denhardt, D. T., Hamilton, R. T., Parfett, C. L. J., Edwards, D. R., Saint Pierre, R., Waterhouse, P., and Nilsen-Hamilton, M.,** Close relationship of the major excreted protein of transformed murine fibroblasts to thiol-dependent cathepsins, *Cancer Res.,* 26, 163, 1986.

272. **Sporn, M. B., Roberts, A. B., Shull, J. H., Smith, J. M., and Ward, J. M.,** Polypeptide TGFs isolated from bovine sources and used for wound healing *in vivo, Science,* 219, 1329, 1983.

273. **Mustoe, T. A., Pierce, G. F., Thomason, A., Gramates, P., Sporn, M. B., and Deuel, T. F.,** Accelerated healing of incisional wounds in rats induced by transforming growth factor-beta, *Science,* 237, 1333, 1987.

274. **Fine, A. and Goldstein, R. H.,** The effect of transforming growth factor-beta on cell proliferation and collagen formation by lung fibroblasts, *J. Biol. Chem.,* 262, 3897, 1987.

275. **Chen, J.-K., Hoshi, H., and McKeehan, W. L.,** TGF type beta specifically stimulates synthesis of proteoglycan in human adult arterial smooth muscle cells, *Proc. Natl. Acad. Sci. U.S.A.,* 84, 5287, 1987.

276. **Anzano, M. A., Roberts, A. B., Meyers, C. A., Komoriya, A., Lamb, C., Smith, J. M., and Sporn, M. B.,** Synergistic interaction of two classes of TGFs from murine sarcoma cells, *Cancer Res.,* 42, 4776, 1982.

277. **Holley, R. W., Böhlen, P., Fava, R., Baldwin, J. H., Kleeman, G., and Armour, R.,** Purification of kidney epithelial cell growth inhibitors, *Proc. Natl. Acad. Sci. U.S.A.,* 77, 5989, 1980.

278. **Holley, R. W., Armour, R., Baldwin, J. H., and Greenfield, S.,** Preparation and properties of a growth inhibitor produced by kidney epithelial cells, *Cell Biol. Int. Rep.,* 7, 525, 1983.

279. **Tucker, R. F., Shipley, G. D., Moses, H. L., and Holley, R. W.,** Growth inhibitor from BSC-1 cells closely related to platelet type beta transforming growth factor, *Science,* 226, 705, 1984.

280. **Sefton, B. M. and Rubin, H.,** Release from density-dependent growth inhibition by proteolytic enzymes, *Nature (London),* 227, 843, 1970.

281. **Chen, L. B. and Buchanan, J. M.,** Mitogenic activity of blood components. I. Thrombin and prothrombin, *Proc. Natl. Acad. Sci. U.S.A.,* 72, 131, 1975.

282. **Moonan, G., Grau-Wagemans, M.-P., Selak, I., Lefebvre, Ph. P., Rogister, B., Vassalli, J. D., and Belin, D.,** Plasminogen activator is a mitogen for astrocytes in developing cerebellum, *Dev. Brain Res.,* 20, 41, 1985.

283. **Bar-Shavit, R., Kahn, A. J., Mann, K. G., and Wilner, G. D.,** Identification of a thrombin sequence with growth factor activity on macrophages, *Proc. Natl. Acad. Sci. U.S.A.,* 83, 976, 1986.

284. **Glenn, K. C., Carney, D. H., Fenton, J. W., II, and Cunningham, D. D.,** Thrombin active site regions required for fibroblast receptor binding and initiation of cell division, *J. Biol. Chem.,* 255, 6609, 1980.

285. **Low, D. A., Scott, R. W., Baker, J. B., and Cunningham, D. D.,** Cells regulate their mitogenic response to thrombin through release of protease nexin, *Nature (London),* 298, 476, 1982.

286. **Westermark, B. and Blomquist, E.,** Stimulation of fibroblast migration by epidermal growth factor, *Cell Biol. Int. Rep.,* 4, 649, 1980.

287. **Blay, J. and Brown, K. D.,** Epidermal growth factor promotes the chemotactic migration of cultured rat intestinal epithelial cells, *J. Cell. Physiol.,* 124, 107, 1985.

288. **Boyle, M. D. P., Lawman, M. J. P., Gee, A. P., and Young, M.,** Nerve growth factor, a chemotactic factor for polymorphonuclear leukocytes *in vivo, J. Immunol.,* 134, 564, 1985.

289. **Grotendorst, G. R., Seppa, H. E. J., Kleinman, H. K., and Martin, G. R.,** Attachment of smooth muscle cells to collagen and their migration toward platelet-derived growth factor, *Proc. Natl. Acad. Sci. U.S.A.,* 78, 3669, 1981.

290. **McCauslan, B. R., Bender, V., Reilly, W., and Moss, B. A.,** New functions of epidermal growth factor: stimulation of capillary endothelial cell migration and matrix dependent proliferation, *Cell Biol. Int. Rep.,* 9, 175, 1985.

291. **Postlethwaite, A. E., Keski-Oja, J., Moses, H. L., and Kang, A. H.,** Stimulation of the chemotactic migration of human fibroblasts by transforming growth factor type beta, *J. Exp. Med.,* 165, 251, 1987.

292. **Sherman, M. J.,** Studies on the temporal correlation between secretion of plasminogen activator and stages of early mouse embryogenesis, *Oncodev. Biol. Med.,* 1, 7, 1980.

293. **Martin, O. and Arias, F.,** Plasminogen activator production by trophoblast cells *in vitro:* effect of steroid hormones and protein synthesis inhibitors, *Am. J. Obstet. Gynecol.,* 142, 402, 1982.

294. **Lee, L.-S. and Weinstein, I. B.,** Epidermal growth factor, like phorbol esters, induces plasminogen activator in HeLa cells, *Nature (London),* 274, 696, 1978.

295. **Lin, H.-S. and Gordon, S.,** Secretion of plasminogen activator by marrow-derived mononuclear phagocytes and its enhancement by colony-stimulating factor, *J. Exp. Med.,* 150, 231, 1979.

296. **Mira-y-Lopez, R., Joseph-Silverstein, J., Rifkin, D., and Ossowski, L.,** Identification of a pituitary factor responsible for enhancement of plasminogen activator activity in breast tumor cells, *Proc. Natl. Acad. Sci. U.S.A.,* 83, 7780, 1986.

297. **Chiang, and Nilsen-Hamilton, M.,** Unpublished data.

298. **Ebner, K.,** Personal communication.

299. **Wilder, E. L. and Linzer, D. I. H.,** Participation of multiple factors, including proliferin, in the inhibition of myogenic differentiation, *Mol. Cell. Biol.,* 9, 430, 1989.

Chapter 9

THE EXPRESSION OF ONCOGENES IN MAMMALIAN EMBRYOGENESIS

Ann Anderson Kiessling and Geoffrey M. Cooper

TABLE OF CONTENTS

I. INTRODUCTION

Similarities between embryonic and malignant cells have been described frequently. Continual cell division, the expression of some embryonic proteins in tumor cells, and the undifferentiated phenotypes of embryonic and tumor cells are the bases of the comparisons. For these reasons, genes known to be involved in neoplastic transformation seem likely candidates for expression in embryonic cells. In particular, the role of oncogenes in the regulation of growth and differentiation of cells suggests that their normal cellular counterparts, the proto-oncogenes, might function as regulatory genes during development. The preimplantation period of embryo development is particularly well-suited for studies of defined cell commitment events. Sensitive methodologies now make it feasible to examine proto-oncogene expression before and after the initial cell differentiation events that occur during preimplantation embryogenesis.

Three programs of gene expression are unique to early embryogenesis. The first is the gamete program, operational during meiosis and the initial steps of fertilization. The second is the cleavage program that begins at the zygote or two-cell stage and directs the series of cell divisions without cell commitment that give rise to the mass of totipotent cells that comprise the morula. The third is the program of commitment at the late morula/early blastocyst stage that gives rise to trophoblast and inner cell mass (ICM) cells, marking the first differentiation event of embryogenesis. Thus, within a handful of cell cycles and a similar number of days, differentiated, haploid gemetes give rise to undifferentiated, diploid stem cells that then enter a new program of cell differentiation. This chapter will focus on proto-oncogene expression in cells during peri-fertilization events, including meiosis.

II. ONCOGENES AND PROTO-ONCOGENES

Cellular oncogenes have been identified by three approaches: (1) hybridization with retroviral oncogenes, (2) transformation of transfected cells in culture, and (3) detection of genes that are frequently altered in neoplastic cells by DNA rearrangement or amplification. Together, these approaches have identified over 60 cellular genes which, as activated oncogenes, can induce at least some aspects of neoplastic transformation. The oncogenes which have been characterized so far encode proteins which can be divided into five functional groups (Table 1): growth factors, plasma membrane proteins with tyrosine kinase activity, plasma membrane guanine nucleotide binding proteins, cytoplasmic proteins associated with serine/threonine kinase activity, and nuclear proteins. It is generally thought that these proteins function as signal transducing regulatory molecules, the activity of which can lead to cell proliferation (see a recent review[1]).

Oncogenes are altered versions of normal cellular genes called proto-oncogenes. The alteration of a normal gene to produce an oncogene or oncogene protein involves one or more of at least three different mechanisms. One mechanism is a change in the regulation of gene expression (e.g., by a gene rearrangement), which results in protein expression not under normal cellular control. A second mechanism of activation is a point mutation in the gene that leads to a single amino acid substitution in the protein product, rendering its activity independent of normal cellular controls. The third mechanism of activation is DNA rearrangements that yield a recombinant fusion protein product, which is also independent of normal cellular controls.

The notion that proto-oncogenes play important roles in normal cell physiology is supported by their evolutionary conservation throughout vertebrates and, in some cases, in insects and yeast. Five proto-oncogenes have been identified as normal cellular genes by DNA sequence analyses. Thus, the sis, erbB, fms, erbA, and jun oncogenes are derived from normal cell genes encoding platelet-derived growth factor, epidermal growth factor receptor, macrophage colony stimulating factor receptor, thyroid hormone receptor, and transcription factor AP-1, respec-

TABLE 1
Oncogene Functional Groups

Function	Oncogene
Growth factor	sis, hst, int-2, fgf-5
Tyrosine kinase	erb B, erb B-2, fms, src, abl, fes, yes, ros, fgr, met, ret, trk, pim, kit
Guanine nucleotide-binding	rasH, rasK, rasN
Serine/threonine kinase	mos, raf
Nuclear	c-myc, N-myc, L-myc, myb, fos, p53, erb A, jun
Other	crk, mas, dbl

Note: The oncogenes listed are representative examples of the functional groups indicated. "Other" refers to recently identified oncogenes whose functions appear distinct from the 5 major groups.

tively.[2-8] However, the normal physiological roles of other proto-oncogenes remain unknown. While some proto-oncogenes are expressed in many types of proliferating cells, others display more restricted patterns of expression, suggesting that they function in specific developmental pathways. For example, the src and ras proto-oncogenes are expressed at high levels in terminally differentiated neurons and can induce neuronal differentiation of PC 12 pheochromocytoma cells in culture, suggesting the possibility that these genes may normally function in neuronal differentiation.[9-11]

As methodologies have become more sensitive, the investigation of proto-oncogene function during development has begun. Proto-oncogenes that play a role in embryonic development could function at defined stages to bring about a specific commitment process, or they could function constitutively, as, for example, to support the rapid cell divisions characteristic of embryonic cells.

III. FERTILIZATION AND CLEAVAGE

The processes of fertilization and early cleavage have been well described morphologically for several species. The cellular program directing them begins during meiosis of each of the gametes. Although, in both sexes, meiosis gives rise to gametes with haploid chromosomes that have had the opportunity to undergo cross-over rearrangements, functionally the process is sex specific. Male germ cells enter prophase, undergo DNA replication to become 4n, and transit through distinct morphological stages termed leptotene, zygotene, and pachytene during the time of genetic recombination and chromosome pairing. The reduction divisions that give rise to 4 haploid spermatids from each pachytene germ cell occur relatively quickly, apparently without a recognizable interval between Metaphase I and Metaphase II. In contrast, oocytes enter meiotic prophase in the female fetus, become 4n, and maintain meiotic arrest for decades. Resumption of oocyte maturation proceeds to Metaphase II, where a second arrest occurs that is interrupted only by fertilization or activation. Therefore, whereas some aspects of the molecular control of meiosis must be common to gametes of both sexes, the midmeiosis arrest at Metaphase II is peculiar to oocytes.

Transcription occurs in prophase germ cells of both sexes, but whether or not it continues on the haploid genome is not clear. It is assumed that nuclear formation does not occur following Metaphase I in spermatids but does occur following Metaphase II. RNAs specific to the post-meiotic stages of spermiogenesis have been isolated,[12] and their absence in pachytene stage spermatocytes indicates they are transcribed on the haploid genome following Metaphase II. Therefore, it is possible that the haploid spermatids express transcripts that are operational in sperm maturation, and possibly (although there is no evidence for it) in fertilization and early cleavage. In contrast, the only nucleus that forms following meiosis in the oocyte is the female pronucleus of the activated or fertilized zygote. If one assumes that transcription does not occur on condensed metaphase chromosomes, the oocyte transcripts essential for maintenance of Metaphase II as well as for fertilization and early cleavage must be accumulated prior to breakdown of the germinal vesicle. This is supported by the studies of RNA synthesis in oocytes that have revealed synthesis of all classes of RNAs in the growing oocyte and a marked decrease in detectable RNA synthesis in the antral follicle oocyte.[13]

Evidence from several lines of investigation indicates that fertilization and early cleavage rely heavily, if not exclusively, on the molecular program established in the oocyte during meiosis. Although postmeiotic sperm do contain stage-specific RNAs,[12] these transcripts are not detected in mature sperm and there is no evidence that sperm RNA plays a role in fertilization or zygote formation. The bulk of information is in keeping with the generally held belief that maternal messenger RNAs direct these processes.

Most mammalian ova are ovulated at a stage intermediate between meiosis one and meiosis two and stay arrested at Metaphase II until fertilization occurs. The first opportunity for transcription to resume occurs following decondensation of the chromosomes in the male and female pronuclei. Studies directed at determining the role of new transcription in development have utilized transcriptional inhibitors to determine if development of zygotes to the two-cell stage was affected.[4] Although the first cleavage was not affected, subsequent cleavages are blocked. This is interpreted as indicating that RNA synthesis is not necessary for the first cleavage but is necessary for subsequent cleavages. Therefore, work to date indicates that embryonic gene expression resumes in the uncommitted blastomeres of the two-cell embryo following the loss of the bulk of maternal mRNAs. Approximately 50% of the bulk RNA synthesized by the growing oocyte is lost in the mature ovum that has extruded a polar body and is arrested at Metaphase II. A further decrease occurs immediately following fertilization and only approximately 10% remains by the mid to late two-cell stage. It is possible that this dramatic loss of RNA is the principal mechanism by which the cell program of the differentiated gamete is forced to give way to the cell program of the undifferentiated two-cell embryo. The uncommitted cell program is operational during the early cleavages and is then replaced or augmented by cell programs of commitment.

IV. PROTO-ONCOGENE EXPRESSION IN MAMMALIAN GERM CELLS

The proto-oncogene, abl, which is transcribed in a number of somatic cells, is found in postmeiotic spermatids as a unique transcript.[15] Several other oncogenes which are transcribed in somatic cells are also expressed in male germ cells. Raf, rasH, rasK, and rasN are expressed throughout spermatogenesis;[16,17] pim and int-1 are expressed postmeiotically[16,18] and c-myc, fos, and jun are expressed in Type B spermatogonia.[17] The proto-oncogene, mos, which is not transcribed in most somatic cells, also appears as a unique transcript in spermatids.[19-21]

The mos proto-oncogene is also the first oncogene to be described in mammalian oocytes.

TABLE 2
Oncogene Expression During Embryonic Development

Stage	Species	Oncogene
Maternal	Mouse	mos
	Drosophila	rel, src, raf
	Xenopus	myc, Vg1, FGF
Embryonic	Mouse	rasH, rasK, abl, int-1, int-2
	Drosophila	int-1
	Xenopus	FGF
Extra-embryonic	Mouse	fms, fos

This proto-oncogene is transcribed during mouse oocyte growth and mos RNA accumulates to high levels, on the order of 100,000 copies per cell, in fully grown oocytes arrested at the diplotene stage of meiotic prophase.[19,20] The specific expression of mos in both male and female germ cells suggests the possibility that mos functions in a process common to both male and female germ cell development, presumably meiosis.

The product of the mos proto-oncogene is a protein of 37 kDa and is a member of the serine/threonine protein kinase family. A number of proteins are specifically modified by phosphorylation in ovulated and fertilized eggs, and protein phosphorylation is thought to play a regulatory role, allowing rapid activation and deactivation of proteins possibly essential for meiosis and early cleavages.[22-27] It is therefore tempting to consider the hypothesis that mos encodes a regulatory kinase which functions in these early developmental events.

To begin to define the function of the mos RNA accumulated in mouse oocytes, its stability and polyadenylation during oocyte maturation, fertilization, and early cleavage was investigated.[28-30] These studies indicated that mos RNA is retained in mature ovulated eggs and zygotes but is degraded at the two-cell embryo stage, consistent with the fate of bulk maternal RNA in the mouse. In addition, mos RNA was found to be posttranscriptionally polyadenylated in concert with oocyte maturation.[30] These results suggest that mos functions as a maternal mRNA which is likely translated during oocyte maturation or fertilization.

To investigate the role of mos in oocyte meiosis or early embryo development, germinal vesicle stage oocytes were micro-injected with anti-sense oligonucleotides.[45] Injected oocytes undergo germinal vesicle breakdown and polar body extrusion. However, within 24 to 48 h, the phenotype of injected oocytes is similar to that described for oocytes exposed to protein synthesis inhibitors:[31] reformation of nuclei and possible cleavage to two cells. These results suggest a role for the mos proto-oncogene in oocyte maturation.

V. PROTO-ONCOGENE EXPRESSION IN MOUSE EMBRYOS

Proto-oncogene expression in preimplantation embryos has not yet been investigated. However, expression of a few proto-oncogenes has been measured in the postimplantation mouse embryo (Table 2). Abl, ras H, and ras K are expressed in embryonic and extraembryonic tissues throughout postimplantation development; extraembryonic membranes express fms and fos at high levels.[32,33] Fms encodes the tyrosine kinase which is the receptor for macrophage colony stimulating factor.[7] Fos encodes a nuclear protein which functions as a transcriptional regulator in conjunction with jun (transcription factor AP-1).[8] Int-1 is specifically expressed in the developing central nervous system of 8 to 13 d mouse embryos and int-2 RNA is detected specifically in 7 to 8 d embryos at the egg cylinder stage, suggesting that these proto-oncogenes function in postimplantation development.[18,34,35] Int-1 encodes a secreted protein and int-2 encodes a protein which is a member of the fibroblast growth factor family.

VI. PROTO-ONCOGENE EXPRESSION IN *XENOPUS* AND *DROSOPHILA*

Analysis of transcription of proto-oncogenes in oocytes of other species (Table 2) indicated high levels of expression of myc in *Xenopus* and of src and raf in *Drosophila*.[36-39] Furthermore, transcripts encoding proteins related to transforming growth factor β (Vg-1) and to fibroblast growth factor (FGF) have been identified as maternal mRNAs in *Xenopus* oocytes.[40-41] Analysis of development mutants in *Drosophila* has also demonstrated that the segment development gene wingless is *Drosophila* Int-1 and that the maternal effect gene dorsal is the *Drosophila* homolog of the rel oncogene.[42-44] Thus, in both *Xenopus* and *Drosophila*, genes related to growth factors and oncogenes function in early embryonic development.

VII. SUMMARY

The investigation of proto-oncogene expression and function during gamete maturation, fertilization, and embryogenesis is itself embryonic and fecund. Embryogenesis involves sequential restrictions in developmental potential of cells. Commitment events always proceed along one of two possible pathways. Once committed, the cells enter into fixed patterns of differentiation. This step-by-step developmental progression suggests theoretical models that involve sequential gene regulation. An example of such controlled gene expression would be the synthesis of specific transcriptional regulators that will function during the next cell cycle. The functions of oncogenes in signal transducing pathways which ultimately affect programs of gene expression suggest their potential activities in embryogenesis. The expression of mos as a mouse maternal RNA and the activity of other proto-oncogenes and growth factors in early *Xenopus* and *Drosophila* development suggest that proto-oncogenes will play important roles in early development of mammalian embryos.

REFERENCES

1. **Bishop, J. M.,** The molecular genetics of cancer, *Science*, 235, 305, 1987.
2. **Doolittle, R. F., Hunkapiller, M. W., Hood, L. E., Devare, S. G., Robbins, K. C., Aaronson, S. A., and Antoniades, H. N.,** Simian sarcoma virus onc gene, v-*sis*, is derived from the gene (or genes) encoding a platelet-derived growth factor, *Science*, 275, 221, 1983.
3. **Ullrich, A., Schlessinger, J., and Waterfield, M. D.,** Close similarity of epidermal growth factor receptor and v-erb-B oncogene protein sequences, *Nature (London)*, 307, 521, 1984.
4. **Waterfield, M. D., Scrace, G. T., Whittle, N., Stroobant, P., Johnsson, A., Wasteson, A., Westermark, B., Heldin, C. H., Huang, J. S., and Deuel, T. F.,** Platelet-derived growth factor is structurally related to the putative transforming protein p28[sis] of simian sarcoma virus, *Nature (London)*, 304, 35, 1983.
5. **Weinberger, C., Thompson, C. C., Ong, E. S., Lebo, R., Gruol, D. J., and Evans, R. M.,** The c-erb-A gene encodes a thyroid hormone receptor, *Nature (London)*, 324, 641, 1986.
6. **Sap, J., Munoz, A., Damm, K., Goldberg, Y., Ghysdael, J., Leutz, A., Beug, H., and Vennstrom, B.,** The c-erb-A protein is a high-affinity receptor for thyroid hormone, *Nature (London)*, 625, 324, 1986.
7. **Sherr, C. J., Rettenmier, C. W., Sacca, R., Roussel, M. F., Look, A. T., and Stanley, E. R.,** The c-fms proto-oncogene product is related to the receptor for the mononuclear phagocyte growth factor, *Cell*, 41, 665, 1985.
8. **Vogt, P. K. and Tjian, R.,** Jun: a transcriptional regulator turned oncogenic, *Oncogene*, 3, 3, 1988.

9. **Brugge, J. S., Cotton, P. C., Queral, A. E., Barret, J. N., Nonner, D., and Keane, R. W.,** Neurones express high levels of a structurally modified activated form of pp60^{c-src}, *Nature (London)*, 316, 554, 1985.

10. **Bar-Sagi, D. and Feramisco, J. R.,** Microinjection of the ras oncogene protein into PC12 cells induces morphological differentiation, *Cell*, 82, 841, 1985.

11. **Furth, M. E., Aldrich, T. H., and Cardo, C. C.,** Expression of ras proto-oncogene proteins in normal human tissues, *Oncogene*, 1, 47, 1987.

12. **Hecht, N. B.,** Regulation of gene expression during mammalian spermatogenesis, in *Experimental Approaches to Mammalian Embryonic Development*, Rossant, J. and Pederson, R. A., Eds., Cambridge University Press, Cambridge, 1986, 151.

13. **Wassarman, P. M.,** Oogenesis: synthetic events in the developing mammalian egg, in *Mechanism and Control of Animal Fertilization*, Hartman, J. F., Ed., Academic Press, New York, 1983.

14. **Braude, P., Pelham, H., Flach, G., and Lobatto, R.,** Post-transcriptional control in the early mouse embryo, *Nature (London)*, 282, 102, 1979.

15. **Ponzetto, C. and Wolgemuth, D. J.,** Haploid expression of a unique c-abl transcript in the mouse male germ line, *Mol. Cell Biol.*, 5, 1791, 1985.

16. **Sorrentino, V., McKinney, M. D., Giorgi, M., Geremia, R., and Fleissner, E.,** Expression of cellular protooncogenes in the mouse male germ line: a distinctive 2.4-kilobase pim-1 transcript is expressed in haploid postmeiotic cells, *Proc. Natl. Acad. Sci. U.S.A.*, 85, 2191, 1988.

17. **Wolfes, H. R., Kogawa, K., Millette, C. F., and Cooper, G. M.,** Protooncogene expression in spermatogenesis: specific expression of nuclear protoocogenes prior to entry into meiotic prophase. Submitted.

18. **Shackleford, G. M. and Varmus, H. E.,** Expression of the proto-oncogene int-1 is restricted to postmeiotic male germ cells and the neural tube of midgestational embryos, *Cell*, 50, 89, 1987.

19. **Goldman, D. S., Kiessling, A. A., Millette, C. F., and Cooper, G. M.,** Expression of c-mos RNA in germ cells of male and female mice, *Proc. Natl. Acad. Sci. U.S.A.*, 84, 4509, 1987.

20. **Mutter, G. L. and Wolgemuth, D. J.,** Distinct developmental patterns of c-mos protooncogene expression in female and male mouse germ cells, *Proc. Natl. Acad. Sci. U.S.A.*, 84, 5301, 1987.

21. **Propst, F., Rosenberg, M., Iyer, A., Kaul, K., and Vande Woude, G. F.,** C-mos protooncogene RNA transcripts in mouse tissues: structural features, developmental regulation and localization in specific cell types, *Mol. Cell Biol.*, 7, 1629, 1987.

22. **Van Blerkom, J.,** Intrinsic and extrinsic patterns of molecular differentiation during oogenesis, embryogenesis and organogenesis in mammals, in *Cellular and Molecular Aspects of Implantation*, Glasser, S. R. and Bullock, D. W., Eds., Plenum Press, New York, 1981.

23. **Pratt, H. P., Bolton, V. N., and Gudgeon, K. A.,** The legacy from the oocyte and its role in controlling early development of the mouse embryo, *Ciba Found. Symp.*, 98, 197, 1983.

24. **Howlett, S. K., Bolton, V. N., and Gudgeon, K. A.,** Sequence and regulation of morphological and molecular events during the first cell cycle of mouse embryogenesis, *J. Embryol. Exp. Morphol.*, 87, 175, 1985.

25. **Bornslaeger, E. A., Mattei, P., and Schultz, R. M.,** Involvement of cAMP-dependent protein kinase and protein phosphorylation in regulation of mouse oocyte maturation, *Dev. Biol.*, 114, 453, 1986.

26. **Poueymirou, W. T. and Schultz, R. M.,** Differential effects of activators of cAMP-dependent protein kinase and protein kinase C on cleavage of one-cell mouse embryos and protein synthesis and phosphorylation in one- and two-cell embryos, *Dev. Biol.*, 121, 489, 1987.

27. **Howlett, S. D.,** A set of proteins showing cell cycle dependent modification in the early mouse embryo, *Cell*, 45, 387, 1986.

28. **Keshet, E., Rosenberg, M. P., Mercer, J. A., Propst, F., Vande Woude, G. F., Jenkins, N. A., and Copeland, N. G.,** Developmental regulation of ovarian-specific Mos expression, *Oncogene*, 2, 235, 1988.

29. **Mutter, G. L., Grills, G. S., and Wolgemuth, D. J.,** Evidence for the involvement of the proto-oncogene c-mos in mammalian meiotic maturation and possibly very early embryogenesis, *EMBO J.*, 7, 683, 1988.

30. **Goldman, D. S., Kiessling, A. A., and Cooper, G. M.,** Post-transcriptional processing suggests that c-mos functions as a maternal message in mouse eggs, *Oncogene*, 3, 159, 1988.

31. **Clarke, H. J. and Masui, Y.,** The induction of reversible and irreversible chromosome decondensation by protein synthesis inhibition during meiotic maturation of mouse oocytes, *Dev. Biol.*, 97, 291, 1983.

32. **Muller, R., Slamon, D. J., Tremblay, J. M., Cline, M. J., and Verma, I. M.,** Differential expression of cellular oncogenes during pre- and postnatal development of the mouse, *Nature (London)*, 299, 640, 1982.

33. **Muller, R., Verma, I. M., and Adamson E. D.,** Expression of c-onc genes: c-fos transcripts accumulate to high levels during development of mouse placenta, yolk sac and amnion, *EMBO J.*, 2, 679, 1983.

34. **Wilkinson, D. G., Bailes, J. A., and McMahon, A. P.,** Expression of the proto-oncogene int-1 is restricted to specific neural cells in the developing mouse embryo, *Cell*, 50, 79, 1987.

35. **Jakobovits, A., Shackleford, G. M., Varmus, H. E., and Martin, G. R.,** Two proto-oncogenes implicated in mammary carcinogenesis, int-1 and int-2, are independently regulated during mouse development, *Proc. Natl. Acad. Sci. U.S.A.*, 83, 7806, 1986.

36. **Taylor, M. V., Gusse, M., Evan, G. I., Dathan, N., and Mechali, M.,** *Xenopus* myc proto-oncogene during development: expression as a stable maternal mRNA uncoupled from cell division, *EMBO J.,* 5, 3563, 1986.

37. **Godeau, F., Persson, H., Gray, H. E., and Pardee, A. B.,** C-myc expression is dissociated from DNA synthesis and cell division in *Xenopus* oocyte and early embryonic development, *EMBO J.,* 5, 3571, 1986.

38. **Gregory, R. J., Kammermeyer, K. L., Vincent, W. S., III, and Wadsworth, S. G.,** Primary sequence and developmental expression of a novel *Drosophila* melanogaster src gene, *Mol. Cell. Biol.,* 7, 2119, 1987.

39. **Mark, G. E., McIntyre, R. J., Digan, M. E., Ambrosio, L., and Perrimon, N.,** *Drosophila melanogaster* homologs of the raf oncogene, *Mol. Cell Biol.,* 7, 2134, 1987.

40. **Weeks, D. L. and Melton, D. A.,** A maternal mRNA localized to the vegetal sphere in *Xenophus* eggs codes for a growth factor related to TGF-beta, *Cell,* 51, 861, 1987.

41. **Kimelman, D. and Kirschner, M.,** Synergistic induction of mesoderm by FGF and TGF-beta and the induction of an mRNA coding for FGF in the early *Xenopus* embryo, *Cell,* 51, 869, 1987.

42. **Rijsewijk, F., Schuermann, M., Wagenaar, E., Parren, P., Weigel, D., and Nusse, R.,** The *Drosophila* homolog of the mouse mammary oncogene int-1 is identical to the segment polarity gene wingless, *Cell,* 50, 649, 1987.

43. **Cabrera, C. V., Alonso, M. D., Johnston, P., Phillips, R. G., and Lawrence, P. A.,** Phenocopies induced with antisense RNA identify the wingless gene, *Cell,* 50, 659, 1987.

44. **Steward, R.,** Dorsal, an embryonic polarity gene in the *Drosophila,* is homologene to the vertebrate proto-oncogene c-rel, *Science,* 238, 692, 1987.

45. **O'Keefe, S., Wolfes, H., Kiessling, A. A., and Cooper, G.,** Microinjection of antisense c-mos oligonucleotides prevents metaphase II arrest in the maturing mouse egg, submitted.

INDEX

A

Actin, 146
Adenine nucleotides, 29
Adenosine diphosphate (ADP), 23, 29
Adenosine diphosphate, ATP ratio to, 29, 30
Adenosine monophosphate (AMP), 23, 29
Adenosine monophosphate, cyclic, see Cyclic AMP
Adenosine triphosphate (ATP), 23, 29, 42
  ADP ratio to, 29, 30
  binding site of, 36, 56
  radiolabeled, 37
Adenosine triphosphate (ATP)-ADP translocase, 146
Adenosine triphosphate (ATP)-ADP translocase
    proteins, 147
Adenylate cyclase, 34, 38
ADP, see Adenosine diphosphate
Affinity labeling, 81
aFibroblast growth factor, 137
Aggregation chimeras, 8
Agnatha
  insulin in, 52—54
  insulin receptors in, 58
Agouti locus, 10
AKR-2B, 141
Albumin, see also specific types
  bovine serum, 99, 103
  messenger RNA of, 79
Alleles, 3
Allosteric effectors, carbohydrate metabolism and,
    29—30
Amino acids, 52, 54, 56, 57, see also specific types
    of insulin, 48
AMP, see Adenosine monophosphate
Amphibians, see also specific types
  embryonic studies of, 63
  insulin in, 54
  insulin receptors in, 59—63
  oocyte maturation studies of, 59—63
Anabolic effects, 108, see also specific types
Androgenetic embryos, 4
Androgenetic zygotes, 7
Aneuploidies, 9
Annelida, 50—51
Annion yolk sac mesoderm, 103
Antibodies, 77—79, 127, see also specific types
Anti-insulin receptor antiserum, 103
Antiserum, see also specific types
  anti-insulin receptor, 103
  RNA in, 127
Arthropods, insulin in, 51—52
Assays, see also specific types
  fluorometric, 21—23, 29
  luciferase-based, 29
  ultramicrofluorometric, 23, 29
ATP, see Adenosine triphosphate
Autocrine, 106, 120
Autocrine growth control, 128

Autocrine growth factor, 153
Autoradiography, 81, 95
Aves
  insulin in, 54
  insulin receptors in, 63

B

Beta cells, 74
bFibroblast growth factor, 127, 137
B-glucuronidase, 3
Biodosimeters, 9
Biogenic factors, 8—10, see also specific types
Biologically active platelet-derived growth factors,
    117
Bioluminescence methods, 21, 23
Blastocoel cavities, 95
Blastocoel fluid, 24
Blastocysts, 2, 5, 11, 29, 95
  formation of, 27
  morula-early, 94
  outgrowths of, 64, 98
Blastomeres, 7, 8
B-microglobulin, 3
Bone, 152, 153
Bovine fetal pancreata, 55
Bovine serum albumin (BSA), 99, 103
Brain, 52, 64
Brown adipose tissue, 104
BSA, see Bovine serum albumin
BSC-1 growth inhibitor, 151

C

Cachectin (tumor necrosis factor alpha), 145
Calcium, 40
Calcium-independent enzymes, 42, see also specific
    types
cAMP, see Cyclic AMP
Carbohydrate metabolism, 19—31
  allosteric effectors and, 29—30
  enzyme analysis and, 20—24
  glucose uptake and, 24—29
  insulin and, 48
  metabolite analysis and, 20—24
  pyruvate uptake and, 24—29
Carbon dioxide, 27
Carbon sources, 20
Carcinoma, embryonal, see Embryonal carcinoma
Cartilage, 103, 153
Cartilage-inducing factor, 151
Casein kinase II, 42
Cathepsin L (major excreting protein), 146, 152, 154
cDNA probe, 105
Cell biology, 94
Cell-cell interactions, see Intercellular interactions
Cell death, 146
Cell density quiescence, 149

Metalloprotease inhibition, 146, 152
Metaphase I, 169
Metaphase II, 169, 170
Methylation of DNA, 6
Microglobulin, 3
Microvilli, 99
Midgestational embryos, 108
Migration of cells, 153—155
MIP, see Molluscan insulin-related peptide
Mitochondrial DNA (mtDNA), 8
Mitochondrial inheritance, 8
Mitogen-regulated protein (MRP), 146—149, 151
Molecular biology, 94
Molluscs, see also specific types
  insulin in, 51—52
  insulin-related peptide (MIP) in, 52
Morphometry, 99
Morula-stage, 95
MPF, see Maturation promoting factor
mRNA, see Messenger RNA
MRP, see Mitogen-regulated protein
MRP/PLF, see Mitogen-regulated protein
MSA, see Multiplication-stimulating activity
mtDNA, see Mitochondrial DNA
Multiplication
  of cells, 92, 106
  -stimulating activity (MSA), 106
Muscle, skeletal, 143, see also other specific types
Mutations, 4, 9, see also specific types
Myoblast differentiation, 79

N

NADH, 21
NADPH, 21
Natural fertilization, 4—6
Nerve growth factor (NGF), 137
Nervous system, 50, 57
Neurotransmitters, 34, see also specific types
N gene, see Notch gene
NGF, see Nerve growth factor
Nonmammals, homologues in, 138—139
Nonproliferative functions of growth factors, 145—146
Northern analysis, 104, 121
Notch gene, 128
NRK cells, 126
Nuclear DNA, 10
Nuclear proteins, 168, see also specific types
Nuclear transfer experiments, 6
Nutritional studies, 20

O

OC-15 cells, 115, 138, 144
Occupied insulin receptors, 99
Oligosyndactyly (Os), 10
Oncogenes, 139, see also specific types
  expression of, 10, 167—172
  prot-, see Proto-oncogenes
  viral, 10

Oncogenically transformed cells, 139
Ontogeny
  of insulin-like growth factor receptors, 72
  of insulin production, 56
  of insulin receptors, 72
  of insulin secretion, 56
Oocytes, 2, 54, 58, 59, 95, 169
  amphibian, 59—63
  maturation studies of, 59—63
  meiotic maturation of, 61
  metabolites in, 22
  morphological differences with sperm, 4
Oogenesis, 20
Organogenesis, 63, 103
Os, see Oligosyndactyly
Osteichthyes, 59
OTT-6050, 106
Ovaries, 57
Oviducts, 20, 136

P

PAI-1, 146
Palate, 142
Pancreas, 74
Pancreas, bovine fetal, 55
Pancreatic immunoreactive insulin, 54
Paracrine growth control, 116, 128
Paracrine growth control, embryonal carcinoma cells
  and, 124—125
Parietal endoderm, 103, 115
Paternal alleles, 3
Paternal disomy 11, 5
PC-13 cells, 106, 115, 117, 138
PDGF, see Platelet-derived growth factor
Pentose phosphate pathway, 27
Peptide hormones, 34
C-Peptides, 48
PFK, see 6-Phosphofructokinase
Phloretin, 25
Phosphatidylinositol, 38—41, 60
Phosphatidylinositol 4,5-bisphosphate, 38
6-Phosphofructokinase (PFK), 29, 30
Phospholipase A2-inhibitory proteins, 42
Phospholipase C, 38
Phospholipids, total, 79, see also specific types
Phosphoprotein phosphatase, 36
Phosphorylation, see also specific types
  defined, 34
  insulin-stimulated, 104
  protein tyrosine kinase, 60
Phosphorylation-dephosphorylation reactions, 34, 42
Photoaffinity labeling, 104
Physiological effects of insulin, 48
Placenta, 64, 142, 148
  chorioallantoic, 103
  lactogens in, 151
Plasma membranes, 99
Plasminogen activator, inhibitors of, 146
Plasminogen activators, 115, 152, 154, see also
  specific types